The Anti-Nuclear Movement

Edited by Paul F. Kisak

Contents

Chapter 1

Anti-nuclear movement

120,000 people attended an anti-nuclear protest in Bonn, West Germany, on October 14, 1979, following the Three Mile Island accident.[1]

Anti-Nuclear Power Plant Rally following the Fukushima Daiichi nuclear disaster on 19 September 2011 at Meiji Shrine complex in Tokyo, Japan.

Anti-nuclear demonstration in Colmar, north-eastern France, on October 3, 2009.

The **anti-nuclear movement** is a social movement that opposes various nuclear technologies. Some direct action groups, environmental groups, and professional organisations[2][3] have identified themselves with the movement at the local, national, and international level. Major anti-nuclear groups include Campaign for Nuclear Disarmament, Friends of the Earth, Greenpeace, International Physicians for the Prevention of Nuclear War, and the

Nuclear Information and Resource Service. The initial objective of the movement was nuclear disarmament, though since the late 1960s opposition has included the use of nuclear power. Many anti-nuclear groups oppose both nuclear power and nuclear weapons. The formation of green parties in the 1970s and 1980s was often a direct result of anti-nuclear politics.[4]

Scientists and diplomats have debated nuclear weapons policy since before the atomic bombings of Hiroshima and Nagasaki in 1945.[5] The public became concerned about nuclear weapons testing from about 1954, following extensive nuclear testing in the Pacific. In 1963, many countries ratified the Partial Test Ban Treaty which prohibited atmospheric nuclear testing.[6]

Some local opposition to nuclear power emerged in the early 1960s,[7] and in the late 1960s some members of the scientific community began to express their concerns.[8] In the early 1970s, there were large protests about a proposed nuclear power plant in Wyhl, West Germany. The project was cancelled in 1975 and anti-nuclear success at Wyhl inspired opposition to nuclear power in other parts of Europe and North America.[9][10] Nuclear power became an issue of major public protest in the 1970s.[11]

A protest against nuclear power occurred in July 1977 in Bilbao, Spain, with up to 200,000 people in attendance. Following the Three Mile Island accident in 1979, an anti-nuclear protest was held in New York City, involving 200,000 people. In 1981, Germany's largest anti-nuclear power demonstration took place to protest against the Brokdorf Nuclear Power Plant west of Hamburg; some 100,000 people came face to face with 10,000 police officers. The largest protest was held on June 12, 1982, when one million people demonstrated in New York City against nuclear weapons. A 1983 nuclear weapons protest in West Berlin had about 600,000 participants. In May 1986, following the Chernobyl disaster, an estimated 150,000 to 200,000 people marched in Rome to protest against the Italian nuclear program. In the US, public opposition preceded the shutdown of the Shoreham, Yankee Rowe, Millstone 1, Rancho Seco, Maine Yankee, and many other nuclear power plants.

For many years after the 1986 Chernobyl disaster nuclear power was off the policy agenda in most countries, and the anti-nuclear power movement seemed to have won its case. Some anti-nuclear groups disbanded. In the 2000s (decade), however, following public relations activities by the nuclear industry,[12][13][14][15][16] advances in nuclear reactor designs, and concerns about climate change, nuclear power issues came back into energy policy discussions in some countries. The 2011 Japanese nuclear accidents subsequently undermined the nuclear power industry's proposed renaissance and revived nuclear opposition worldwide, putting governments on the defensive.[17] As of 2016, countries such as Australia, Austria, Denmark, Greece, Malaysia, New Zealand, and Norway have no nuclear power stations and remain opposed to nuclear power.[18][19] Germany, Italy, Spain, Sweden and Switzerland are phasing-out nuclear power.[19][20][21][22] Globally, more nuclear power reactors have closed than opened in recent years.[21]

Women Strike for Peace during the Cuban Missile Crisis

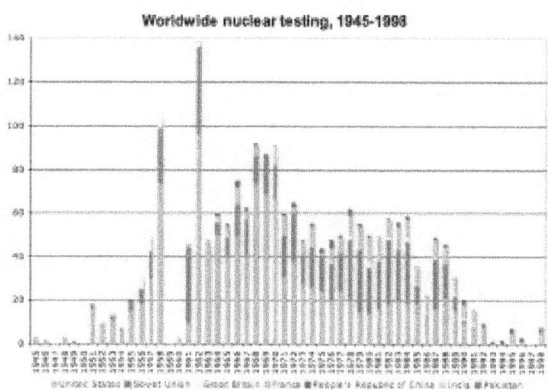

Worldwide nuclear testing totals, 1945-1998.

1.1 History and issues

1.1.1 Roots of the movement

Main article: History of the anti-nuclear movement
 The application of nuclear technology, as a source of energy and as an instrument of war, has been controversial.[23][24][25][26][27][28] These issues are discussed in nuclear weapons debate, nuclear power debate, and uranium mining debate.

Scientists and diplomats have debated nuclear weapons policy since before the Atomic bombings of Hiroshima and Nagasaki in 1945.[5] The public became concerned about nuclear weapons testing from about 1954, following extensive nuclear testing in the Pacific. In 1961, at the height of the Cold War, about 50,000 women brought together by Women Strike for Peace marched in 60 cities in the United States to demonstrate against nuclear weapons.[29][30] In 1963, many countries ratified the Partial Test Ban Treaty which prohibited atmospheric nuclear testing.[6]

Some local opposition to nuclear power emerged in the early 1960s,[7] and in the late 1960s some members of the scientific community began to express their concerns.[8] In the early 1970s, there were large protests about a proposed nuclear power plant in Wyhl, Germany. The project was cancelled in 1975 and anti-nuclear success at Wyhl inspired opposition to nuclear power in other parts of Europe and North America.[9][10] Nuclear power became an issue of major public protest in the 1970s.[11]

1.1.2 Anti-nuclear perspectives

See also: Lists of nuclear disasters and radioactive incidents

Concerns about nuclear weapons

The 18,000 km² expanse of the Semipalatinsk Test Site (indicated in red), which covers an area the size of Wales. The Soviet Union conducted 456 nuclear tests at Semipalatinsk from 1949 until 1989 with little regard for their effect on the local people or environment. The full impact of radiation exposure was hidden for many years by Soviet authorities and has only come to light since the test site closed in 1991.[31]

See also: Nuclear ethics and Uranium mining § Environment

From an anti-nuclear point of view, there is a threat to modern civilization from global nuclear war by accidental or deliberate nuclear strike.[32] Some climate scientists estimate that a war between two countries that resulted in 100 Hiroshima-size atomic explosions would cause significant loss of life, in the tens of millions from climatic effects alone. Soot thrown up into the atmosphere could blanket the earth, causing food chain disruption in what is termed a nuclear winter.[33][34]

Many anti-nuclear weapons groups cite the 1996 Advisory Opinion of the International Court of Justice, *Legality of the Threat or Use of Nuclear Weapons*, in which it found that 'the threat or use of nuclear weapons would generally be contrary to the rules of international law applicable in armed conflict'.[35]

Ridding the world of nuclear weapons has been a cause for pacifists for decades. But more recently mainstream politi-

cians and retired military leaders have advocated nuclear disarmament. In January 2007 an article in the *Wall Street Journal*, authored by Henry Kissinger, Bill Perry, George Shultz and Sam Nunn.[36] These men were veterans of the cold-war who believed in using nuclear weapons for deterrence. But they now reversed their previous position and asserted that instead of making the world safer, nuclear weapons had become a source of extreme concern.[37]

During the era of nuclear weapons testing many local communities were affected, and some are still affected by uranium mining, and radioactive waste disposal.[32]

Concerns about nuclear power

Following the 2011 Japanese Fukushima nuclear disaster, authorities shut down the nation's 54 nuclear power plants. As of 2013, the Fukushima site remains highly radioactive, with some 160,000 evacuees still living in temporary housing, and some land will be unfarmable for centuries. The difficult cleanup job will take 40 or more years, and cost tens of billions of dollars.[38][39]

The abandoned city of Prypiat, Ukraine, following the 1986 Chernobyl disaster. The Chernobyl nuclear power plant is in the background.

See also: Lists of nuclear disasters and radioactive incidents and Nuclear safety and security

President Jimmy Carter leaving the Three Mile Island accident for Middletown, Pennsylvania, April 1, 1979.

There are large variations in peoples' understanding of the issues surrounding nuclear power, including the technology itself, its deployment, climate change, and energy security. There is a wide spectrum of views and concerns over nuclear power [40] and it remains a controversial area of public policy. [41]

Many studies have shown that the public "perceives nuclear power as a very risky technology" and, around the world, nuclear energy has declined in popularity since the Fukushima Daiichi nuclear disaster. [42] [43] [44] Anti-nuclear critics see nuclear power as a dangerous, expensive way to boil water to generate electricity. [45] Opponents of nuclear power have raised a number of related concerns: [46]

- Nuclear accidents: a concern that the core of a nuclear power plant could overheat and melt down, releasing radioactivity.

- Radioactive waste disposal: a concern that nuclear power results in large amounts of radioactive waste, some of which remains dangerous for very long periods.

- Nuclear proliferation: a concern that some types of nuclear reactor designs use and/or produce fissile material which could be used in nuclear weapons.

- High cost: a concern that nuclear power plants are very expensive.

- Attacks on nuclear plants: a concern that nuclear facilities could be targeted by terrorists or criminals.

- Curtailed civil liberties: a concern that the risk of nuclear accidents, proliferation and terrorism may be used to justify restraints on citizen rights.

Of these concerns, nuclear accidents and disposal of long-lived radioactive waste have probably had the greatest public impact worldwide. [46] Anti-nuclear campaigners point to the 2011 Fukushima nuclear emergency as proof that nuclear power can never be 100% safe. [47]

In his book *Global Fission: The Battle Over Nuclear Power*, Jim Falk explores connections between technological concerns and political concerns. Falk suggests that concerns of citizen groups or individuals who oppose nuclear power have often focused initially on the "range of physical hazards which accompany the technology" and leads to a "concern over the political relations of the nuclear industry" A more neutral observer might observe that this is nothing more than a conspiracy theory. Baruch Fischhoff, a social science professor said that many people really do not trust the nuclear industry. [48] Wade Allison, a physics professor actually says "radiation is safe & all nations should embrace nuclear technology" [49]

M.V. Ramana says that "distrust of the social institutions that manage nuclear energy is widespread", and a 2001 survey by the European Commission found that "only 10.1 percent of Europeans trusted the nuclear industry". This public distrust is periodically reinforced by nuclear safety violations, or through ineffectiveness or corruption of the nuclear regulatory authorities. Once lost, says Ramana, trust is extremely difficult to regain. [50]

Faced with public antipathy, the nuclear industry has "tried a variety of strategies to persuade the public to accept nuclear power", including the publication of numerous "fact sheets" that discuss issues of public concern. M.V. Ramana says that none of these strategies have been very successful. [51] Nuclear proponents have tried to regain public support by offering newer, safer, reactor designs. These designs include those that incorporate passive safety and Small Modular Reactors. While these reactor designs "are intended to inspire trust, they may have an unintended effect: creating distrust of older reactors that lack the touted safety features". [52]

Since 2000 the nuclear industry has undertaken an international media and lobbying campaign to promote nuclear power as a solution to the greenhouse effect and climate change. [53] Nuclear power, the industry says, emits no or negligible amounts of carbon dioxide. Anti-nuclear groups respond by saying that only reactor operation is free of carbon dioxide emissions. All other stages of the nuclear fuel chain – mining, milling, transport, fuel fabrication, enrichment, reactor construction, decommissioning and waste management – use fossil fuels and hence emit carbon dioxide. [53] [54] [55]

In 2011, a French court fined Électricité de France (EDF) €1.5m and jailed two senior employees for spying on Greenpeace, including hacking into Greenpeace's computer systems. Greenpeace was awarded €500,000 in damages.[56][57]

There is a wide range of published energy-related studies which conclude that energy efficiency programs and renewable power technologies are a better energy option than nuclear power plants. This diverse range of studies come from many different sources, across the political spectrum, and from various academic disciplines, which suggests that there is a consensus among many independent, non-partisan energy experts that nuclear power plants are a poor way to produce electrical power.[58]

Other technologies

Protest against ITER in France, 2009.

The nuclear fusion project International Thermonuclear Experimental Reactor is constructing the world's largest and most advanced experimental tokamak nuclear fusion reactor in the south of France. A collaboration between the European Union (EU), India, Japan, China, Russia, South Korea and the United States, the project aims to make a transition from experimental studies of plasma physics to electricity-producing fusion power plants. In 2005, Greenpeace International issued a press statement criticizing government funding of the ITER, believing the money should have been diverted to renewable energy sources and claiming that fusion energy would result in nuclear waste and nuclear weapons proliferation issues.[59] A French association including about 700 anti-nuclear groups, Sortir du nucléaire (Get Out of Nuclear Energy), claimed that ITER was a hazard because scientists did not yet know how to manipulate the high-energy deuterium and tritium hydrogen isotopes used in the fusion process.[60] According to most anti-nuclear groups, nuclear fusion power "remains a distant dream".[61] The World Nuclear Association says that fusion "presents so far insurmountable scientific and engineering challenges".[62] Construction of the ITER facility began in 2007, but the project has run into many delays and

budget overruns. The facility is now not expected to begin operations until the year 2027 – 11 years after initially anticipated.[63]

Anti-nuclear groups advocate reduced reliance on reactor-produced medical radioisotopes, through the use of alternative radioisotope production and alternative clinical technologies.[64] Cyclotrons are being increasingly used to produce medical radioisotopes to the point where nuclear reactors are no longer needed to make the most common medical isotopes.[65]

1.1.3 Nuclear-free alternatives

See also: 100% renewable energy, Soft energy path, Renewable energy commercialisation, Non-nuclear future, and Clean Tech Nation

Anti-nuclear groups say that reliance on nuclear energy

Three renewable energy sources: solar energy, wind power, and hydroelectricity.

The 150 MW Andasol Solar Power Station is a commercial parabolic trough solar thermal power plant, located in Spain. The Andasol plant uses tanks of molten salt to store solar energy so that it can continue generating electricity even when the sun isn't shining.[66]

Photovoltaic SUDI shade is an autonomous and mobile station in France that provides energy for electric vehicles using solar energy.

can be reduced by adopting energy conservation and energy efficiency measures. Energy efficiency can reduce energy consumption while providing the same level of energy "services".[67] Renewable energy flows involve natural phenomena such as sunlight, wind, tides, plant growth, and geothermal heat, as the International Energy Agency explains:[68]

> Renewable energy is derived from natural processes that are replenished constantly. In its various forms, it derives directly from the sun, or from heat generated deep within the earth. Included in the definition is electricity and heat generated from solar, wind, ocean, hydropower, biomass, geothermal resources, and biofuels and hydrogen derived from renewable resources.

Anti-nuclear groups also favour the use of renewable energy, such as hydro, wind power, solar power, geothermal energy and biofuel.[69] According to the International Energy Agency renewable energy technologies are essential contributors to the energy supply portfolio, as they contribute to world energy security and provide opportunities for mitigating greenhouse gases.[70] Fossil fuels are being replaced by clean, climate-stabilizing, non-depletable sources of energy. According to Lester R. Brown:

> ...the transition from coal, oil, and gas to wind, solar, and geothermal energy is well under way. In the old economy, energy was produced by burning something —oil, coal, or natural gas —leading to the carbon emissions that have come to define our economy. The new energy economy harnesses the energy in wind, the energy coming from the sun, and heat from within the earth itself.[71]

In 2014 global wind power capacity expanded 16% to 369,553 MW.[72] Yearly wind energy production is also growing rapidly and has reached around 4% of worldwide electricity usage,[73] 11.4% in the EU,[74] and it is widely used in Asia, and the United States. In 2014, worldwide installed photovoltaics capacity increased to 177 gigawatts (GW), sufficient to supply 1 percent of global electricity demands.[75] Solar thermal energy stations operate in the USA and Spain, and as of 2016, the largest of these is the 392 MW Ivanpah Solar Electric Generating System in California.[76][77] The world's largest geothermal power installation is The Geysers in California, with a rated capacity of 750 MW. Brazil has one of the largest renewable energy programs in the world, involving production of ethanol fuel from sugar cane, and ethanol now provides 18% of the country's automotive fuel. Ethanol fuel is also widely available in the USA.

Greenpeace advocates a reduction of fossil fuels by 50% by 2050 as well as phasing out nuclear power, contending that innovative technologies can increase energy efficiency, and suggests that by 2050 most electricity will come from renewable sources.[69] The International Energy Agency estimates that nearly 50% of global electricity supplies will need to come from renewable energy sources in order to halve carbon dioxide emissions by 2050 and minimise climate change impacts.[78]

Mark Z. Jacobson, a Stanford professor, says producing all new energy with wind power, solar power, and hydropower by 2030 is feasible and existing energy supply arrangements could be replaced by 2050. Barriers to implementing the renewable energy plan are seen to be "primarily social and political, not technological or economic". Jacobson says that energy costs with a wind, solar, water system should be similar to today's energy costs.[79]

1.2 Anti-nuclear organizations

See also: List of anti-nuclear groups, List of anti-nuclear power groups, and List of anti-nuclear groups in the United States

The anti-nuclear movement is a social movement which operates at the local, national, and international level. Various types of groups have identified themselves with the movement:[3]

- direct action groups, such as the Clamshell Alliance and Shad Alliance;
- environmental groups, such as Friends of the Earth and Greenpeace;
- consumer protection groups, such as Ralph Nader's Critical Mass;

Members of Nevada Desert Experience hold a prayer vigil during the Easter period of 1982 at the entrance to the Nevada Test Site.

- professional organisations,*[2] such as International Physicians for the Prevention of Nuclear War; and

- political parties such as European Free Alliance.

Anti-nuclear groups have undertaken public protests and acts of civil disobedience which have included occupations of nuclear plant sites. Other salient strategies have included lobbying, petitioning government authorities, influencing public policy through referendum campaigns and involvement in elections. Anti-nuclear groups have also tried to influence policy implementation through litigation and by participating in licensing proceedings.*[80]

Anti-nuclear power organisations have emerged in every country that has had a nuclear power programme. Protest movements against nuclear power first emerged in the USA, at the local level, and spread quickly to Europe and the rest of the world. National nuclear campaigns emerged in the late 1970s. Fuelled by the Three Mile Island accident and the Chernobyl disaster, the anti-nuclear power movement mobilised political and economic forces which for some years "made nuclear energy untenable in many countries" .*[81] In the 1970s and 1980s, the formation of green parties was often a direct result of anti-nuclear politics (e.g., in Germany and Sweden).*[4]

Some of these anti-nuclear power organisations are reported to have developed considerable expertise on nuclear power and energy issues.*[82] In 1992, the chairman of the Nuclear Regulatory Commission said that "his agency had been pushed in the right direction on safety issues because of the pleas and protests of nuclear watchdog groups" .*[83]

1.2.1 International organizations

- European Nuclear Disarmament, which held annual conventions in the 1980s involving thousands of anti-nuclear weapons activists mostly from Western Europe but also from Eastern Europe, the United States, and Australia.*[84]

- Friends of the Earth International, a network of environmental organizations in 77 countries.*[85] Since 2014, however, FOE (UK) has softened its stance; the fierce opposition against nuclear reactors has shifted into a more pragmatic opposition, which still opposes the construction of new nuclear (fission) reactors, but doesn't campaign against closing down the existing ones any more.*[86]

- Global Zero, an international non-partisan group of 300 world leaders dedicated to achieving the elimination of nuclear weapons.*[87]

- Global Initiative to Combat Nuclear Terrorism, an international partnership of 83 nations.

- Greenpeace International, a non-governmental environmental organization*[88] with offices in over 41 countries and headquarters in Amsterdam, Netherlands.*[89]

- International Campaign to Abolish Nuclear Weapons

- International Network of Engineers and Scientists for Global Responsibility

- International Physicians for the Prevention of Nuclear War, which had affiliates in 41 nations in 1985, representing 135,000 physicians;*[84] IPPNW was awarded the UNESCO Peace Education Prize in 1984 and the Nobel Peace Prize in 1985.*[90]

- Nuclear Information and Resource Service

- OPANAL

- Parliamentarians for Nuclear Non-Proliferation and Disarmament, a global network of over 700 parliamentarians from more than 75 countries working to prevent nuclear proliferation.*[91]

- Pax Christi International, a Catholic group which took a "sharply anti-nuclear stand" .*[84]

- Ploughshares Fund

- Pugwash Conferences on Science and World Affairs

- Socialist International, the world body of social democratic parties.*[92]

- Sōka Gakkai, a peace-orientated Buddhist organisation, which held anti-nuclear exhibitions in Japanese
  cities during the late 1970s, and gathered 10 million
  signatures on petitions calling for the abolition of nuclear weapons.*[92]*[93]

- United Nations Office for Disarmament Affairs

- World Disarmament Campaign*[92]

- World Information Service on Energy, basedin Amsterdam, The Netherlands

- World Union for Protection of Life

1.2.2   Other groups

National and local anti-nuclear groups are listed at Anti-
nuclear groups in the United States and List of anti-nuclear
groups.

1.2.3   Symbols

1.3   Activities

1.3.1   Large protests

Main article: Anti-nuclear protests

In 1971, the town of Wyhl, in Germany, was a proposed
site for a nuclear power station. In the years that followed,
public opposition steadily mounted, and there were large
protests. Television coverage of police dragging away farmers and their wives helped to turn nuclear power into a major
issue. In 1975, an administrative court withdrew the construction licence for the plant.*[9]*[10]*[95] The Wyhl experience encouraged the formation of citizen action groups
near other planned nuclear sites.*[9]

In 1972, the nuclear disarmament movement maintained a
presence in the Pacific, largely in response to French nuclear testing there. New Zealand activists sailed boats into
the test zone, interrupting the testing program.*[96]*[97]
In Australia, thousands of people joined protest marches
in Adelaide, Melbourne, Brisbane, and Sydney. Scientists
issued statements demanding an end to the nuclear tests.

Demonstration against French nuclear testing in 1995 in Paris.

Demonstration in Lyon, France in the 1980s against nuclear tests

In Fiji, anti-nuclear activists formed an Against Testing on
Mururoa organization.*[97]

In the Basque Country (Spain and France), a strong anti-
nuclear movement emerged in 1973, which ultimately
led to the abandonment of most of the planned nuclear
power projects.*[98] On July 14, 1977, in Bilbao, between
150,000 and 200,000 people protested against the Lemoniz
Nuclear Power Plant. This has been called the "biggest
ever anti-nuclear demonstration" .*[99]

In France, there were mass protests in the early 1970s,
organized at nearly every planned nuclear site in France.
Between 1975 and 1977, some 175,000 people protested
against nuclear power in ten demonstrations.*[1] In 1977
there was a massive demonstration at the Superphénix
breeder reactor in Creys-Malvillein which culminated in violence.*[100]

In West Germany, between February 1975 and April 1979,
some 280,000 people were involved in seven demonstra-

On 12 December 1982, 30,000 women held hands around the 6 miles (9.7 km) perimeter of the base, in protest against the decision to site American cruise missiles there

tions at nuclear sites. Several site occupations were also attempted. Following the Three Mile Island accident in 1979, some 120,000 people attended a demonstration against nuclear power in Bonn.[1]

In the Philippines, there were many protests in the late 1970s and 1980s against the proposed Bataan Nuclear Power Plant, which was built but never operated.[101]

In 1981, Germany's largest anti-nuclear power demonstration protested against the construction of the Brokdorf Nuclear Power Plant west of Hamburg. Some 100,000 people came face to face with 10,000 police officers.[95][102][103]

In the late 1970s and early 1980s, the revival of the nuclear arms race, triggered a new wave of protests about nuclear weapons. Older organizations such as the Federation of Atomic Scientists revived, and newer organizations appeared, including the Nuclear Weapons Freeze Campaign and Physicians for Social Responsibility.[104] In the UK, on 1 April 1983, about 70,000 people linked arms to form a 14-mile-long human chain between three nuclear weapons centres in Berkshire.[105]

On Palm Sunday 1982, 100,000 Australians participated in anti-nuclear rallies in the nation's largest cities. Growing year by year, the rallies drew 350,000 participants in 1985.[97]

In May 1986, following the Chernobyl disaster, clashes between anti-nuclear protesters and West German police were common. More than 400 people were injured in mid-May at a nuclear-waste reprocessing plant being built near Wackersdorf.[106] Also in May 1986, an estimated 150,000 to 200,000 people marched in Rome to protest against the Italian nuclear program, and 50,000 marched in Milan.[107] Hundreds of people walked from Los Angeles to Washington, D.C. in 1986 in what is referred to as the Great Peace March for Global Nuclear Disarmament. The march took nine months to traverse 3,700 miles (6,000 km), advancing approximately fifteen miles per day.[108]

The anti-nuclear organisation "Nevada Semipalatinsk" was formed in 1989 and was one of the first major anti-nuclear groups in the former Soviet Union. It attracted thousands of people to its protests and campaigns which eventually led to the closure of the nuclear test site in north-east Kazakhstan, in 1991.[109][110][111][112]

The World Uranium Hearing was held in Salzburg, Austria in September 1992. Anti-nuclear speakers from all continents, including indigenous speakers and scientists, testified to the health and environmental problems of uranium mining and processing, nuclear power, nuclear weapons, nuclear tests, and radioactive waste disposal. People who spoke at the 1992 Hearing included: Thomas Banyacya, Katsumi Furitsu, Manuel Pino and Floyd Red Crow Westerman.[113][114]

1.3.2 Protests in the United States

Main article: Anti-nuclear protests in the United States
There were many **anti-nuclear protests in the United States** which captured national public attention during the 1970s and 1980s. These included the well-known Clamshell Alliance protests at Seabrook Station Nuclear Power Plant and the Abalone Alliance protests at Diablo Canyon Nuclear Power Plant, where thousands of protesters were arrested. Other large protests followed the 1979 Three Mile Island accident.[115]

A large anti-nuclear demonstration was held in May 1979 in Washington D.C., when 65,000 people including the Governor of California, attended a march and rally against nuclear power.[116] In New York City on September 23, 1979, almost 200,000 people attended a protest against nuclear power.[117] Anti-nuclear power protests preceded the shutdown of the Shoreham, Yankee Rowe, Millstone I, Rancho Seco, Maine Yankee, and about a dozen other nuclear power plants.[118]

On June 12, 1982, one million people demonstrated in New York City's Central Park against nuclear weapons and for an end to the cold war arms race. It was the largest anti-

Anti-nuclear protest in 1979 following the Three Mile Island Accident.

April 2011 OREPA rally at the Y-12 weapons plant entrance

nuclear protest and the largest political demonstration in American history.[119][120] International Day of Nuclear Disarmament protests were held on June 20, 1983 at

50 sites across the United States.[121][122] In 1986, hundreds of people walked from Los Angeles to Washington DC in the Great Peace March for Global Nuclear Disarmament.[123] There were many Nevada Desert Experience protests and peace camps at the Nevada Test Site during the 1980s and 1990s.[124][125]

On May 1, 2005, 40,000 anti-nuclear/anti-war protesters marched past the United Nations in New York, 60 years after the atomic bombings of Hiroshima and Nagasaki.[126][127] This was the largest anti-nuclear rally in the U.S. for several decades.[97] In the 2000s there were protests about, and campaigns against, several new nuclear reactor proposals in the United States.[128][129][130] In 2013, four aging, uncompetitive, reactors were permanently closed: San Onofre 2 and 3 in California, Crystal River 3 in Florida, and Kewaunee in Wisconsin.[131][132] Vermont Yankee, in Vernon, is scheduled to close in 2014, following many protests. Protesters in New York State are seeking to close Indian Point Energy Center, in Buchanan, 30 miles from New York City.[132]

1.3.3 Recent developments

For many years after the 1986 Chernobyl disaster nuclear power was off the policy agenda in most countries, and the anti-nuclear power movement seemed to have won its case. Some anti-nuclear groups disbanded. In the 2000s (decade), however, following public relations activities by the nuclear industry,[14][15][16][133] advances in nuclear reactor designs, and concerns about climate change, nuclear power issues came back into energy policy discussions in some countries. The 2011 Japanese nuclear accidents subsequently undermined the nuclear power industry's proposed come back.[17]

2004-2006

In January 2004, up to 15,000 anti-nuclear protesters marched in Paris against a new generation of nuclear reactors, the European Pressurised Water Reactor (EPWR).[134]

On May 1, 2005, 40,000 anti-nuclear/anti-war protesters marched past the United Nations in New York, 60 years after the atomic bombings of Hiroshima and Nagasaki.[126][127] This was the largest anti-nuclear rally in the U.S. for several decades.[135] In Britain, there were many protests about the government's proposal to replace the aging Trident weapons system with a newer model. The largest protest had 100,000 participants and, according to polls, 59 percent of the public opposed the move.[135]

2007-2009

A scene from the 2007 Stop EPR (European Pressurised Reactor) protest in Toulouse, France.

Start of anti-nuclear march from Geneva to Brussels, 2009

Anti-nuclear protest near nuclear waste disposal centre at Gorleben in Northern Germany, on 8 November 2008.

Anti-nuclear march from London to Geneva, 2008

On March 17, 2007 simultaneous protests, organised by *Sortir du nucléaire*, were staged in five French towns to protest construction of EPR plants; Rennes, Lyon, Toulouse, Lille, and Strasbourg.[136][137]

In June 2007, 4,000 local residents, students and anti-nuclear activists took to the streets in the city of Kudus in Indonesia's Central Java, calling on the Government to abandon plans to build a nuclear power plant there.[138]

In February 2008, a group of concerned scientists and engineers called for the closure of the Kashiwazaki-Kariwa Nuclear Power Plant in Japan.[139][140]

The International Conference on Nuclear Disarmament took place in Oslo in February 2008, and was organized by The Government of Norway, the Nuclear Threat Initiative and the Hoover Institute. The Conference was entitled *Achieving the Vision of a World Free of Nuclear Weapons* and had the purpose of building consensus between nuclear weapon states and non-nuclear weapon states in relation to the Nuclear Non-proliferation Treaty.[141]

During a weekend in October 2008, some 15,000 people disrupted the transport of radioactive nuclear waste from France to a dump in Germany. This was one of the largest such protests in many years and, according to *Der Spiegel*, it signals a revival of the anti-nuclear movement in Germany.[142][143][144] In 2009, the coalition of green parties in the European parliament, who are unanimous in their anti-nuclear position, increased their presence in the parliament from 5.5% to 7.1% (52 seats).[145]

In October 2008 in the United Kingdom, more than 30 people were arrested during one of the largest anti-nuclear protests at the Atomic Weapons Establishment at Aldermaston for 10 years. The demonstration marked the start of the UN World Disarmament Week and involved about 400 people.[146]

In 2008 and 2009, there have been protests about, and criticism of, several new nuclear reactor proposals in the United States.[128][129][130] There have also been some objections to license renewals for existing nuclear plants.[147][148]

A convoy of 350 farm tractors and 50,000 protesters took part in an anti-nuclear rally in Berlin on September 5, 2009. The marchers demanded that Germany close all nuclear plants by 2020 and close the Gorleben radioactive dump.[149][150] Gorleben is the focus of the anti-nuclear movement in Germany, which has tried to derail train transports of waste and to destroy or block the approach roads to the site. Two above-ground storage units house 3,500 containers of radioactive sludge and thousands of tonnes of spent fuel rods.[151]

2010

KETTENreAKTION! in Uetersen, Germany

On April 21, 2010, a dozen environmental organizations called on the United States Nuclear Regulatory Commission to investigate possible limitations in the AP1000 reactor design. These groups appealed to three federal agencies to suspend the licensing process because they believed containment in the new design is weaker than existing reactors.[152]

On April 24, 2010, about 120,000 people built a human chain (KETTENreAKTION!) between the nuclear plants at Krümmel and Brunsbüttel. In this way they were demonstrating against the plans of the German government to extend the life of nuclear power reactors.[153]

In May 2010, some 25,000 people, including members of peace organizations and 1945 atomic bomb survivors, marched for about two kilometers from downtown New York to the United Nations headquarters, calling for the elimination of nuclear weapons.[154] In September 2010, German government policy shifted back toward nuclear energy, and this generated some new anti-nuclear sentiment in Berlin and beyond.[155] On September 18, 2010, tens of thousands of Germans surrounded Chancellor Angela Merkel's office in an anti-nuclear demonstration that organisers said was the biggest of its kind since the 1986 Chernobyl disaster.[156] In October 2010, tens of thousands of people protested in Munich against the nuclear power policy of Angela Merkel's coalition government. The action was the largest anti-nuclear event in Bavaria for more than two decades.[157] In November 2010, there were

violent protests against a train carrying reprocessed nuclear waste in Germany. Tens of thousands of protesters gathered in Dannenberg to signal their opposition to the cargo. Around 16,000 police were mobilised to deal with the protests.[158][159]

In December 2010, some 10,000 people (mainly fishermen, farmers and their families) turned out to oppose the Jaitapur Nuclear Power Project in the Maharashtra state of India, amid a heavy police presence.[160]

In December 2010, five anti-nuclear weapons activists, including octogenarians and Jesuit priests, were convicted of conspiracy and trespass in Tacoma, USA. They cut fences at Naval Base Kitsap-Bangor in 2009 to protest submarine nuclear weapons, and reached an area near where Trident nuclear warheads are stored in bunkers. Members of the group could face up to 10 years in prison.[161]

2011

Anti-nuclear demonstration in Munich, Germany, March 2011.

In January 2011, five Japanese young people held a hunger strike for more than a week, outside the Prefectural Government offices in Yamaguchi City, to protest against the planned Kaminoseki Nuclear Power Plant near the environmentally sensitive Seto Inland Sea.[162]

Following the Fukushima Daiichi nuclear disaster, anti-nuclear opposition intensified in Germany. On 12 March 2011, 60,000 Germans formed a 45-km human chain from Stuttgart to the Neckarwestheim power plant.[163] On 14 March, 110,000 people protested in 450 other German towns, with opinion polls indicating 80% of Germans opposed the government's extension of nuclear power.[164] On March 15, 2011, Angela Merkel said that seven nuclear power plants which went online before 1980 would be closed and the time would be used to study speedier renewable energy commercialization.[165]

In March 2011, around 2,000 anti-nuclear protesters demonstrated in Taiwan for an immediate halt to the construction of the island's fourth nuclear power plant. The

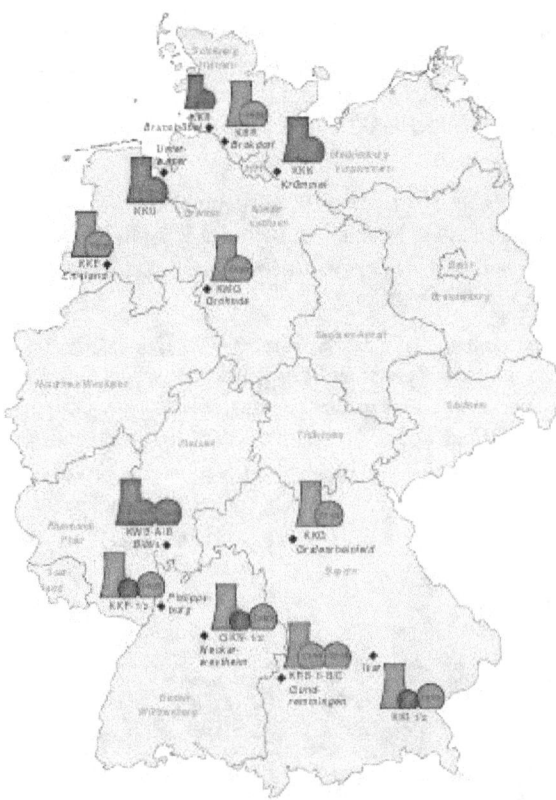

Eight of the seventeen operating reactors in Germany were permanently shut down following the March 2011 Fukushima nuclear disaster.

Buddhist monks of Nipponzan-Myōhōji protest against nuclear power near the Diet of Japan in Tokyo on April 5, 2011.

Human chain against nuclear plant in Turkey on 17 April 2011

Castor demonstration in Dannenberg, Germany, November 2011.

protesters were also opposed to plans to extend the lifespan of three existing nuclear plants.[166]

In March 2011, more than 200,000 people took part in anti-nuclear protests in four large German cities, on the eve of state elections. Organisers called it the largest anti-nuclear demonstration the country has seen.[167][168] Thousands of Germans demanding an end to the use of nuclear power took part in nationwide demonstrations on 2 April 2011. About 7,000 people took part in anti-nuclear protests in Bremen. About 3,000 people protested outside RWE's headquarters in Essen.[169]

Citing the Fukushima nuclear disaster, environmental activists at a U.N. meeting in April 2011 "urged bolder steps to tap renewable energy so the world doesn't have to choose between the dangers of nuclear power and the ravages of climate change".[170]

In mid-April, 17,000 people protested at two demonstrations in Tokyo against nuclear power.[171]

In India, environmentalists, local farmers and fishermen have been protesting for months over the planned Jaitapur Nuclear Power Project six-reactor complex, 420 km south of Mumbai. If built, it would be one of the world's largest nuclear power complexes. Protests have escalated following Japan's Fukushima nuclear disaster and during two days of violent rallies in April 2011, a local man was killed and dozens were injured.[172]

In May 2011, some 20,000 people turned out for Switzerland's largest anti-nuclear power demonstration in 25 years. Demonstrators marched peacefully near the Beznau Nuclear Power Plant, the oldest in Switzerland, which started operating 40 years ago.[173][174] Days after the anti-nuclear rally, Cabinet decided to ban the building of new nuclear power reactors. The country's five existing reactors would be allowed to continue operating, but "would not be replaced at the end of their life span".[22]

In May 2011, 5,000 people joined a carnival-like anti-nuclear protest in Taipei City. This was part of a nationwide "No Nuke Action" protest, urging the government to stop construction of a Fourth Nuclear Plant and pursue a more sustainable energy policy.[175]

On World Environment Day in June 2011, environmental groups demonstrated against Taiwan's nuclear power policy. The Taiwan Environmental Protection Union, together with 13 environmental groups and legislators, gathered in Taipei and protested against the nation's three operating nuclear power plants and the construction of a fourth plant.[176]

Three months after the Fukushima nuclear disaster, thousands of anti-nuclear protesters marched in Japan. Company workers, students, and parents with children rallied across Japan, "venting their anger at the government's handling of the crisis, carrying flags bearing the words 'No Nukes!' and 'No More Fukushima'." [177]

In August 2011, about 2,500 people including farmers and fishermen marched in Tokyo. They are suffering heavy losses following the Fukushima nuclear disaster, and called for prompt compensation from plant operator TEPCO and the government.[178]

In September 2011, anti-nuclear protesters, marching to the beat of drums, "took to the streets of Tokyo and other cities to mark six months since the March earthquake and tsunami and vent their anger at the government's handling of the nuclear crisis set off by meltdowns at the Fukushima power plant".[179] Protesters called for a complete shutdown of Japanese nuclear power plants and demanded a shift in government policy toward alternative sources of energy. Among the protestors were four young men who started a 10-day hunger strike to bring about change in Japan's nuclear policy.[179]

Tens of thousands of people marched in central Tokyo in September 2011, chanting "Sayonara nuclear power" and waving banners, to call on Japan's government to abandon atomic energy in the wake of the Fukushima nuclear disaster. Author Kenzaburō Ōe and musician Ryuichi Sakamoto were among the event's supporters.[180]

Since the March 2011 Japanese Fukushima nuclear disaster, "populations around proposed Indian NPP sites have launched protests that are now finding resonance around the country, raising questions about atomic energy as a clean and safe alternative to fossil fuels".[181] Assurances by Prime Minister Manmohan Singh that all safety measures

will be implemented, have not been heeded, and there have thus been mass protests against the French-backed 9900 MW Jaitapur Nuclear Power Project in Maharashtra and the 2000 MW Koodankulam Nuclear Power Plant in Tamil Nadu. The state government of West Bengal state has also refused permission to a proposed 6000 MW facility where six Russian reactors were to be built.[181] A Public Interest Litigation (PIL) has also been filed against the government's civil nuclear program at the apex Supreme Court. The PIL specifically asks for the "staying of all proposed nuclear power plants till satisfactory safety measures and cost-benefit analyses are completed by independent agencies".[181][182]

Michael Banach, the current Vatican representative to the International Atomic Energy Agency, told a conference in Vienna in September 2011 that the Japanese nuclear disaster created new concerns about the safety of nuclear plants globally. Auxiliary bishop of Osaka Michael Goro Matsuura said this serious nuclear power incident should be a lesson for Japan and other countries to abandon nuclear projects. He called on the worldwide Christian solidarity to provide wide support for this anti-nuclear campaign. Statements from bishops' conferences in Korea and the Philippines called on their governments to abandon atomic power. Nobel laureate Kenzaburō Ōe has said Japan should decide quickly to abandon its nuclear reactors.[183]

In the UK, in October 2011, more than 200 protesters blockaded the Hinkley Point C nuclear power station site. Members of the Stop New Nuclear alliance barred access to the site in protest at EDF Energy's plans to build two new reactors on the site.[184]

2012

Protest at Neckarwestheim, Germany, 11 March 2012.

In January 2012, 22 South Korean women's' groups appealed for a nuclear free future, saying they believe nuclear weapons and power reactors "threaten our lives, the lives of our families and all living creatures" . The women said they feel an enormous sense of crisis after the Fukushima nuclear disaster in March 2011, which demonstrated the destructive power of radiation in the disruption of human lives, environmental pollution, and food contamination.[185]

Thousands of demonstrators took to the streets of Yokohama, Japan, on January 14–15, 2012, to show their support for a nuclear power-free world. The demonstration showed that organized opposition to nuclear power has gained momentum following the Fukushima nuclear disaster. The most immediate demand of the demonstrators was for the protection of rights, including basic human rights such as health care, for those affected by the Fukushima accident.[186]

In January 2012, three hundred anti-nuclear protestors marched against plans to build a new nuclear power station at Wylfa in the UK. The march was organised by Pobl Atal Wylfa B, Greenpeace and Cymdeithas yr Iaith, which are supporting a farmer who is in dispute with Horizon.[187]

On the anniversary of the 11 March earthquake and tsunami, protesters across Japan called for the abolishment of nuclear power and nuclear reactors.[188] In Koriyama, Fukushima, 16,000 people called for the end of nuclear power. In Shizuoka Prefecture, 1,100 people appealed for the scrapping of the Hamaoka Nuclear Power Plant. In Tsuruga, Fukui, 1,200 people marched in the streets of the city of Tsuruga, the home of the Monju fast-breeder reactor prototype and other nuclear reactors. In Nagasaki and Hiroshima, anti-nuclear protesters and atomic-bomb survivors marched together and demanded that Japan should end its nuclear dependency.[188]

Austrian Chancellor Werner Faymann expects anti-nuclear petition drives to start in at least six European Union countries in 2012 in an effort to have the EU abandon nuclear power. Under the EU's Lisbon Treaty, petitions that attract at least one million signatures can seek legislative proposals from the European Commission, which would pave the way for anti-nuclear activists to garner support.[189]

In March 2012, about 2,000 people staged an anti-nuclear protest in Taiwan's capital following the massive tsunami that hit Japan one year ago. The protesters rallied in Taipei to renew calls for a nuclear-free island. They "want the government to scrap a plan to operate a newly constructed nuclear power plant - the fourth in densely populated Taiwan". Scores of aboriginal protesters "demanded the removal of 100,000 barrels of nuclear waste stored on their Orchid Island".[190]

In March 2012, hundreds of anti-nuclear demonstrators converged on the Australian headquarters of global mining giants BHP Billiton and Rio Tinto. The 500-strong march through southern Melbourne called for an end to uranium mining in Australia, and included speeches and performances by representatives of the expatriate Japanese community as well as Australia's Indigenous communities, who are concerned about the effects of uranium mining near tribal lands. There were also events in Sydney.[191]

In March 2012, South Korean environmental groups held a rally in Seoul to oppose nuclear power. Over 5,000 people attended, and the turnout was one of the largest in recent memory for an anti-nuclear rally. The demonstration demanded that President Lee Myung Bak abandon his policy of promoting nuclear power.[192]

In March 2012, police said they had arrested nearly 200 anti-nuclear activists who were protesting the restart of work at the long-stalled Indian Kudankulam nuclear power plant.[193]

In June 2012, tens of thousands of Japanese protesters participated in anti-nuclear power rallies in Tokyo and Osaka, over the government's decision to restart the first idled reactors since the Fukushima disaster, at Oi Nuclear Power Plant in Fukui Prefecture.[194]

2013

Anti-nuclear protesters in Taipei

Thousands of protesters marched in Tokyo on March 11, 2013 calling on the government to reject nuclear power.[195]

In March 2013, 68,000 Taiwanese protested across major cities against nuclear power and the island's fourth nuclear plant, which is under construction. Taiwan's three existing nuclear plants are near the ocean, and prone to geological fractures, under the island.[196]

In April 2013, thousands of Scottish campaigners, MSPs, and union leaders, rallied against nuclear weapons. The Scrap Trident Coalition wants to see an end to nuclear weapons, and says saved monies should be used for health, education and welfare initiatives. There was also a block-ade of the Faslane Naval Base, where Trident missiles are stored.[197]

2014

Anti-nuclear protesters shot with water cannons in Taiwan

In March 2014, around 130,000 Taiwanese marched for an anti-nuclear protest around Taiwan. They demanded that the government remove nuclear power plants in Tai-wan. The march came ahead of the 3rd anniversary of Fukushima disaster. Around 50,000 people marched in Taipei while another three separate events were held around other Taiwanese cities attended by around 30,000 peo-ple.[198][199] Among the participants are the organiza-tions from Green Citizen Action's Alliance, Homemakers United Foundation, Taiwan Association for Human Rights and Taiwan Environmental Protection Union.[200] Fac-ing on-going opposition and a host of delays, construction of the Lungmen Nuclear Power Plant was halted in April 2014.[201]

1.3.4 Casualties

Casualties during anti-nuclear protests include:

- On 9 December 1982, Norman Mayer, an American anti-nuclear weapons activist was shot and killed by the United States Park Police after threatening to blow up the Washington Monument, Washington, D.C., unless a national dialogue on the threat of nuclear weapons was seriously undertaken.

- On 10 July 1985, the flagship of Greenpeace, Rainbow Warrior, was sunk by French agents in New Zealand waters, and a Greenpeace photographer was killed.

Anti-nuclear demonstrations near Gorleben, Lower Saxony, Ger-many, 8 May 1996.

The ship was involved in protests against nuclear weapons testing at Mururoa Atoll. The French Gov-ernment initially denied any involvement with the sinking but eventually admitted its guilt in October 1985. Two French agents pleaded guilty to charges of manslaughter and the French Government paid $7 million in damages.[202]

- In 1990, two pylons holding high voltage power lines connecting the French and Italian grid were blown up by Italian eco-terrorists, and the attack is believed to have been directly in opposition against the Super-phénix.[203]

- In 2004, activist Sébastien Briat, who had tied himself to train tracks in front of a shipment of reprocessed nuclear waste, was run over by the wheels of the train. The event happened in Avricourt, France and the fuel (totaling 12 containers) was from a German plant, on its way to be reprocessed.[204]

1.4 Impact

1.4.1 Impact on popular culture

See also: List of films about nuclear issues

Beginning in the 1950s, anti-nuclear ideas received cov-erage in the popular media with novels such as *Fail-Safe* and feature films such as *Godzilla (1954),Dr. Strangelove or: How I Learned to Stop Worrying and Love the Bomb* (1964), *The China Syndrome* (1979), *Silkwood* (1983), and *The Rainbow Warrior* (1992).

Dr. Strangelove explored   "what might happen within the Pentagon ... if some maniac Air Force general should sud-

Montage of film stills from the International Uranium Film Festival.

denly order a nuclear attack on the Soviet Union". One reviewer called the movie "one of the cleverest and most incisive satiric thrusts at the awkwardness and folly of the military that has ever been on the screen". .[205]

The China Syndrome has been described as a "gripping 1979 drama about the dangers of nuclear power" which had an extra impact when the real-life accident at the Three Mile Island nuclear plant occurred several weeks after the film opened. Jane Fonda plays a TV reporter who witnesses a near-meltdown (the "China syndrome" of the title) at a local nuclear plant, which was averted by a quick-thinking engineer, played by Jack Lemmon. The plot suggests that corporate greed and cost-cutting "have led to potentially deadly faults in the plant's construction". .[206]

Silkwood was inspired by the true-life story of Karen Silkwood, who died in a suspicious car accident while investigating alleged wrongdoing at the Kerr-McGee plutonium plant where she worked.[25]

Dark Circle is a 1982 American documentary film that focuses on the connections between the nuclear weapons and the nuclear power industries, with a strong emphasis on the individual human and protracted U.S. environmental costs involved. A clear point made by the film is that while only two bombs were dropped on Japan, many hundreds were exploded in the United States. The film won the Grand Prize for documentary at the Sundance Film Festival and received a national Emmy Award for "Outstanding individual achievement in news and documentary." [207] For the

opening scenes and about half of its length, the film focuses on the Rocky Flats Plant and its plutonium contamination of the area's environment.

Ashes to Honey (ミツバチの羽音と地球の回転 *Mitsubashi no haoto to chikyū no kaiten*), (literally "*Humming of Bees and Rotation of the Earth*") is a Japanese documentary directed by Hitomi Kamanaka and released in 2010.[208] It is the third in Kamanaka's trilogy of films on the problems of nuclear power and radiation, preceded by *Hibakusha at the End of the World* (also known as *Radiation: A Slow Death*) and *Rokkasho Rhapsody*.[209]

Nuclear Tipping Point is a 2010 documentary film produced by the Nuclear Threat Initiative. It features interviews with four American government officials who were in office during the Cold War period, but are now advocating for the elimination of nuclear weapons. They are: Henry Kissinger, George Shultz, Sam Nunn, and William Perry.[210]

Musicians United for Safe Energy (MUSE) was a musical group founded in 1979 by Jackson Browne, Graham Nash, Bonnie Raitt, and John Hall, following the Three Mile Island nuclear accident. The group organized a series of five *No Nukes* concerts held at Madison Square Garden in New York City in September 1979. On September 23, 1979, almost 200,000 people attended a large anti-nuclear rally staged by MUSE on the then-empty north end of the Battery Park City landfill in New York.[117] The album *No Nukes*, and a film, also titled *No Nukes*, were both released in 1980 to document the performances.

In 2007, Bonnie Raitt, Graham Nash, and Jackson Browne, as part of the No Nukes group, recorded a music video of the Buffalo Springfield song "For What It's Worth".[211][212]

1.4.2 Impact on policy

pair of billboards in Davis, California advertising its nuclear-free policy.

See also: Nuclear energy policy, Nuclear power by country.

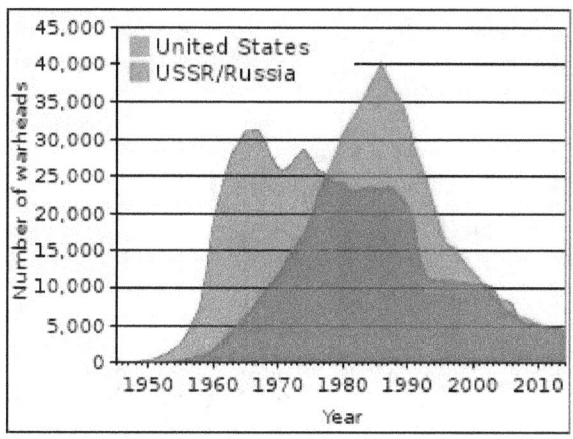

U.S. and USSR/Russian nuclear weapons stockpiles, 1945–2005.

Nuclear free zone, List of canceled nuclear plants in the United States, and Anti-nuclear movement in Australia

The *Bulletin of the Atomic Scientists* is a nontechnical on-line magazine that has been published continuously since 1945, when it was founded by former Manhattan Project physicists after the atomic bombings of Hiroshima and Nagasaki. The *Bulletin's* primary aim is to inform the public about nuclear policy debates while advocating for the international control of nuclear weapons. One of the driving forces behind the creation of the *Bulletin* was the amount of public interest surrounding atomic energy at the dawn of the atomic age. In 1945 the public interest in atomic warfare and weaponry inspired contributors to the *Bulletin* to attempt to inform those interested about the dangers and destruction that atomic war could bring about.[213] In the 1950s, the *Bulletin* was involved in the formation of the Pugwash Conferences on Science and World Affairs, annual conferences of scientists concerned about nuclear proliferation.

Historian Lawrence S. Wittner has argued that anti-nuclear sentiment and activism led directly to government policy shifts about nuclear weapons. Public opinion influenced policymakers by limiting their options and also by forcing them to follow certain policies over others. Wittner credits public pressure and anti-nuclear activism with "Truman's decision to explore the Baruch Plan, Eisenhower's efforts towards a nuclear test ban and the 1958 testing moratorium, and Kennedy's signing of the Partial Test Ban Treaty".[214]

In terms of nuclear power, *Forbes* magazine, in the September 1975 issue, reported that "the anti-nuclear coalition has been remarkably successful ... [and] has certainly slowed the expansion of nuclear power." [27] California has banned the approval of new nuclear reactors since the late 1970s because of concerns over waste disposal,[215]

and some other U.S. states have a moratorium on construction of nuclear power plants.[216] Between 1975 and 1980, a total of 63 nuclear units were canceled in the USA. Anti-nuclear activities were among the reasons, but the primary motivations were the overestimation of future demand for electricity and steadily increasing capital costs, which made the economics of new plants unfavorable.[217]

The proliferation of nuclear weapons became a presidential priority issue for the Carter Administration in the late 1970s.[218] To deal with proliferation problems, President Carter promoted stronger international control over nuclear technology, including nuclear reactor technology. Although a strong supporter of nuclear power generally, Carter turned against the breeder reactor because the plutonium it produced could be diverted into nuclear weapons.[218]

For many years after the 1986 Chernobyl disaster nuclear power was off the policy agenda in most countries. In recent years, intense public relations activities by the nuclear industry, increasing evidence of climate change and failures to address it, have brought nuclear power issues back to the forefront of policy discussion in the nuclear renaissance countries.[53][219] But some countries are not prepared to expand nuclear power and are still divesting themselves of their nuclear legacy, through nuclear power phase-out legislation.[219]

Under the *New Zealand Nuclear Free Zone, Disarmament, and Arms Control Act 1987*, all territorial sea and land of New Zealand is declared a nuclear free zone. Nuclear-powered and nuclear-armed ships are prohibited from entering the country's territorial waters. Dumping of foreign radioactive waste and development of nuclear weapons in the country is outlawed.[220] Despite common misconception, this act does not make nuclear power plants illegal, nor does it make radioactive medical treatments produced in overseas reactors illegal.[221] A 2008 survey shows that 19% of New Zealanders favour nuclear power as the best energy source, while 77% prefer wind power as the best energy source.[222]

On 26 February 1990, FW de Klerk issued orders to terminate the country's nuclear weapons programme, which until then had been a state secret.[223] South Africa becomes the first country in the world to voluntary give-up its nuclear weapons programme.

Ireland, in 1999, had no plans to change its non-nuclear stance and pursue nuclear power in the future.[224]

In the United States, the Navajo Nation forbids uranium mining and processing in its land.[225]

In the United States, a 2007 University of Maryland survey showed that 73 percent of the public surveyed favours the elimination of all nuclear weapons, 64 percent support removing all nuclear weapons from high alert, and 59 percent support reducing U.S. and Russian nuclear stockpiles to 400 weapons each. Given the unpopularity of nuclear weapons, U.S. politicians have been wary of supporting new nuclear programs. Republican-dominated congresses "have defeated the Bush administration's plan to build so-called 'bunker-busters' and 'mini-nukes'." [135]

The Megatons to Megawatts Program converts weapons-grade material from nuclear warheads into fuel for nuclear power plants. [226]

Thirty-one countries operate nuclear power plants. [227] Nine nations possess nuclear weapons: [228]

> Today, some 26,000 nuclear weapons remain in the arsenals of the nine nuclear powers, with thousands on hair-trigger alert. Although U.S., Russian, and British nuclear arsenals are shrinking in size, those in the four Asian nuclear nations —China, India, Pakistan, and North Korea—are growing, in large part because of tensions among them. This Asian arms race also has possibilities of bringing Japan into the nuclear club. [97]

During Barack Obama's successful U.S. presidential election campaign, he advocated the abolition of nuclear weapons. Since his election he has reiterated this goal in several major policy addresses. [97] In 2010, the Obama administration negotiated a new weapons accord with Russia for a reduction of the maximum number of deployed nuclear weapons on each side from 2,200 to between 1,500 and 1,675—a reduction of some 30 percent. In addition, President Obama has committed $15 billion over the next five years to improving the safety of the nuclear weapons stockpile. [229]

Following the 2011 Japanese nuclear accidents, the Italian government put a one-year moratorium on plans to revive nuclear power. [230] On 11—12 June 2011, Italian voters passed a referendum to cancel plans for new reactors. Over 94% of the electorate voted in favor of the construction ban, with 55% of the eligible voters participating, making the vote binding. [231]

German Chancellor Angela Merkel's coalition announced on May 30, 2011, that Germany's 17 nuclear power stations will be shut down by 2022, in a policy reversal following Japan's Fukushima I nuclear accidents and anti-nuclear protests within Germany. Seven of the German power stations were closed temporarily in March, and they will remain off-line and be permanently decommissioned. An eighth was already off line, and will stay so. [232]

As of 2011, countries such as Australia, Austria, Denmark, Greece, Ireland, Italy, Latvia, Liechtenstein, Luxembourg, Malta, Portugal, Israel, Malaysia,

New Zealand, and Norway remain opposed to nuclear power. [18] [19] Germany and Switzerland are phasing-out nuclear power. [19] [22]

1.4.3 Public opinion surveys on nuclear issues

Main article: Public opinion on nuclear issues

In 2005, the International Atomic Energy Agency presented the results of a series of public opinion surveys in the *Global Public Opinion on Nuclear Issues* report. [233] Majorities of respondents in 14 of the 18 countries surveyed believe that the risk of terrorist acts involving radioactive materials at nuclear facilities is high, because of insufficient protection. While majorities of citizens generally support the continued use of existing nuclear power reactors, most people do not favour the building of new nuclear plants, and 25% of respondents feel that all nuclear power plants should be closed down. [233] Stressing the climate change benefits of nuclear energy positively influences 10% of people to be more supportive of expanding the role of nuclear power in the world, but there is still a general reluctance to support the building of more nuclear power plants. [233]

There is little support across the world for building new nuclear reactors, a 2011 poll for the BBC indicates. The global research agency GlobeScan, commissioned by BBC News, polled 23,231 people in 23 countries from July to September 2011, several months after the Fukushima nuclear disaster. In countries with existing nuclear programmes, people are significantly more opposed than they were in 2005, with only the UK and US bucking the trend. Most believe that boosting energy efficiency and renewable energy can meet their needs. [42]

1.5   Criticism

See also: List of pro-nuclear environmentalists and List of nuclear power groups

Attempts to reach political agreement on effective policies for climate change continue, and pro-nuclear environmentalists seek to reverse the traditionally anti-nuclear attitudes of environmentalists. Filmmaker Rob Stone's *Pandora's Promise* (2013) is a good example of this trend. [235]

Some environmentalists criticise the anti-nuclear movement for under-stating the environmental costs of fossil fuels and non-nuclear alternatives, and overstating the environmental costs of nuclear energy. [236] [237] Of the numerous nuclear experts who have offered their expertise in addressing controversies, Bernard Cohen, Professor Emer-

Stewart Brand at a 2010 debate, "Does the world need nuclear energy?"[234]

itus of Physics at the University of Pittsburgh, is likely the most frequently cited. In his extensive writings he examines the safety issues in detail. He is best known for comparing nuclear safety to the relative safety of a wide range of other phenomena.[238][239]

Anti-nuclear activists are accused of encouraging radiophobic emotions among the public. *The War Against the Atom* (Basic Books, 1982) Samuel MacCracken of Boston University argued that in 1982, 50,000 deaths per year could be attributed directly to non-nuclear power plants, if fuel production and transportation, as well as pollution, were taken into account. He argued that if non-nuclear plants were judged by the same standards as nuclear ones, each US non-nuclear power plant could be held responsible for about 100 deaths per year.[240]

The Nuclear Energy Institute[241] (NEI) is the main lobby group for companies doing nuclear work in the USA, while most countries that employ nuclear energy have a national industry group. The World Nuclear Association is the only global trade body. In seeking to counteract the arguments of nuclear opponents, it points to independent studies that quantify the costs and benefits of nuclear energy and compares them to the costs and benefits of alternatives. NEI sponsors studies of its own, but it also references studies performed for the World Health Organisation,[242] for the International Energy Agency,[243] and by university researchers.[244]

Critics of the anti-nuclear movement point to independent studies that show that the capital resources required for renewable energy sources are higher than those required for nuclear power.[243]

Some people, including former opponents of nuclear energy, criticise the movement on the basis of the claim that nuclear power is necessary for reducing carbon dioxide emissions. These individuals include James Lovelock,[236] originator of the Gaia hypothesis, Patrick Moore,[237] a co-founder of Greenpeace and former director of Greenpeace International, George Monbiot and Stewart Brand, creator of the Whole Earth Catalog.[245][246] Lovelock goes further to refute claims about the danger of nuclear energy and its waste products.[247] In a January 2008 interview, Moore said that "It wasn't until after I'd left Greenpeace and the climate change issue started coming to the forefront that I started rethinking energy policy in general and realised that I had been incorrect in my analysis of nuclear as being some kind of evil plot." [248]

Some anti-nuclear organisations have acknowledged that their positions are subject to review.[249] Nuclear-energy opponents take the position that militant environmentalist organisations have not changed their views:

> While some environmentalists, in the interests of reducing the CO_2 emissions associated with burning carbon-based fuels, have switched from anti- to pro-nuclear power in recent years, it is clear that many —if not most —of the militant environmentalist organizations remain adamantly opposed to the expansion of nuclear power. Many even propose decommissioning and dismantling the existing nuclear power electrical plants.[250]

In April 2007, Dan Becker, Director of Global Warming for the Sierra Club, declared, "Switching from dirty coal plants to dangerous nuclear power is like giving up smoking cigarettes and taking up crack."[251] James Lovelock criticizes holders of such a view: "Opposition to nuclear energy is based on irrational fear fed by Hollywood-style fiction, the Green lobbies and the media." ". . .I am a Green and I entreat my friends in the movement to drop their wrongheaded objection to nuclear energy."[252]

George Monbiot, an English writer known for his environmental and political activism, once expressed deep antipathy to the nuclear industry.[253] He finally rejected his later neutral position regarding nuclear power in March 2011. Although he "still loathe[s] the liars who run the nuclear industry",[254] Monbiot now advocates its use, having been convinced of its relative safety by what he considers the limited effects of the 2011 Japan tsunami on nuclear reactors in the region." [254] Subsequently, he has harshly condemned the anti-nuclear movement, writing that it "has misled the world about the impacts of radiation on human health ... made [claims] ungrounded in science, unsupportable when challenged and wildly wrong." He singled out Helen Caldicott for, he wrote, making unsourced and inaccurate claims, dismissing contrary evidence as part of a cover-up, and overstating the death toll from the Chernobyl disaster by a factor of more than 140." [255]

1.6 See also

- ⇐ *The Atomic Age* – Wikipedia book
- Doomsday Clock
- Environmental movement
- John Gofman
- International Day against Nuclear Tests
- List of Chernobyl-related articles
- List of nuclear whistleblowers
- List of peace activists
- List of states with nuclear weapons
- Lists of nuclear disasters and radioactive incidents
- Gregory Minor
- Nuclear-free zone
- Nuclear organizations (Wikipedia category)
- Nuclear power phase-out
- Nuclear safety
- Nuclear-Free Future Award
- Nuclear weapons in popular culture
- *Pandora's Promise*
- The Bomb (film)
- The Ribbon International
- Uranium
- Vulnerability of nuclear plants to attack
- World Association of Nuclear Operators

1.7 Notes and references

[1] Herbert P. Kitschelt. Political Opportunity and Political Protest: Anti-Nuclear Movements in Four Democracies *British Journal of Political Science*, Vol. 16, No. 1, 1986, p. 71.

[2] Fox Butterfield. Professional Groups Flocking to Antinuclear Drive, *The New York Times*, March 27, 1982.

[3] William A. Gamson and Andre Modigliani. Media Coverage and Public Opinion on Nuclear Power Archived 24 March 2012 at the Wayback Machine., *American Journal of Sociology*, Vol. 95, No. 1, July 1989, p. 7.

[4] John Barry and E. Gene Frankland. *International Encyclopedia of Environmental Politics*, 2001, p. 24.

[5] Jerry Brown and Rinaldo Brutoco (1997). *Profiles in Power: The Anti-nuclear Movement and the Dawn of the Solar Age*, Twayne Publishers, pp. 191-192.

[6] Wolfgang Rudig (1990). *Anti-nuclear Movements: A World Survey of Opposition to Nuclear Energy*, Longman, p. 54-55.

[7] Paula Garb. Review of Critical Masses, *Journal of Political Ecology*, Vol 6, 1999.

[8] Wolfgang Rudig (1990). *Anti-nuclear Movements: A World Survey of Opposition to Nuclear Energy*, Longman, p. 52.

[9] Stephen Mills and Roger Williams (1986). Public Acceptance of New Technologies Routledge, pp. 375-376.

[10] Robert Gottlieb (2005). Forcing the Spring: The Transformation of the American Environmental Movement, Revised Edition, Island Press, USA, p. 237.

[11] Jim Falk (1982). *Global Fission: The Battle Over Nuclear Power*, Oxford University Press, pp. 95-96.

[12] Leo Hickman (28 November 2012). "Nuclear lobbyists wined and dined senior civil servants, documents show". *The Guardian*.

[13] Diane Farseta (September 1, 2008). "The Campaign to Sell Nuclear". *Bulletin of the Atomic Scientists*. pp. 38–56.

[14] Jonathan Leake. "The Nuclear Charm Offensive" *New Statesman*, 23 May 2005.

[15] Union of Concerned Scientists. Nuclear Industry Spent Hundreds of Millions of Dollars Over the Last Decade to Sell Public, Congress on New Reactors, New Investigation Finds Archived 27 November 2013 at the Wayback Machine. News Center, February 1, 2010.

[16] Nuclear group spent $460,000 lobbying in 4Q Archived 23 October 2012 at the Wayback Machine. *Business Week*, March 19, 2010.

[17] "Japan crisis rouses anti-nuclear passions globally". *Washington Post*. March 16, 2011.

[18] "Nuclear power: When the steam clears". *The Economist*. March 24, 2011.

[19] Duroyan Fertl (June 5, 2011). "Germany: Nuclear power to be phased out by 2022". *Green Left*.

[20] Erika Simpson and Ian Fairlie, Dealing with nuclear waste is so difficult that phasing out nuclear power would be the best option, Lfpress, February 26, 2016.

[21] "Difference Engine: The nuke that might have been". *The Economist*. Nov 11, 2013.

[22] James Kanter (May 25, 2011). "Switzerland Decides on Nuclear Phase-Out". *New York Times*.

[23] "Sunday Dialogue: Nuclear Energy, Pro and Con". *New York Times*. February 25, 2012.

[24] Union-Tribune Editorial Board (March 27, 2011). "The nuclear controversy". *Union-Tribune*.

[25] Robert Benford. The Anti-nuclear Movement (book review) *American Journal of Sociology*, Vol. 89, No. 6, (May 1984), pp. 1456-1458.

[26] James J. MacKenzie. Review of The Nuclear Power Controversy by Arthur W. Murphy *The Quarterly Review of Biology*, Vol. 52, No. 4 (Dec., 1977), pp. 467-468.

[27] Walker, J. Samuel (2004). *Three Mile Island: A Nuclear Crisis in Historical Perspective* (Berkeley: University of California Press), pp. 10-11.

[28] Jim Falk (1982). *Global Fission: The Battle Over Nuclear Power*, Oxford University Press.

[29] Woo, Elaine (January 30, 2011). "Dagmar Wilson dies at 94: organizer of women's disarmament protesters". *Los Angeles Times*.

[30] Hevesi, Dennis (January 23, 2011). "Dagmar Wilson, Anti-Nuclear Leader, Dies at 94". *The New York Times*.

[31] Togzhan Kassenova (28 September 2009). "The lasting toll of Semipalatinsk's nuclear testing". *Bulletin of the Atomic Scientists*.

[32] Frida Berrigan. The New Anti-Nuclear Movement *Foreign Policy in Focus*, April 16, 2010.

[33] Philip Yam. Nuclear Exchange, *Scientific American*, June 2010, p. 24.

[34] Alan Robock and Owen Brian Toon. Local Nuclear War, Global Suffering, *Scientific American*, January 2010, p. 74-81.

[35] For the full text of the Advisory Opinion

[36] Hugh Gusterson (30 March 2012). "The new abolitionists". *Bulletin of the Atomic Scientists*.

[37] "Nuclear endgame: The growing appeal of zero". *The Economist*. June 16, 2011.

[38] Richard Schiffman (12 March 2013). "Two years on, America hasn't learned lessons of Fukushima nuclear disaster". *The Guardian*.

[39] Martin Fackler (June 1, 2011). "Report Finds Japan Underestimated Tsunami Danger". *New York Times*.

[40] Sustainable Development Commission. Public engagement and nuclear power

[41] Sustainable Development Commission. Is Nuclear the Answer? p. 12.

[42] Richard Black (25 November 2011). "Nuclear power 'gets little public support worldwide'". *BBC News*.

[43] M.V. Ramana (July 2011). "Nuclear power and the public". *Bulletin of the Atomic Scientists*. p. 44.

[44] Mark Cooper (July 2011). "The implications of Fukushima: The US perspective". *Bulletin of the Atomic Scientists*. p. 9.

[45] Helen Caldicott (2006). *Nuclear Power is Not the Answer to Global Warming or Anything Else*, Melbourne University Press. ISBN 0-522-85251-3, p. xvii

[46] Brian Martin. Opposing nuclear power: past and present, *Social Alternatives*, Vol. 26, No. 2, Second Quarter 2007, pp. 43-47.

[47] Bibi van der Zee (22 March 2011). "Japan nuclear crisis puts UK public off new power stations". *The Guardian*.

[48] Matthew L. Wald. Edging Back to Nuclear Power *New York Times*, April 21, 2010.

[49] Prof. Wade Allison. "Why radiation is safe & all nations should embrace nuclear technology".

[50] M. V. Ramana (July 2011). "Nuclear power and the public". *Bulletin of the Atomic Scientists*. pp. 47–48.

[51] M.V. Ramana (July 2011). "Nuclear power and the public". *Bulletin of the Atomic Scientists*. p. 46.

[52] M. V. Ramana (July 2011). "Nuclear power and the public". *Bulletin of the Atomic Scientists*. p. 48.

[53] Mark Diesendorf. Is nuclear energy a possible solution to global warming? Archived 22 July 2012 at the Wayback Machine.

[54] Kurt Kleiner. Nuclear energy: assessing the emissions *Nature Reports*, Vol. 2, October 2008, pp. 130-131.

[55] Mark Diesendorf (2007). *Greenhouse Solutions with Sustainable Energy*, University of New South Wales Press, p. 252.

[56] Richard Black (10 November 2011). "EDF fined for spying on Greenpeace nuclear campaign". BBC. Retrieved 11 November 2011.

[57] Hanna Gersmann (10 November 2011). "EDF fined € 1.5m for spying on Greenpeace". *The Guardian*. Retrieved 11 November 2011.

[58] *Contesting the Future of Nuclear Power, Non-Nuclear Futures, Reaction Time.*

[59] Nuclear fusion reactor project in France: an expensive and senseless nuclear stupidity|Greenpeace International

[60] France Wins Nuclear Fusion Plant|Germany|Deutsche Welle|28 June 2005

[61] Jim Green (2012). "New Reactor Types - pebble bed, thorium, plutonium, fusion". *Friends of the Earth*.

[62] World Nuclear Association (2005). "Nuclear Fusion Power".

[63] W Wayt Gibbs (30 December 2013). "Triple-threat method sparks hope for fusion". *Nature*.

[64] Jim Green. "Medical radioisotope supply options for Australia". *Friends of the Earth*.

[65] Robert F. Service (20 February 2012). "Nuclear Reactors Not Needed to Make the Most Common Medical Isotope". *Science Now*.

[66] Edwin Cartlidge (18 November 2011). "Saving for a rainy day". *Science (Vol 334)*: 922–924.

[67] Greenpeace International and European Renewable Energy Council (January 2007). *Energy Revolution: A Sustainable World Energy Outlook*, p. 7.

[68] IEA Renewable Energy Working Party (2002). *Renewable Energy... into the mainstream*, p. 9.

[69] Greenpeace International and European Renewable Energy Council (January 2007). *Energy Revolution: A Sustainable World Energy Outlook*

[70] International Energy Agency (2007). *Renewables in global energy supply: An IEA facts sheet* (PDF) OECD, 34 pages.

[71] Lester R. Brown. *Plan B 4.0: Mobilizing to Save Civilization*, Earth Policy Institute, 2009, p. 135.

[72] "GWEC Global Wind Statistics 2014" (PDF). GWEC. 10 February 2015.

[73] The World Wind Energy Association (2014). *2014 Half-year Report*. WWEA. pp. 1–8.

[74] Wind in power: 2015 European statistics- EWEA

[75] Tam Hunt (9 March 2015). "The Solar Singularity Is Nigh". *Greentech Media*. Retrieved 29 April 2015.

[76] "World largest solar thermal plant syncs to the grid". Spectrum.ieee.org. Retrieved 28 November 2014.

[77] "World's Largest Solar Thermal Power Project at Ivanpah Achieves Commercial Operation", NRG press release, February 13, 2014.

[78] International Energy Agency. IEA urges governments to adopt effective policies based on key design principles to accelerate the exploitation of the large potential for renewable energy 29 September 2008.

[79] Mark A. Delucchi and Mark Z. Jacobson (2011). "Providing all global energy with wind, water, and solar power, Part II: Reliability, system and transmission costs, and policies" (PDF). *Energy Policy*. Elsevier Ltd. pp. 1170–1190.

[80] Herbert P. Kitschelt. Political Opportunity and Political Protest: Anti-Nuclear Movements in Four Democracies *British Journal of Political Science*, Vol. 16, No. 1, 1986, p. 67.

[81] Wolfgang Rudig (1990). *Anti-nuclear Movements: A World Survey of Opposition to Nuclear Energy*, Longman, p. 1.

[82] Lutz Mez, Mycle Schneider and Steve Thomas (Eds.) (2009). *International Perspectives of Energy Policy and the Role of Nuclear Power*, Multi-Science Publishing Co. Ltd. p. 279.

[83] Matthew L. Wald. Nuclear Agency's Chief Praises Watchdog Groups, *The New York Times*, June 23, 1992.

[84] Lawrence S. Wittner (2009). *Confronting the Bomb: A Short History of the World Nuclear Disarmament Movement*, Stanford University Press, pp. 164-165.

[85] "About Friends of the Earth International". Friends of the Earth International. Archived from the original on 4 May 2009. Retrieved 2009-06-25.

[86] FOE shift on nuclear power

[87] http://www.globalzero.org?name=2.htm&id=2

[88] United Nations, Department of Public Information, Non-Governmental Organizations

[89] Greenpeace International: Greenpeace worldwide

[90] Profile from Helix Magazine

[91] Henry Mhara (Oct 17, 2011). "Coltart elected anti-nuclear organisation president". *News Day*.

[92] Lawrence S. Wittner (2009). *Confronting the Bomb: A Short History of the World Nuclear Disarmament Movement*, Stanford University Press, p. 128.

[93] Lawrence S. Wittner (2009). *Confronting the Bomb: A Short History of the World Nuclear Disarmament Movement*, Stanford University Press, p. 125.

[94] World's best-known protest symbol turns 50, *BBC News*, 20 March 2008.

[95] Nuclear Power in Germany: A Chronology

[96] Paul Lewis. David McTaggart, a Builder of Greenpeace, Dies at 69 *The New York Times*, March 24, 2001.

[97] Lawrence S. Wittner. Nuclear Disarmament Activism in Asia and the Pacific, 1971-1996 *The Asia-Pacific Journal*, Vol. 25-5-09, June 22, 2009.

[98] Lutz Mez, Mycle Schneider and Steve Thomas (Eds.) (2009). *International Perspectives of Energy Policy and the Role of Nuclear Power*, Multi-Science Publishing Co. Ltd. p. 371.

[99] Wolfgang Rudig (1990). *Anti-nuclear Movements: A World Survey of Opposition to Nuclear Energy*, Longman, p. 138.

[100] Dorothy Nelkin and Michael Pollak (1982). *The Atom Besieged: Antinuclear Movements in France and Germany*, ASIN: B0011LXE0A. p. 3.

[101] Yok-shiu F. Lee and Alvin Y. So (1999). Asia's Environmental Movements: Comparative Perspectives M.E. Sharpe, pp. 160-161.

[102] West Germans Clash at Site of A-Plant *New York Times*, March 1, 1981 p. 17.

[103] Violence Mars West German Protest *New York Times*, March 1, 1981 p. 17

[104] Lawrence S. Wittner. "Disarmament movement lessons from yesteryear". *Bulletin of the Atomic Scientists*, 27 July 2009.

[105] Paul Brown, Shyama Perera and Martin Wainwright. Protest by CND stretches 14 miles *The Guardian*, 2 April 1983.

[106] John Greenwald. Energy and Now, the Political Fallout, *TIME*, June 2, 1986.

[107] Marco Giugni (2004). Social protest and policy change p. 55.

[108] Hundreds of Marchers Hit Washington in Finale of Nation-wide Peace March *Gainesville Sun*, November 16, 1986.

[109] "Semipalatinsk: 60 years later (collection of articles)". *Bulletin of the Atomic Scientists*, September 2009. Retrieved 2009-10-01. External link in |publisher= (help)

[110] World: Asia-Pacific: Kazakh anti-nuclear movement celebrates tenth anniversary *BBC News*, February 28, 1999.

[111] Matthew Chance. Inside the nuclear underworld: Deformity and fear *CNN.com*, August 31, 2007.

[112] Protests Stop Devastating Nuclear Tests: The Nevada-Semipalatinsk Anti-Nuclear Movement in Kazakhstan

[113] Nuclear-Free Future Award. "World Uranium Hearing, a Look Back".

[114] Nuclear-Free Future Award. "The Declaration of Salzberg"

[115] Giugni, Marco (2004). *Social Protest and Policy Change: Ecology, Antinuclear, and Peace Movements* p. 44.

[116] Giugni, Marco (2004). *Social Protest and Policy Change: Ecology, Antinuclear, and Peace Movements* p. 45.

[117] Herman, Robin (September 24, 1979). "Nearly 200,000 Rally to Protest Nuclear Energy". *New York Times*. p. B1.

[118] Williams, Estha. Nuke Fight Nears Decisive Moment *Valley Advocate*, August 28, 2008.

[119] Jonathan Schell. The Spirit of June 12 *The Nation*, July 2, 2007.

[120] 1982 - a million people march in New York City Archived 16 June 2010 at the Wayback Machine.

[121] Harvey Klehr. Far Left of Center: The American Radical Left Today Transaction Publishers, 1988, p. 150.

[122] 1,400 Anti-nuclear protesters arrested *Miami Herald*, June 21, 1983.

[123] Hundreds of Marchers Hit Washington in Finale of Nation-waide Peace March *Gainesville Sun*, November 16, 1986.

[124] Robert Lindsey. 438 Protesters are Arrested at Nevada Nuclear Test Site *New York Times*, February 6, 1987.

[125] 493 Arrested at Nevada Nuclear Test Site *New York Times*, April 20, 1992.

[126] Lance Murdoch. Pictures: New York MayDay anti-nuke/war march Archived 28 July 2011 at the Wayback Machine. *IndyMedia*, 2 may 2005.

[127] Anti-Nuke Protests in New York *Fox News*, May 2, 2005.

[128] Protest against nuclear reactor *Chicago Tribune*, October 16, 2008.

[129] Southeast Climate Convergence occupies nuclear facility *Indymedia UK*, August 8, 2008.

[130] Anti-Nuclear Renaissance: A Powerful but Partial and Tentative Victory Over Atomic Energy

[131] Mark Cooper (18 June 2013). "Nuclear aging: Not so graceful". *Bulletin of the Atomic Scientists*.

[132] Matthew Wald (June 14, 2013). "Nuclear Plants, Old and Uncompetitive, Are Closing Earlier Than Expected". *New York Times*.

[133] Diane Farseta (September 1, 2008). "The Campaign to Sell Nuclear". *Bulletin of the Atomic Scientists*. **64** (4): 38–56. doi:10.2968/064004009.

[134] Thousands march in Paris anti-nuclear protest *ABC News*, January 18, 2004.

[135] Lawrence S. Wittner. A rebirth of the anti-nuclear weapons movement? Portents of an anti-nuclear upsurge *Bulletin of the Atomic Scientists*, 7 December 2007.

[136] "French protests over EPR". Nuclear Engineering International. 2007-04-03.

[137] "France hit by anti-nuclear protests". Evening Echo. 2007-04-03.

[138] "Thousands protest against Indonesian nuclear plant". ABC News. June 12, 2007.

[139] Japan Nuclear Plant Not Safe to Restart After Quake, Group Says

[140] Close Kashiwazaki-Kariwa Nuclear Power Plant

[141] "International Conference on Nuclear Disarmament". February 2008.

[142] The Renaissance of the Anti-Nuclear Movement Spiegel Online, 11/10/2008.

[143] Anti-Nuclear Protest Reawakens: Nuclear Waste Reaches German Storage Site Amid Fierce Protests Spiegel Online, 11/11/2008.

[144] Simon Sturdee. Police break up German nuclear protest The Age, November 11, 2008.

[145] Green boost in European elections may trigger nuclear fight, Nature, 9 June 2009.

[146] More than 30 arrests at Aldermaston anti-nuclear protest The Guardian, 28 October 2008.

[147] Maryann Spoto. Nuclear license renewal sparks protest Star-Ledger. June 02, 2009.

[148] Anti-nuclear protesters reach capitol Rutland Herald, January 14, 2010.

[149] Eric Kirschbaum. Anti-nuclear rally enlivens German campaign Reuters, September 5, 2009.

[150] 50,000 join anti-nuclear power march in Berlin The Local, September 5, 2009.

[151] Roger Boyes. German nuclear programme threatened by old mine housing waste The Times, January 22, 2010.

[152] "Groups say new Vogtle Reactors need study". August Chronicle. Retrieved 2010-04-24.

[153] "German nuclear protesters form 75-mile human chain". Reuters. 2010-04-25. Archived from the original on 27 April 2010. Retrieved 2010-04-25.

[154] A-bomb survivors join 25,000-strong anti-nuclear march through New York Mainichi Daily News, May 4, 2010.

[155] James Norman and Dave Sweeney. Germany's 'hot autumn' of nuclear discontent Sydney Morning Herald, September 14, 2010.

[156] Dave Graham. Thousands of Germans attend anti-nuclear protest National Post, September 18, 2010.

[157] Tens of thousands take part in Munich anti-nuclear protest Deutsche Welle, 9 October 2010.

[158] Rachael Brown. Violent protests against nuclear waste train ABC News, November 8, 2010.

[159] Atomic waste train back on move after anti-nuclear blockade Deutsche Welle, 5 November 2010.

[160] Indians protest against nuclear plant (December 4, 2010) World News Australia.

[161] Valdes, Manuel (December 13, 2010). Anti-nuclear weapon protesters convicted in Tacoma The Washington Post.

[162] "Five Japanese in Hunger Strike Against Kaminoseki Nuclear Power Plant". January 29, 2011.

[163] Stamp, David (14 March 2011). "Germany suspends deal to extend nuclear plants' life". Reuters. Retrieved 15 March 2011.

[164] Knight, Ben (15 March 2011). "Merkel shuts down seven nuclear reactors". Deutsche Welle. Retrieved 15 March 2011.

[165] James Kanter and Judy Dempsey (March 15, 2011). "Germany Shuts 7 Plants as Europe Plans Safety Tests". New York Times.

[166] "Over 2,000 rally against nuclear plants in Taiwan". AFP. March 20, 2011.

[167] "Anti-nuclear Germans protest on eve of state vote". Reuters. March 26, 2011.

[168] Judy Dempsey (March 27, 2011). "Merkel Loses Key German State on Nuclear Fears". New York Times.

[169] "Thousands of Germans protest against nuclear power". Bloomberg Businessweek. 2 April 2011. Archived from the original on 8 May 2011.

[170] "Activists call for renewable energy at UN meeting". The Associated Press. 4 April 2011.

[171] Krista Mahr (April 11, 2011). "What Does Fukushima's Level 7 Status Mean?". Time.

[172] Amanda Hodge (April 21, 2011). "Fisherman shot dead in Indian nuke protest". The Australian.

[173] "Biggest anti-nuclear Swiss protests in 25 years". Bloomberg Businessweek. 22 May 2011. Archived from the original on 26 October 2012.

[174] "Anti-nuclear protests attract 20,000". Swissinfo. May 22, 2011.

[175] Lee I-Chia (May 1, 2011). "Anti-nuclear rally draws legions". Taipei Times.

[176] Lee I-Chia (June 5, 2011). "Conservationists protest against nuclear policies". Tapai Times.

[177] Antoni Slodkowski (June 15, 2011). "Japan anti-nuclear protesters rally after quake". *Reuters*.

[178] "Fukushima farmers, fishermen protest over nuclear crisis". *Mainichi Daily News*. 13 August 2011. Archived from the original on 2 September 2011.

[179] Olivier Fabre (11 September 2011). "Japan anti-nuclear protests mark 6 months since quake". *Reuters*.

[180] "Thousands march against nuclear power in Tokyo". *USA Today*. September 2011.

[181] Siddharth Srivastava (27 October 2011). "India's Rising Nuclear Safety Concerns". *Asia Sentinel*.

[182] Ranjit Devraj (25 October 2011). "Prospects Dim for India's Nuclear Power Expansion as Grassroots Uprising Spreads". *Inside Climate News*.

[183] Mari Yamaguchi (September 2011). "Kenzaburo Oe, Nobel Winner Urges Japan To Abandon Nuclear Power". *Huffington Post*.

[184] "Hinkley Point power station blockaded by anti-nuclear protesters". *The Guardian*. 3 October 2011.

[185] ""We want a nuclear-free peaceful world" say South Korea's women". *Women News Network*. January 13, 2012.

[186] "Protesting nuclear power". *The Japan Times*. January 22, 2012.

[187] Elgan Hearn (January 25, 2012). "Hundreds protest against nuclear power station plans". *Online Mail*.

[188] The Mainichi Shimbun (12 March 2012) Antinuclear protests held across Japan on anniversary of disaster

[189] "Austria expects EU anti-nuclear campaign this year". *Reuters*. Mar 12, 2012.

[190] "About 2,000 Taiwanese stage anti-nuclear protest". *Straits Times*. 11 March 2011.

[191] Phil Mercer (11 March 2012). "Australian Rallies Remember Fukushima Disaster". *VOA News*. Archived from the original on 12 March 2012.

[192] "Antinuclear rally held in Seoul on eve of Japan quake anniversary". *Mainichi Daily*. March 11, 2012.

[193] "Nearly 200 arrested in India nuclear protest". *France24*. 20 March 2012.

[194] "Oi prompts domestic, U.S. antinuclear rallies". *The Japan Times*. June 24, 2012.

[195] Thousands in Japan anti-nuclear protest two years after Fukushima Reuters

[196] Yu-Huay Sun (Mar 11, 2013). "Taiwan Anti-Nuclear Protests May Derail $8.9 Billion Power Plant". *Bloomberg*.

[197] "Thousands of anti-nuclear protesters attend Glasgow march against Trident". *Daily Record*. 13 April 2013.

[198] http://www.dw.de/anti-nuclear-protests-in-taiwan-draw-a-17483190

[199] http://www.voanews.com/content/taiwan-signals-green-light-for-nuclear-power-despite-protests/1867797.html

[200] http://focustaiwan.tw/news/asoc/201403040006.aspx

[201] "Taiwan to halt construction of fourth nuclear power plant". Reuters. 28 April 2014. Retrieved 28 April 2014.

[202] Newtan, Samuel Upton (2007). *Nuclear War 1 and Other Major Nuclear Disasters of the 20th Century*. AuthorHouse. p. 96.

[203] WISE Paris. The threat of nuclear terrorism:from analysis to precautionary measures. 10 December 2001.

[204] Indymedia UK. Activist Killed in Anti-nuke Protest.

[205] Bosley Crowther. Movie Review: Dr. Strangelove (1964) *The New York Times*, January 31, 1964.

[206] The China Syndrome (1979) *The New York Times*.

[207] *Dark Circle*, DVD release date March 27, 2007, Directors: Judy Irving, Chris Beaver, Ruth Landy. ISBN 0-7670-9304-6.

[208] "Mitsubashi no haoto to chikyū no kaiten". *Cinema Today* (in Japanese). Retrieved 1 December 2012.

[209] "Mitsubashi no haoto to chikyū no kaiten Kawanaka Hitomi". *Eiga Geijutsu* (in Japanese). Retrieved 1 December 2012.

[210] "Documentary Advances Nuclear Free Movement". NPR. Retrieved 2010-06-10.

[211] "For What It's Worth." No Nukes Reunite After Thirty Years

[212] Musicians Act to Stop New Atomic Reactors

[213] Boyer, Paul S. (1985). *By the Bomb's Early Light*. Pantheon. p. 70. ISBN 9780394528786.

[214] Woodrow Wilson International Center for Scholars. Confronting the Bomb: A Short History of the World Nuclear Disarmament Movement

[215] Jim Doyle. Nuclear power industry sees opening for revival *San Francisco Chronicle*, March 9, 2009.

[216] Minnesota House says no to new nuclear power plants Archived 5 May 2009 at the Wayback Machine. *StarTribune.com*, April 30, 2009.

[217] Rebecca A. McNerney (1998). The Changing Structure of the Electric Power Industry p. 110.

[218] William A. Gamson and Andre Modigliani. Media Coverage and Public Opinion on Nuclear Power. *American Journal of Sociology*, Vol. 95, No. 1, July 1989, p. 15.

[219] Research and Markets: International Perspectives on Energy Policy and the Role of Nuclear Power *Reuters*, May 6, 2009.

[220] New Zealand Nuclear Free Zone, Disarmament, and Arms Control Act

[221] "Nuclear Energy Prospects in New Zealand". World Nuclear Association. April 2009. Archived from the original on 3 January 2010. Retrieved 2009-12-09.

[222] "Nuclear power backed by 19%". *Television New Zealand*. 7 April 2008. Retrieved 16 September 2011.

[223] http://people.reed.edu/~{}ahm/Courses/Stan-PS-314-2009-Q1_PNP/Syllabus/EReadings/Albright1994South.pdf

[224] Electricity Regulation Act, 1999

[225] Navajo Nation outlaws uranium mining

[226] "Archived copy". Archived from the original on 8 July 2011. Retrieved 2012-09-15. *Bulletin of atomic scientists*

[227] Mycle Schneider, Steve Thomas, Antony Froggatt, Doug Koplow (August 2009). The World Nuclear Industry Status Report, German Federal Ministry of Environment, Nature Conservation and Reactor Safety, p. 6.

[228] Ralph Summy. Confronting the Bomb (book review). *Social Alternatives*, Vol. 28, No. 3, 2009, p. 64.

[229] Jeremy Bernstein. Nukes for Sale *The New York Review of Books*, April 14, 2010.

[230] "Italy puts 1 year moratorium on nuclear". *Businessweek*. March 23, 2011.

[231] "Italy Nuclear Referendum Results". 13 June 2011. Archived from the original on 25 March 2012.

[232] Annika Breidthardt (May 30, 2011). "German government wants nuclear exit by 2022 at latest". *Reuters*.

[233] International Atomic Energy Agency (2005). Global Public Opinion on Nuclear Issues and the IAEA: Final Report from 18 Countries pp. 6-7.

[234] "Stewart Brand + Mark Z. Jacobson: Debate: Does the world need nuclear energy?". *TED*. February 2010. Retrieved 21 October 2013.

[235] van Munster R. Sylvest C. Pro-Nuclear Environmentalism: Should We Learn to Stop Worrying and Love Nuclear Energy? *Technology and Culture*, 2015 Oct 56(4):789-811. doi: 10.1353/tech.2015.0107.

[236] James Lovelock: Nuclear power is the only green solution

[237] Going Nuclear

[238] Bernard Cohen

[239] The Nuclear Energy Option

[240] Samuel MacCracken, *The War Against the Atom*, 1982. Basic Books, pp. 60-61

[241] Nuclear Energy Institute website

[242] Fourth Ministerial Conference on Environment and Health: Budapest, Hungary, 23–25 June 2004

[243] Executive Summary

[244] Ari Rabl and Mona. Dreicer. Health and Environmental Impacts of Energy Systems. *International Journal of Global Energy Issues*, vol.18(2/3/4), 113-150 (2002)

[245] Environmental Heresies

[246] An Early Environmentalist, Embracing New 'Heresies'

[247] James Lovelock

[248]

[249] Some rethinking nuke opposition USA Today

[250] William F. Jasper. NGO Demonstrators: No to Coal, No to Oil, No to Nuclear *New American*, 16 December 2009.

[251] http://www.newstrib.com/featured-series/energy-series/Articles/A_7-20-2007_1_4.pdf

[252] The Independent, 24 May 2004

[253] George Monbiot "The nuclear winter draws near". *The Guardian*, 30 March 2000

[254] Monbiot, George (21 March 2011). "Why Fukushima made me stop worrying and love nuclear power". *The Guardian*. Retrieved 22 March 2011.

[255] Monbiot, George (4 April 2011). "Evidence Meltdown". *The Guardian*. Archived from the original on 9 April 2011. Retrieved 17 April 2011.

1.8 Bibliography

See also: List of books about nuclear issues and List of films about nuclear issues

- Brown, Jerry and Rinaldo Brutoco (1997). *Profiles in Power: The Anti-nuclear Movement and the Dawn of the Solar Age*, Twayne Publishers.

- Byrne, John and Steven M. Hoffman (1996). *Governing the Atom: The Politics of Risk*, Transaction Publishers.

- Clarfield, Gerald H. and William M. Wiecek (1984). *Nuclear America: Military and Civilian Nuclear Power in the United States 1940-1980*, Harper & Row.

- Cooke, Stephanie (2009). *In Mortal Hands: A Cautionary History of the Nuclear Age*, Black Inc.

- Cragin, Susan (2007). *Nuclear Nebraska: The Remarkable Story of the Little County That Couldn' t Be Bought*, AMACOM.

- Dickerson, Carrie B. and Patricia Lemon (1995). *Black Fox: Aunt Carrie's War Against the Black Fox Nuclear Power Plant*, Council Oak Publishing Company, ISBN 1-57178-009-2

- Diesendorf, Mark (2009). *Climate Action: A Campaign Manual for Greenhouse Solutions*, University of New South Wales Press.

- Diesendorf, Mark (2007). *Greenhouse Solutions with Sustainable Energy*, University of New South Wales Press.

- Elliott, David (2007). *Nuclear or Not? Does Nuclear Power Have a Place in a Sustainable Energy Future?*, Palgrave.

- Falk, Jim (1982). *Global Fission: The Battle Over Nuclear Power*, Oxford University Press.

- Fradkin, Philip L. (2004). *Fallout: An American Nuclear Tragedy*, University of Arizona Press.

- Giugni, Marco (2004). *Social Protest and Policy Change: Ecology, Antinuclear, and Peace Movements in Comparative Perspective*, Rowman and Littlefield.

- Lovins, Amory B. (1977). *Soft Energy Paths: Towards a Durable Peace*, Friends of the Earth International, ISBN 0-06-090653-7

- Lovins, Amory B. and John H. Price (1975). *Non-Nuclear Futures: The Case for an Ethical Energy Strategy*, Ballinger Publishing Company, 1975, ISBN 0-88410-602-0

- Lowe, Ian (2007). *Reaction Time: Climate Change and the Nuclear Option*, Quarterly Essay.

- McCafferty, David P. (1991). *The Politics of Nuclear Power: A History of the Shoreham Power Plant*, Kluwer.

- Natti, Susanna and Bonnie Acker (1979). *No Nukes: Everyone's Guide to Nuclear Power*, South End Press.

- Newtan, Samuel Upton (2007). *Nuclear War 1 and Other Major Nuclear Disasters of the 20th Century*, AuthorHouse.

- Ondaatje, Elizabeth H. (c1988). *Trends in Antinuclear Protests in the United States, 1984-1987*, Rand Corporation.

- Parkinson, Alan (2007). *Maralinga: Australia' s Nuclear Waste Cover-up*, ABC Books.

- Pernick, Ron and Clint Wilder (2012). *Clean Tech Nation: How the U.S. Can Lead in the New Global Economy*.

- Peterson, Christian (2003). *Ronald Reagan and Antinuclear Movements in the United States and Western Europe, 1981-1987*, Edwin Mellen Press.

- Price, Jerome (1982). *The Antinuclear Movement*, Twayne Publishers.

- Rudig, Wolfgang (1990). *Anti-nuclear Movements: A World Survey of Opposition to Nuclear Energy*, Longman.

- Schneider, Mycle, Steve Thomas, Antony Froggatt, Doug Koplow (August 2009). *The World Nuclear Industry Status Report*, German Federal Ministry of Environment, Nature Conservation and Reactor Safety.

- Smith, Jennifer (Editor), (2002). *The Antinuclear Movement*, Cengage Gale.

- Sovacool, Benjamin K. (2011). *Contesting the Future of Nuclear Power: A Critical Global Assessment of Atomic Energy*, World Scientific.

- Surbrug, Robert (2009). *Beyond Vietnam: The Politics of Protest in Massachusetts, 1974-1990*, University of Massachusetts Press.

- Walker, J. Samuel (2004). *Three Mile Island: A Nuclear Crisis in Historical Perspective*, University of California Press.

- Wellock, Thomas R. (1998). *Critical Masses: Opposition to Nuclear Power in California, 1958-1978*, The University of Wisconsin Press, ISBN 0-299-15850-0

- Wills, John (2006). *Conservation Fallout: Nuclear Protest at Diablo Canyon*, University of Nevada Press.

- Wittner, Lawrence S. (2009). *Confronting the Bomb: A Short History of the World Nuclear Disarmament Movement*, Stanford University Press.

1.9 External links

- The M and S Collection at the Library of Congress contains anti-nuclear movement materials.

Chapter 2

Anti-nuclear movement in the United States

A sign pointing to an old fallout shelter in New York City.

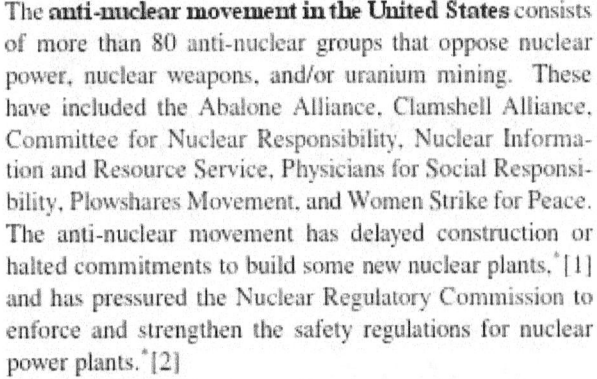

Anti-nuclear poster from the 1970s American movement

The **anti-nuclear movement in the United States** consists of more than 80 anti-nuclear groups that oppose nuclear power, nuclear weapons, and/or uranium mining. These have included the Abalone Alliance, Clamshell Alliance, Committee for Nuclear Responsibility, Nuclear Information and Resource Service, Physicians for Social Responsibility, Plowshares Movement, and Women Strike for Peace. The anti-nuclear movement has delayed construction or halted commitments to build some new nuclear plants,[1] and has pressured the Nuclear Regulatory Commission to enforce and strengthen the safety regulations for nuclear power plants.[2]

Anti-nuclear protests reached a peak in the 1970s and 1980s and grew out of the environmental movement.[3] Campaigns that captured national public attention involved the Calvert Cliffs Nuclear Power Plant, Seabrook Station

Nuclear Power Plant (by the Clamshell Alliance), Diablo Canyon Power Plant, Shoreham Nuclear Power Plant, and Three Mile Island.[1]

Beginning in the 1980s, many anti-nuclear power activists began shifting their interest, by joining a rapidly growing nuclear freeze movement, and the primary concern about nuclear hazards in the US changed from the problems of nuclear power plants to the prospects of nuclear war.[4] On June 12, 1982, one million people demonstrated in New York City's Central Park against nuclear weapons and for an end to the cold war arms race. It was the largest anti-nuclear protest and the largest political demonstration in American history.[5][6] International Day of Nuclear Disarmament protests were held on June 20, 1983, at 50 sites across the United States.[7][8] There were many Nevada Desert Experience protests and peace camps at the Nevada Test Site during the 1980s and 1990s.[9][10]

More recent campaigning by anti-nuclear groups has related to several nuclear power plants including the Enrico Fermi Nuclear Power Plant.*[11]*[12] Indian Point Energy Center,*[13] Oyster Creek Nuclear Generating Station.*[14] Pilgrim Nuclear Generating Station.*[15] Salem Nuclear Power Plant,*[16] and Vermont Yankee Nuclear Power Plant.*[17] There have also been campaigns relating to the Y-12 Nuclear Weapons Plant,*[18] the Idaho National Laboratory,*[19] Yucca Mountain nuclear waste repository proposal,*[20] the Hanford Site, the Nevada Test Site,*[21] Lawrence Livermore National Laboratory,*[22] and transportation of nuclear waste from the Los Alamos National Laboratory.*[23]

Some scientists and engineers have expressed reservations about nuclear power, including: Barry Commoner, S. David Freeman, John Gofman, Arnold Gundersen, Mark Z. Jacobson, Amory Lovins, Arjun Makhijani, Gregory Minor, M.V. Ramana, Joseph Romm and Benjamin K. Sovacool. Scientists who have opposed nuclear weapons include Paul M. Doty, Hermann Joseph Muller, Linus Pauling, Eugene Rabinowitch, M.V. Ramana and Frank N. von Hippel.

2.1   Emergence of the movement

See also: Nuclear weapons and the United States

2.1.1   Emergence of the anti-nuclear weapons movement

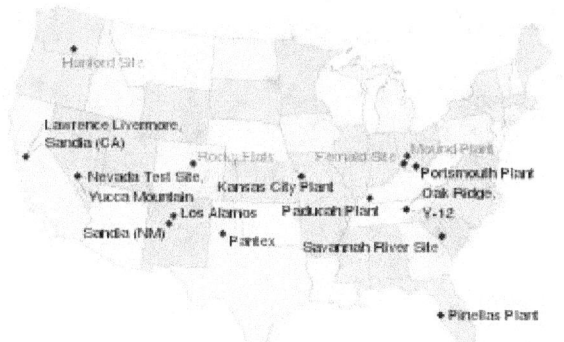

Map of major U.S. nuclear weapons infrastructure sites during the Cold War and into the present. Places with grayed-out names are no longer functioning and are in various stages of environmental remediation.

On November 1, 1961, at the height of the Cold War, about 50,000 women brought together by Women Strike for Peace marched in 60 cities in the United States to demonstrate

Women Strike for Peace during the Cuban Missile Crisis in 1962.

against nuclear weapons. It was the largest national women's peace protest of the 20th century.*[24]

This view of downtown Las Vegas shows a mushroom cloud in the background. Scenes such as this were typical during the 1950s. From 1951 to 1962 the government conducted 100 atmospheric tests at the nearby Nevada Test Site.

The nuclear debate initially was about nuclear weapons policy, and began within the scientific community. Scientific concern about the adverse health effects arising from atmospheric nuclear weapons testing first emerged in 1954.*[25] Professional associations such as the Federation of Atomic

Scientists and the Pugwash Conference on Science and World Affairs were involved.[26] The National Committee for a Sane Nuclear Policy was formed in November 1957, and surveys showed rising public uneasiness about the nuclear arms race---especially atmospheric nuclear weapons tests that sent radioactive fallout around the globe.[27] In 1962, Linus Pauling won the Nobel Peace Prize for his work to stop the atmospheric testing of nuclear weapons, and the "Ban the Bomb" movement spread throughout the United States.[26]

Between 1945 and 1992, the United States maintained a program of vigorous nuclear weapons testing. A total of 1,054 nuclear tests and two nuclear attacks were conducted, with over 900 of them at the Nevada Test Site, and ten on miscellaneous sites in the United States (Alaska, Colorado, Mississippi, and New Mexico).[28] Until November 1962, the vast majority of the U.S. tests were above-ground; after the acceptance of the Partial Test Ban Treaty all testing was relegated underground, in order to prevent the dispersion of nuclear fallout.

The U.S. program of atmospheric nuclear testing exposed some people to the hazards of fallout. Since the Radiation Exposure Compensation Act of 1990, more than $1.38 billion in compensation has been approved. The money is going to people who took part in the tests, notably at the Nevada Test Site, and to others exposed to the radiation.[29][30]

All-Atomic Comics *(1976)*
"The arguments against nuclear power are not as well-known [···ꞏ but] raise very serious questions."

2.1.2 Emergence of the anti-nuclear power movement

See also: Nuclear power debate
Unexpectedly high costs in the nuclear weapons program, along with competition with the Soviet Union and a desire to spread democracy through the world, created "...pressure on federal officials to develop a civilian nuclear power industry that could help justify the government's considerable expenditures."[31]

The Atomic Energy Act of 1954 encouraged private corporations to build nuclear reactors and a significant learning phase followed with many early partial core meltdowns and accidents at experimental reactors and research facilities.[32] This led to the introduction of the Price-Anderson Act in 1957, which was, "...an implicit admission that nuclear power provided risks that producers were unwilling to assume without federal backing."[32] The Price-Anderson Act "...shields nuclear utilities, vendors and suppliers against liability claims in the event of a catastrophic accident by imposing an upper limit on private sector liability." Without such protection, private companies were unwilling to become involved. No other technology in the history of American industry has enjoyed such continuing

President Jimmy Carter leaving Three Mile Island for Middletown, Pennsylvania, April 1, 1979

blanket protection.[31]

The first U.S. reactor to face public opposition was Fermi 1 in 1957. It was built approximately 30 miles from Detroit and there was opposition from the United Auto Workers Union.[33]

Pacific Gas & Electric planned to build the first commercially viable nuclear power plant in the US at Bodega Bay, north of San Francisco. The proposal was controversial and conflict with local citizens began in 1958.*[34] The proposed plant site was close to the San Andreas fault and the region's environmentally sensitive fishing and dairy industries. The Sierra Club became actively involved in the controversy.*[35] The conflict ended in 1964, with the forced abandonment of plans for the Bodega Bay power plant. Historian Thomas Wellock traces the birth of the anti-nuclear movement in the United States to the controversy over Bodega Bay.*[34] Attempts to build a nuclear power plant in Malibu were similar to those at Bodega Bay and were also abandoned.*[34]

A small military test reactor exploded at the Stationary Low-Power Reactor Number One in Idaho Falls in January 1961, causing 3 fatalities.*[36] This was caused by a combination of dangerous reactor design plus either sabotage, operator error by experienced operators.*[37] A further partial meltdown at the Enrico Fermi Nuclear Generating Station in Michigan in 1966.*[32]

In his 1963 book *Change, Hope and the Bomb*, David E. Lilienthal criticized nuclear developments, particularly the nuclear industry's failure to address the nuclear waste question. He argued that it would be "...particularly irresponsible to go ahead with the construction of full scale nuclear power plants without a safe method of nuclear waste disposal having been demonstrated." However, Lilienthal stopped short of a blanket rejection of nuclear power. His view was that a more cautious approach was necessary.*[38]

Samuel Walker, in his book *Three Mile Island: A Nuclear Crisis in Historical Perspective*, explains that the growth of the nuclear industry in the U.S. occurred as the environmental movement was being formed. Environmentalists saw the advantages of nuclear power in reducing air pollution, but became critical of nuclear technology on other grounds.*[39] The view that nuclear power was better for the environment than conventional fuels was partially undermined in the late 1960s when major controversy erupted over the effects of waste heat from nuclear plants on water quality. The nuclear industry "...gradually and reluctantly took action to reduce thermal pollution by building cooling towers or ponds for plants on inland waterways." *[39]

Several scientists, including John Gofman and Arthur Tamplin, challenged the prevailing view that the small amounts of radioactivity released by nuclear power plants during normal operation were not a problem. They argued "...that the routine releases were a severe threat to public health and could cause tens of thousands of deaths from cancer each year." *[39] Exchange of views about radiation risks caused uneasiness about nuclear power, especially among those unable to evaluate the conflicting claims.*[39]

The large size of nuclear plants ordered during the late 1960s raised new safety questions and created fears of a severe reactor accident that would send large quantities of radioactivity into the environment. In the early 1970s, a highly contentious debate over the performance of emergency core cooling systems in nuclear plants, designed to prevent a core meltdown that could lead to the "China syndrome", received coverage in the popular media and technical journals.*[17]*[40] The emergency core cooling systems controversy opened up whether the AECs first priority was promotion of the nuclear industry or protection of public health and safety.*[41]

By the early 1970s, anti-nuclear activity had increased dramatically in conjunction with concerns about nuclear safety and criticisms of a policy-making process that allowed little voice for these concerns. Initially scattered and organized at the local level, opposition to nuclear power became a national movement by the mid-1970s when such groups as the Sierra Club, Friends of the Earth, Natural Resources Defense Council, Union of Concerned Scientists, and Critical Mass became involved.*[42] With the rise of environmentalism in the 1970s, the anti-nuclear movement grew substantially:*[41]

> In 1975–76, ballot initiatives to control or halt the growth of nuclear power were introduced in eight western states. Although they enjoyed little success at the polls, the controls they sought to impose were sometimes adopted in part by state legislature, most notably in California. Interventions in plant licensing proceedings increased, often focusing on technical issues related to safety. This widespread popular ferment kept the issue before the public and contributed to growing public skepticism about nuclear power.*[41]

Another major area of ongoing concern was nuclear waste management. The absence of a working waste management facility became an important issue by the mid-1970s:

> In 1976, the California Energy Commission announced that it would not approve any more nuclear plants unless the utilities could specify fuel and waste disposal costs, an impossible task without decision on reprocessing, spent fuel storage and waste disposal. By the late 1970s, over thirty states had passed legislation regulating various activities associated with nuclear waste.*[43]

Many technologies and materials associated with the creation of a nuclear power program have a dual-use capability, in that they can be used to make nuclear weapons if a country chooses to do so.*[44] In 1975 over 2,000 prominent scientists signed a Declaration on Nuclear Power, prepared by the Union of Concerned Scientists, warning of the dangers of nuclear proliferation and urging the President and Congress to suspend the exportation of nuclear power to other countries, and reduce domestic construction until major problems were resolved.*[45] Theodore Taylor, a former nuclear weapons designer, explained, "...the ease with which nuclear bombs could be manufactured if fissionable material was available." *[40]

In 1976, four nuclear engineers -three from GE and one from the Nuclear Regulatory Commission- resigned, stating that nuclear power was not as safe as their superiors were claiming.*[46]*[47] These men were engineers who had spent most of their working life building reactors, and their defection galvanized anti-nuclear groups across the country.*[48]*[49] They testified to the Joint Committee on Atomic Energy that:

> "the cumulative effect of all design defects and deficiencies in the design, construction and operations of nuclear power plants makes a nuclear power plant accident, in our opinion, a certain event. The only question is when, and where.*[46]

These issues, together with a series of other environmental, technical, and public health questions, made nuclear power the source of acute controversy. Public support, which was strong in the early 1960s, had been shaken. *Forbes*, in the September 1975 issue, reported that "the anti-nuclear coalition has been remarkably successful ... [and] has certainly slowed the expansion of nuclear power." *[17] By the mid-1970s anti-nuclear activism, fueled by dissenting experts, had moved beyond local protests and politics to gain a wider appeal and influence. Although it lacked a single coordinating organization, and did not have uniform goals, it emerged as a movement sharply focused on opposing nuclear power, and the movement's efforts gained a great deal of national attention.*[17]

On March 28, 1979, equipment failures and operator error contributed to loss of coolant and a partial core meltdown at the Three Mile Island Nuclear Power Plant in Pennsylvania. The World Nuclear Association has stated that cleanup of the damaged nuclear reactor system at TMI-2 took nearly 12 years and cost approximately US $973 million.*[50] Benjamin K. Sovacool, in his 2007 preliminary assessment of major energy accidents, estimated that the TMI accident caused a total of $2.4 billion in property damages.*[51] The health effects of the Three Mile Island accident are widely, but not universally, agreed to be very low level.*[50]*[52] The accident triggered protests around the world.*[53]

The 1979 Three Mile Island accident inspired Perrow's book *Normal Accidents*, where a nuclear accident occurs, resulting from an unanticipated interaction of multiple failures in a complex system. TMI was an example of a normal accident because it was "...unexpected, incomprehensible, uncontrollable and unavoidable." *[54]

Perrow concluded that the failure at Three Mile Island was a consequence of the system's immense complexity. Such modern high-risk systems, he realized, were prone to failures however well they were managed. It was inevitable that they would eventually suffer what he termed a 'normal accident'. Therefore, he suggested, we might do better to contemplate a radical redesign, or if that was not possible, to abandon such technology entirely.*[55]

Nuclear power plants are a complex energy system.*[56]*[57] and opponents of nuclear power have criticized the sophistication and complexity of the technology. Helen Caldicott has said: "... in essence, a nuclear reactor is just a very sophisticated and dangerous way to boil water -- analogous to cutting a pound of butter with a chain saw." *[58] These critics of nuclear power advocate the use of energy conservation, efficient energy use, and appropriate renewable energy technologies to create our energy future.*[59] Amory Lovins, from the Rocky Mountain Institute, has argued that centralized electricity systems with giant power plants are becoming obsolete. In their place are emerging "distributed resources"— smaller, decentralized electricity supply sources (including efficiency) that are cheaper, cleaner, less risky, more flexible, and quicker to deploy. Such technologies are often called "soft energy technologies" and Lovins viewed their impacts as more gentle, pleasant, and manageable than hard energy technologies such as nuclear power.*[60]

Nuclear energy systems have a long stay time. The completion of the sequence of activities related to one commercial nuclear power station, from the start of construction through the safe disposal of its last radioactive waste, may take 100–150 years.*[56]

2.1.3 Emergence of the anti-uranium movement

See also: Uranium mining debate and Uranium mining and the Navajo people

Uranium mining is the process of extraction of uranium ore from the ground. A prominent use of uranium from mining is as fuel for nuclear power plants. After mining uranium ores, they are normally processed by grinding the

The Church Rock uranium mill tailings dam breach. A 20-foot breach in the tailings dam formed around 5:30 am on the morning of July 16, 1979.[61]*

ore materials to a uniform particle size and then treating the ore to extract the uranium by chemical leaching. The milling process commonly yields dry powder-form material consisting of natural uranium, "yellowcake", which is sold on the uranium market as U_3O_8, and uranium mining can use large amounts of water.

The Church Rock uranium mill spill occurred in New Mexico on July 16, 1979, when United Nuclear Corporation's Church Rock uranium mill tailings disposal pond breached its dam.*[61]*[62] Over 1,000 tons of solid radioactive mill waste and 93 million gallons of acidic, radioactive tailings solution flowed into the Puerco River, and contaminants traveled 80 miles (130 km) downstream to Navajo County, Arizona and onto the Navajo Nation.*[61] The accident released more radioactivity than the Three Mile Island accident that occurred four months earlier and was the largest release of radioactive material in U.S. history.*[61]*[63]*[64]*[65] Groundwater near the spill was contaminated and the Puerco rendered unusable by local residents, who were not immediately aware of the toxic danger.*[66]

Despite efforts made in cleaning up uranium sites, significant problems stemming from the legacy of uranium development still exist today on the Navajo Nation and in the states of Utah, Colorado, New Mexico, and Arizona. Hundreds of abandoned mines have not been cleaned up and present environmental and health risks in many communities.*[67] The Environmental Protection Agency estimates that there are 4000 mines with documented uranium production, and another 15,000 locations with uranium occurrences in 14 western states,*[68] most found in the Four Corners area and Wyoming.*[69] The *Uranium Mill Tailings Radiation Control Act* is a United States environmental law that amended the Atomic Energy Act of 1954 and gave the Environmental Protection Agency the authority to es-

tablish health and environmental standards for the stabilization, restoration, and disposal of uranium mill waste.*[70]

Anti-uranium activists in the US include: Thomas Banyacya, Manuel Pino and Floyd Red Crow Westerman.

2.2   Specific groups

Main article: List of anti-nuclear groups in the United States
See also: List of nuclear power groups

Anti-nuclear organizations oppose nuclear power, nuclear weapons, and/or uranium mining. More than eighty anti-nuclear groups operate, or have operated, in the United States. These include:

- Abalone Alliance
- Arms Control Association
- Bailly Alliance
- Beyond Nuclear
- Clamshell Alliance
- Committee for Nuclear Responsibility
- Corporate Accountability International
- Council for a Livable World
- Critical Mass
- Friends of the Earth
- Greenpeace USA
- Mothers for Peace
- Musicians United for Safe Energy
- NAU Against Uranium
- Nevada Desert Experience
- No Nukes group
- Nuclear Age Peace Foundation
- Nuclear Control Institute
- Nuclear Information and Resource Service
- Peace Action
- Physicians for Social Responsibility
- Plowshares Movement

- Public Citizen

- The Seneca Women's Encampment for a Future of Peace and Justice

- Shad Alliance

- Sierra Club

- Three Mile Island Alert

- Women Strike for Peace

Members of Nevada Desert Experience hold a prayer vigil during the Easter period of 1982 at the entrance to the Nevada Test Site.

Logo of the Nuclear Information and Resource Service

Some of the most influential groups in the anti-nuclear movement have had some members who were elite scientists, including Nobel Laureates Linus Pauling and Hermann Joseph Muller. In the United States, these scientists have belonged primarily to three groups: the Union of Concerned Scientists, the Federation of American Scientists, and the Committee for Nuclear Responsibility.[71]

Many American religious organizations have a long record of opposing nuclear weapons. Rejecting the development and use of nuclear weapons is "...one of the most widely shared convictions across faith traditions".[72] In the 1980s religious groups organized large anti-nuclear protests involving hundreds of thousands of people, and specific groups involved included the Southern Baptist Convention, and the Episcopal Church. The Protestant, Catholic, and Jewish communities published explicitly anti-nuclear statements, and in 2000 Muslims also began to take a stance against nuclear weaponry.[72]

The platform adopted by the delegates of the Green Party (United States) at their annual Green Congress May 26–28, 2000, reflecting the majority views of the membership, included the creation of self-reproducing, renewable energy

systems and use of federal investments, purchasing, mandates, and incentives to shut down nuclear power plants and phase out fossil fuels.[73]

Recent campaigning by anti-nuclear groups has related to several nuclear power plants including the Enrico Fermi Nuclear Power Plant,[11][12] Indian Point Energy Center,[13] Oyster Creek Nuclear Generating Station,[14] Pilgrim Nuclear Generating Station,[15] Salem Nuclear Power Plant,[16] and Vermont Yankee Nuclear Power Plant.[17] There have also been campaigns relating to the Y-12 Nuclear Weapons Plant,[18] the Idaho National Laboratory,[19] proposed Yucca Mountain nuclear waste repository,[20] the Hanford Site, the Nevada Test Site,[21] Lawrence Livermore National Laboratory,[22] and transportation of nuclear waste from the Los Alamos National Laboratory.[23]

2.3 Anti-nuclear protests

Main article: Anti-nuclear protests in the United States

On November 1, 1961, at the height of the Cold War, about 50,000 women brought together by Women Strike for Peace marched in 60 cities in the United States to demonstrate against nuclear weapons. It was the largest national

Anti-nuclear protest, US, 1977

White House Peace Vigil, the longest running peace vigil in US history, started by Thomas in 1981.

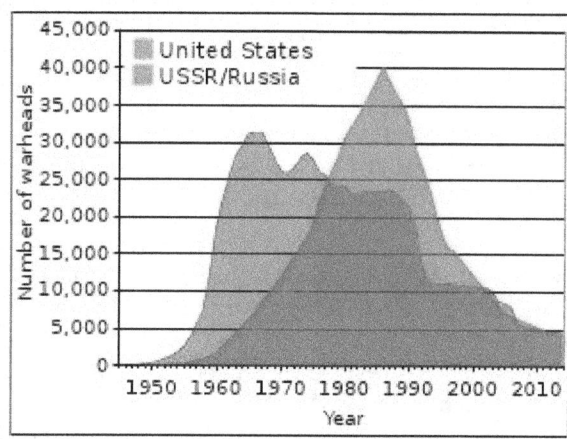

U.S. and USSR/Russian nuclear weapons stockpiles, 1945–2005.

Anti-nuclear protest at Harrisburg in 1979, following the Three Mile Island accident

women's peace protest of the 20th century.[24][74]

There were many anti-nuclear protests in the United States which captured national public attention during the 1970s and 1980s. These included the well-known Clamshell Alliance protests at Seabrook Station Nuclear Power Plant and the Abalone Alliance protests at Diablo Canyon Nuclear

Power Plant, where thousands of protesters were arrested. Other large protests followed the 1979 Three Mile Island accident.[1]

A large anti-nuclear demonstration was held in May 1979 in Washington D.C., when 65,000 people including the Governor of California, attended a march and rally against nuclear power.[75] In New York City on September 23, 1979, almost 200,000 people attended a protest against nuclear power.[76] Anti-nuclear power protests preceded the shutdown of the Shoreham, Yankee Rowe, Millstone I, Rancho Seco, Maine Yankee, and about a dozen other nuclear power plants.[77]

On June 3, 1981, Thomas launched the White House Peace Vigil in Washington, D.C..[78] He was later joined on the vigil by anti-nuclear activists Concepcion Picciotto and Ellen Benjamin.[79]

On June 6, 1982, a crowd of 85,000 gathers at the Rose Bowl in Pasadena, CA for "Peace Sunday: We Have a

April 2011 OREPA rally at the Y-12 nuclear weapons plant entrance

Dream" a rally and concert in support of the United Nations Special Session on Nuclear Disarmament. Performers include Joan Baez, Bob Dylan, Stevie Wonder and Crosby, Stills & Nash.

On June 12, 1982, one million people demonstrated in New York City's Central Park against nuclear weapons and for an end to the cold war arms race. It was the largest anti-nuclear protest and the largest political demonstration in American history.[5][6] International Day of Nuclear Disarmament protests were held on June 20, 1983, at 50 sites across the United States.[7][8] In 1986, hundreds of people walked from Los Angeles to Washington DC in the Great Peace March for Global Nuclear Disarmament.[80] There were many Nevada Desert Experience protests and peace camps at the Nevada Test Site during the 1980s and 1990s.[9][10]

In the 1980s, when fewer nuclear power plants remained in the construction and licensing pipeline, and interest in energy policy as a national issue declined, many anti-nuclear activists switched their focus to nuclear weapons and the arms race.[81] There has also been an *institutionalization* of the anti-nuclear movement,[82] where the anti-nuclear movement carried its contests into less visible, and more specialized institutional areas, such as regulatory and licensing hearings, and legal challenges.[82] At the state level, anti-nuclear groups were also successful in placing several anti-nuclear referendums on the ballot.[83]

On May 1, 2005, 40,000 anti-nuclear/anti-war protesters marched past the United Nations in New York, 60 years after the atomic bombings of Hiroshima and Nagasaki.[84][85] This was the largest anti-nuclear rally in the U.S. for several decades.[86] In 2008, 2009, and

2010, there have been protests about, and campaigns against, several new nuclear reactor proposals in the United States.[87][88][89]

There is an annual protest against U.S. nuclear weapons research at Lawrence Livermore National Laboratory in California and in the 2007 protest, 64 people were arrested.[90] There have been a series of protests at the Nevada Test Site and in the April 2007 Nevada Desert Experience protest, 39 people were cited by police.[91] There have been anti-nuclear protests at Naval Base Kitsap for many years, and several in 2008.[92][93][94] Also in 2008 and 2009, there have been protests about several proposed nuclear reactors.[87][88]

2.4 People with anti-nuclear views

2.4.1 Al Gore

Former vice president Al Gore says he is not anti-nuclear, but has stated that the "...cost of the present generation of reactors is nearly prohibitive."[95] In his 2009 book, *Our Choice*, Gore argues that nuclear power was once "expected to provide virtually unlimited supplies of low-cost electricity", but the reality is that it has been "...an energy source in crisis for the last 30 years."[96] Worldwide growth in nuclear power has slowed in recent years, with no new reactors and an "actual decline in global capacity and output in 2008." In the United States, "...no nuclear power plants ordered after 1972 have been built to completion."[96]

> Of the 253 nuclear power reactors originally ordered in the United States from 1953 to 2008, 48 percent were canceled, 11 percent were prematurely shut down, 14 percent experienced at least a one-year-or-more outage, and 27 percent are operating without having a year-plus outage. Thus, only about one fourth of those ordered, or about half of those completed, are still operating and have proved relatively reliable.[97]

2.4.2 Amory Lovins

In his 2005 book *Winning the Oil Endgame*, Amory Lovins praises nuclear power engineers, but is critical of the nuclear industry:

> No vendor has made money selling power reactors. This is the greatest failure of any enterprise in the industrial history of the world. We don' t mean that as a criticism of nuclear

power's practitioners, on whose skill and devotion we all continue to depend; the impressive operational improvements in U.S. power reactors in recent years deserve great credit. It is simply how technologies and markets evolved, despite the best intentions and immense effort. In nuclear power's heydey, its proponents saw no competitors but central coal-fired power stations. Then, in quick succession, came end-use efficiency, combined-cycle plants, distributed generation (including versions that recovered valuable heat previously wasted), and competitive wind-power. The range of competitors will only continue to expand more and their costs to fall faster than any nuclear technology can match." [98]

In 1988, Lovins argued that improving energy efficiency can simultaneously ameliorate greenhouse warming, reduce acid rain and air pollution, save money, and avoid the problems of nuclear power. Given the urgency of abating global warming, Lovins stated that we cannot afford to invest in nuclear power when those same dollars put into efficiency would displace far more carbon dioxide." [99]

In "Nuclear Power: Climate Fix or Folly," published in 2010, Lovins argued that expanded nuclear power "...does not represent a cost-effective solution to global warming and that investors would shun it were it not for generous government subsidies lubricated by intensive lobbying efforts." * [100]

2.4.3   Joseph Romm

Joseph Romm contends that nuclear power generates about 20 percent of all U.S. electricity, and is a low-carbon source of around-the-clock power, which has received renewed interest in recent years." [101] Yet, Romm says, nuclear power's "own myriad limitations will constrain its growth, especially in the near term", and the limitations include:* [101]

- "Prohibitively high, and escalating, capital costs

- Production bottlenecks in key components needed to build plants

- Very long construction times

- Concerns about uranium supplies and importation issues

- Unresolved problems with availability and security of radioactive waste storage, which has a 100,000 year shelf life

- Large-scale water use and contamination amid shortages

- High electricity prices from new plants" ." [101]

2.4.4   Randall Forsberg

Randall Forsberg (née Watson, 1943–2007) became interested in arms control issues while working at the Stockholm International Peace Research Institute in the late 1960s and early 1970s. In 1974, she returned to the United States, and became a graduate student in international studies at Massachusetts Institute of Technology. In 1979, Forsberg wrote *Call to Halt the Arms Race*, which later was the manifesto of the Nuclear Weapons Freeze Campaign. The document advocated a bilateral halt to the testing, production, deployment and delivery of nuclear weapons." [102]

Forsberg was awarded a doctorate in 1980 and she started the Institute for Defense and Disarmament Studies, which became an important resource for the peace movement and anti-nuclear weapons movement. In 1983 Forsberg was awarded a MacArthur Foundation *genius grant*. In 2005 she became Spitzer Professorship in Political Science at the City College of New York, and died of cancer in 2007 when she was 64 years old." [102]

2.4.5   Christopher Flavin

Many advocates of nuclear power argue that, given the urgency of doing something about climate change quickly, it must be pursued. Christopher Flavin, however, contends that speedy implementation is not one of nuclear power's strong points:* [103]

> Planning, licensing, and constructing even a single nuclear plant typically takes a decade or more, and plants frequently fail to meet completion deadlines. Due to the dearth of orders in recent decades, the world currently has very limited capacity to manufacture many of the critical components of nuclear plants. Rebuilding that capacity will take a decade or more.* [103]

Given the urgency of the climate problem, Flavin emphasizes the rapid commercialization of renewable energy and efficient energy use:

> Improved energy productivity and renewable energy are both available in abundance —and new policies and technologies are rapidly making them more economically competitive with fossil

fuels. In combination, these energy options represent the most robust alternative to the current energy system, capable of providing the diverse array of energy services that a modern economy requires. Given the urgency of the climate problem, that is indeed convenient.*[104]

2.4.6 Other people

Main article: List of anti-nuclear advocates in the United States

Selected other notable individuals who have expressed reservations about nuclear power, nuclear weapons, and/or uranium mining in the US include:*[105]*[106]*[107]

- Larry Bogart
- Helen Caldicott
- Barry Commoner
- Frances Crowe
- Carrie Barefoot Dickerson
- Paul M. Doty
- Jane Fonda
- Randall Forsberg
- Paul Gunter
- John Hall
- Jackie Hudson
- Sam Lovejoy
- Amory Lovins
- Arjun Makhijani
- Gregory Minor
- Hermann Joseph Muller
- Ralph Nader
- Linus Pauling
- Eugene Rabinowitch
- Bonnie Raitt
- Martin Sheen
- Karen Silkwood
- Thomas
- Louie Vitale
- Harvey Wasserman

2.5 Criticism

See also: List of pro-nuclear environmentalists and List of nuclear power groups

In November 2009, *The Washington Post* reported that nu-

Stewart Brand at a 2010 debate, "Does the world need nuclear energy?"[108]

clear power is emerging as "...perhaps the world's most unlikely weapon against climate change, with the backing of even some green activists who once campaigned against it." *[109] The article said that rather than deride the potential for nuclear power, some environmentalists are embracing it, and that presently there is only "muted opposition"— nothing like the protests and plant invasions that helped define the anti-nuclear movement in the United States during the 1970s.*[109]

Patrick Moore, one of the initial founders of Greenpeace, said in a 2008 interview that, "It wasn't until after I'd left Greenpeace and the climate change issue started coming to the forefront that I started rethinking energy policy in general and realized that I had been incorrect in my analysis of nuclear as being some kind of evil plot." *[110] Bernard Cohen, Professor Emeritus of Physics at the University of Pittsburgh, calculates that nuclear power is many times safer than any other form of power generation.*[111]

Critics of the movement point to independent studies showing the capital costs of renewable energy sources are higher than those from nuclear power.*[112] While functioning normally, coal plants, which are the predominate source of electricity in the U.S. (a trend that is expected to continue for some time to come as renewable energy alone cannot cheaply supply constant base load power) shorten nearly 24,000 lives a year in the United States with 2,800 from lung cancer alone. The United States Environmental Protection Agency (EPA) estimates that a range of 13,000 to 34,000 preventable premature deaths could be avoided by the reductions in PM2.5 and ozone produced by coal power plants.[113] In 2008 the World Health Organization (WHO) and other organizations calculated that coal particulates pollution cause approximately one million deaths annually across the world,[114] which is approximately one third of all premature deaths related to all air pollution sources.[115]

Critics argue that the amount of waste generated by nuclear power is very small, as all the high-level nuclear waste from 50+ years of operation of the world's nuclear reactors would fit into a single football field to the depth of five feet.[116] Furthermore, U.S. coal power plants presently create nearly a million tons of low-level radioactive waste per day and therefore release more total radioactivity than the nation's nuclear plants,[117] due to the uranium and thorium found naturally within the coal. Nuclear proponents also point out that cost and the quantity of waste figures for the operation of nuclear power plants are commonly derived from nuclear reactors built using second generation designs, dating from the 1960s. Advanced reactor designs are estimated to be even cheaper to operate and generate less than 1% the amount of waste of current designs, like Integral Fast Reactors or Pebble Bed Reactors.

It is because of these facts that proponents argue that nuclear fission power is the safest means currently available to entirely replace the use of fossil fuels, and pro-nuclear environmentalists argue that a combination of both nuclear energy and renewable energy would be the fastest, safest, and cheapest way forward.[118]

In 2007 Gwyneth Cravens outlined the message of her newest book, *Power to Save the World: The Truth About Nuclear Energy*. It argues for nuclear power as a safe energy source and an essential preventive of global warming. *Pandora's Promise* is a 2013 documentary film, directed by Robert Stone. It presents an argument that nuclear energy, typically feared by environmentalists, is in fact the only feasible way of meeting humanity's growing need for energy while also addressing the serious problem of climate change. The movie features several notable individuals (some of whom were once vehemently opposed to nuclear power, but who now speak in support of it), including: Stewart Brand, Gwyneth Cravens, Mark Lynas, Richard Rhodes and Michael Shellenberger.[119] Anti-nuclear advocate Helen Caldicott appears briefly.[120]

As of 2014, the U.S. nuclear industry has begun a new lob-bying effort, hiring three former senators —Evan Bayh, a Democrat; Judd Gregg, a Republican; and Spencer Abraham, a Republican —as well as William M. Daley, a former staffer to President Obama. The initiative is called Nuclear Matters, and it has begun a newspaper advertising campaign.[121]

2.6   Recent developments

pair of billboards in Davis, California advertising its nuclear-free policy.
See also: List of prospective nuclear units in the United States

As of early 2010, anti-nuclear groups such as Physicians for Social Responsibility, NukeFree.org, and NIRS were actively fighting federal loan guarantees for new nuclear plant construction. In February 2010, several groups coordinated a national call-in day to Congress to attempt to stop $54 billion in federal loan guarantees for new nuclear plants. However, the first such loan guarantee of $8.3 billion was offered to Southern Company that same month.[122]

In January 2010, about 175 anti-nuclear activists participated in a 126-mile walk in an effort to block the re-licensing of Vermont Yankee Nuclear Power Plant.[123] In February 2010, numerous anti-nuclear activists and private citizens gathered in Montpelier, to be at hand as the Vermont Senate voted 26 to 4 against the "Public Good" certificate needed for continued operation of Vermont Yankee past 2012.[124]

In April 2010 a dozen environmental groups (including Friends of the Earth, South Carolina's Sierra Club, Nuclear Watch South, the Southern Alliance for Clean Energy, Georgia Women's Action for New Directions) stated that the proposed AP1000 reactor containment design is "...inherently less safe than current reactors."[125] Arnold Gundersen, a nuclear engineer, authored a 32-page report arguing that the new AP1000 reactors will be vulnerable to leaks caused by corrosion holes. There are plans to construct the Westinghouse AP1000 reactors at seven sites across the southeast, including Plant Vogtle in Burke County, Georgia.[125][126]

In October 2010, Michael Mariotte, executive director of the Nuclear Information and Resource Service anti-nuclear group, predicted that the U.S. nuclear industry will not experience a nuclear renaissance, for the most simple of reasons: "nuclear reactors make no economic sense". The economic slump has driven down electricity demand and the price of competing energy sources, and Congress has failed to pass climate change legislation, making nuclear economics very difficult.[127]

Governor-elect Peter Shumlin is a prominent opponent of the Vermont Yankee Nuclear Power Plant and two days after Shumlin was elected in November 2010, Entergy put the plant up for sale.[128]

2.6.1 Post-Fukushima

Following the 2011 Japanese Fukushima nuclear disaster, authorities shut down the nation's 54 nuclear power plants. As of 2013, the Fukushima site remains highly radioactive, with some 160,000 evacuees still living in temporary housing, and some land will be unfarmable for centuries. The difficult cleanup job will take 40 or more years, and cost tens of billions of dollars.[129][130]

Following the 2011 Japanese nuclear accidents, activists who were involved in the movement's emergence (such as Graham Nash and Paul Gunter), suggest that Japan's nuclear crisis may rekindle an anti-nuclear protest movement in the United States. The aim, they say, is "...not just to block the Obama administration's push for new nuclear construction, but to convince Americans that existing plants pose dangers." [131]

In March 2011, 600 people gathered for a weekend protest outside the Vermont Yankee plant. The demonstration was held to show support for the thousands of Japanese people who are endangered by possible radiation from the Fukushima Daiichi nuclear disaster.[132]

In April 2011, Rochelle Becker, executive director of the Alliance for Nuclear Responsibility said that the United States should review its nuclear accident liability limits, in the light of the economic impacts of the Fukushima disaster.[133]

The New England region has a long history of anti-nuclear activism and 75 people held a State House rally on April 6, 2011, to "protest the region's aging nuclear plants and the increasing stockpile of radioactive spent fuel rods at them." [134] The protest was held shortly before a State House hearing where legislators were scheduled to hear representatives of the region's three nuclear plants – Pilgrim in Plymouth, Vermont Yankee in Vernon, and Seabrook in New Hampshire—talk about the safety of their reactors in the light of the Japanese nuclear crisis. Vermont Yankee and Pilgrim have reactor designs similar to the crippled Japanese nuclear plants.[134]

As of April 2011, a total of 45 groups and individuals from across the nation are formally asking the U.S. Nuclear Regulatory Commission (NRC) to immediately suspend all licensing and other activities at 21 proposed nuclear reactor projects in 15 states until the NRC completes a thorough post-Fukushima reactor crisis examination. The petitioners also are asking the NRC to supplement its own investigation by establishing an independent commission comparable to that set up in the wake of the serious, though less severe, 1979 Three Mile Island accident. The petitioners include Public Citizen, Southern Alliance for Clean Energy, and San Luis Obispo Mothers for Peace.[135][136][137]

Thirty two years after the No Nukes concert in New York, on August 7, 2011, a Musicians United for Safe Energy benefit concert was held Mountain View, California, to raise money for MUSE and for Japanese tsunami/nuclear disaster relief. The show was powered off-grid and artists included Jackson Browne, Bonnie Raitt, John Hall, Graham Nash, David Crosby, Stephen Stills, Kitaro, Jason Mraz, Sweet Honey and the Rock, the Doobie Brothers, Tom Morello, and Jonathan Wilson. In February 2012, the United States Nuclear Regulatory Commission approved the construction of two additional reactors at the Vogtle Electric Generating Plant, the first reactors to be approved in over 30 years since the Three Mile Island accident,[138] but NRC Chairman Gregory Jaczko cast a dissenting vote citing safety concerns stemming from Japan's 2011 Fukushima nuclear disaster, and saying "I cannot support issuing this license as if Fukushima never happened".[139] One week after Southern received the license to begin major construction on the two new reactors, a dozen environmental and anti-nuclear groups are suing to stop the Plant Vogtle expansion project, saying "public safety and environmental problems since Japan's Fukushima Daiichi nuclear reactor accident have not been taken into account".[140]

The nuclear reactors to be built at Vogtle are new AP1000

third generation reactors, which are said to have safety improvements over older power reactors.[138] However, John Ma, a senior structural engineer at the NRC, is concerned that some parts of the AP1000 steel skin are so brittle that the "impact energy" from a plane strike or storm driven projectile could shatter the wall.[141] Edwin Lyman, a senior staff scientist at the Union of Concerned Scientists, is concerned about the strength of the steel containment vessel and the concrete shield building around the AP1000.[141] Arnold Gundersen, a nuclear engineer commissioned by several anti-nuclear groups, released a report which explored a hazard associated with the possible rusting through of the steel liner of the containment structure.[126]

In March 2012, activists protested at San Onofre Nuclear Generating Station to mark the one-year anniversary of the nuclear meltdowns in Fukushima, Japan. Around 200 people rallied in San Onofre State Beach to listen to several speakers, including two Japanese residents who lived through the Fukushima meltdowns. *Residents Organizing for Safe Environment* and several other anti-nuclear energy organizations, organized the event and about 100 activists came in from San Diego.[142]

As of March 2012, 23 aging nuclear power plants continue to operate, including some similar in design to those that melted down in Fukushima, such as Vermont Yankee, and Indian Point 2 just 24 miles north of New York City. Vermont Yankee has reached the end of its projected lifetime operation but, despite strong local opposition, the NRC favored extending its license; however, on August 27, 2013, Entergy (VT Yankee's owner) announced it was decommissioning the plant and that "The station is expected to cease power production after its current fuel cycle and move to safe shutdown in the fourth quarter of 2014." [143] On March 22, 2012, "more than 1,000 people marched to the plant in protest, and about 130 engaging in civil disobedience were arrested". [144]

According to a 2012 Pew Research Center poll, 44 percent of Americans favor and 49 percent oppose the promotion of increased use of nuclear power, while 69 percent favor increasing federal funding for research on wind power, solar power, and hydrogen energy technology.[144][145]

In 2013, four aging, uncompetitive, reactors were permanently closed: San Onofre 2 and 3 in California, Crystal River 3 in Florida, and Kewaunee in Wisconsin.[146][147] Vermont Yankee will close in 2014. New York State is seeking to close Indian Point Energy Center, in Buchanan, 30 miles from New York City.[147]

With reference to the pro-nuclear film *Pandora's Promise*, economics professor, John Quiggin, comments that it presents the environmental rationale for nuclear power, but that reviving nuclear power debates is a distraction, and the main problem with the nuclear option is that it is not economically viable. Quiggin says that we need more efficient energy use and more renewable energy commercialization.[148]

2.7 See also

2.8 References

[1] Giugni, Marco (2004). *Social Protest and Policy Change: Ecology, Antinuclear, and Peace Movements* p. 44.

[2] Jerry Brown and Rinaldo Brutoco (1997). *Profiles in Power: The Anti-nuclear movement and the Dawn of the Solar Age*, p. 198.

[3] Herbert P. Kitschelt. Political Opportunity and Political Protest: Anti-Nuclear Movements in Four Democracies *British Journal of Political Science*, Vol. 16, No. 1, 1986, p. 62.

[4] Lisa Lynch (2012). "'We don't wanna be radiated:' Documentary Film and the Evolving Rhetoric of Nuclear Energy Activism" (PDF). *American Literature Ecocriticism Issue*.

[5] Jonathan Schell. The Spirit of June 12 *The Nation*, July 2, 2007.

[6] 1982 - a million people march in New York City Archived June 16, 2010, at the Wayback Machine.

[7] Harvey Klehr. Far Left of Center: The American Radical Left Today Transaction Publishers, 1988, p. 150.

[8] 1,400 Anti-nuclear protesters arrested *Miami Herald*, June 21, 1983.

[9] Robert Lindsey. 438 Protesters are Arrested at Nevada Nuclear Test Site *New York Times*, February 6, 1987.

[10] 493 Arrested at Nevada Nuclear Test Site *New York Times*, April 20, 1992.

[11] Groups petition against new nuclear plant

[12] Fermi 3 opposition takes legal action to block new nuclear reactor Archived March 30, 2010, at the Wayback Machine.

[13] Hudson River Lovers Fight to Shutter Aging Nuclear Power Plant

[14] Oyster Creek's time is up, residents tell board Archived September 30, 2007, at the Wayback Machine. *Examiner*, June 28, 2007.

[15] Pilgrim Watch (undated). Pilgrim Watch

[16] Unplugsalem.org (undated). UNPLUG Salem

[17] Walker, J. Samuel (2004). *Three Mile Island: A Nuclear Crisis in Historical Perspective* (Berkeley: University of California Press), pp. 10-11.

[18] Stop the Bombs! April 2010 Action Event at Y-12 Nuclear Weapons Complex.

[19] Keep Yellowstone Nuclear Free (2003). Keep Yellowstone Nuclear Free Archived November 22, 2009, at the Wayback Machine.

[20] Sierra Club. (undated). Deadly Nuclear Waste Transport Archived March 8, 2005, at the Wayback Machine.

[21] 22 Arrested in Nuclear Protest *New York Times*, August 10, 1989.

[22] Hundreds Protest at Livermore Lab *The TriValley Herald*, August 11, 2003.

[23] Concerned Citizens for Nuclear Safety (undated). About CCNS

[24] Woo, Elaine (January 30, 2011). "Dagmar Wilson dies at 94: organizer of women's disarmament protesters". *Los Angeles Times*.

[25] Wolfgang Rudig (1990). *Anti-nuclear Movements: A World Survey of Opposition to Nuclear Energy*, Longman, p. 55.

[26] Jerry Brown and Rinaldo Brutoco (1997). *Profiles in Power: The Anti-nuclear Movement and the Dawn of the Solar Age*, Twayne Publishers, pp. 191-192.

[27] Lawrence S. Wittner. Preserving the Golden Rule as a Piece of Anti-Nuclear History *History News Network*, 8 February 2010.

[28] Carey Sublette. Gallery of U.S. Nuclear Tests

[29] What governments offer to victims of nuclear tests *The Associated Press*, March 24, 2009.

[30] Radiation Exposure Compensation System: Claims to Date

[31] John Byrne and Steven M. Hoffman (1996). *Governing the Atom: The Politics of Risk*, Transaction Publishers, p. 136.

[32] Benjamin K. Sovacool. The costs of failure: A preliminary assessment of major energy accidents, 1907–2007, *Energy Policy* 36 (2008), p. 1808.

[33] Michael D. Mehta (2005). Risky business: nuclear power and public protest in Canada Lexington Books, p. 35.

[34] Paula Garb. Review of Critical Masses, *Journal of Political Ecology*, Vol 6, 1999.

[35] Thomas Raymond Wellock (1998). Critical Masses: Opposition to Nuclear Power in California, 1958-1978, The University of Wisconsin Press, pp. 27-28.

[36] McKeown, William (2003). *Idaho Falls: The Untold Story of America's First Nuclear Accident*. ECW Press. ISBN 1-55022-562-6.

[37] Hathaway, William (2006). *Idaho Falls*. Charleston, SC: Arcadia Pub. ISBN 0738548707.

[38] Wolfgang Rudig (1990). *Anti-nuclear Movements: A World Survey of Opposition to Nuclear Energy*, Longman, p. 61.

[39] Walker, J. Samuel (2004). *Three Mile Island: A Nuclear Crisis in Historical Perspective* (Berkeley: University of California Press), p. 10.

[40] Wolfgang Rudig (1990). *Anti-nuclear Movements: A World Survey of Opposition to Nuclear Energy*, Longman, pp. 66-67.

[41] John Byrne and Steven M. Hoffman (1996). *Governing the Atom: The Politics of Risk*, Transaction Publishers, p. 144.

[42] John Byrne and Steven M. Hoffman (1996). *Governing the Atom: The Politics of Risk*, Transaction Publishers, p. 205.

[43] John Byrne and Steven M. Hoffman (1996). *Governing the Atom: The Politics of Risk*, Transaction Publishers, p. 219.

[44] Steven E. Miller & Scott D. Sagan (Fall 2009). "Nuclear power without nuclear proliferation?". *Dædalus*.

[45] Ann Morrissett Davidon (December 1979). "The U.S. Anti-nuclear Movement". *Bulletin of the Atomic Scientists*.

[46] Mark Hertsgaard (1983). *Nuclear Inc. The Men and Money Behind Nuclear Energy*, Pantheon Books, New York, p. 72.

[47] Jim Falk (1982). *Global Fission: The Battle Over Nuclear Power*, Oxford University Press, p. 95.

[48] The San Jose Three *TIME*, Feb. 16, 1976.

[49] "The Struggle over Nuclear Power". Time. 1976-03-08. Retrieved 2016-11-04. (subscription required (help)).

[50] World Nuclear Association. Three Mile Island Accident January 2010.

[51] Benjamin K. Sovacool. The costs of failure: A preliminary assessment of major energy accidents, 1907–2007, *Energy Policy* 36 (2008), p. 1807.

[52] Mangano, Joseph (2004). Three Mile Island: Health study meltdown. *Bulletin of the atomic scientists*, 60(5), pp. 31–35.

[53] Mark Hertsgaard (1983). *Nuclear Inc. The Men and Money Behind Nuclear Energy*, Pantheon Books, New York, p. 95 & 97.

[54] Perrow, C. (1982), 'The President's Commission and the Normal Accident', in Sils, D., Wolf, C. and Shelanski, V. (Eds), *Accident at Three Mile Island: The Human Dimensions*, Westview, Boulder, pp.173–184.

[55] Pidgeon, N. (2011). "In retrospect: Normal Accidents". *Nature*. **477** (7365): 404. doi:10.1038/477404a.

[56] Storm van Leeuwen, Jan (2008). Nuclear power – the energy balance

[57] Wolfgang Rudig (1990). *Anti-nuclear Movements: A World Survey of Opposition to Nuclear Energy*. Longman, p. 53 & p. 61.

[58] Helen Caldicott (2006). *Nuclear power is not the answer to global warming or anything else*. Melbourne University Press. ISBN 0-522-85251-3, p.xvii

[59] Southern Alliance for Clean Energy (undated). Why Nuclear is Risky

[60] Amory B. Lovins (1977). *Soft Energy Paths: Toward a Durable Peace*. Penguin Books.

[61] Pasternak, Judy (2010). *Yellow Dirt: A Poisoned Land and a People Betrayed*. Free Press. p. 149. ISBN 1416594825.

[62] "Navajos mark 20th anniversary of Church Rock spill", *The Daily Courier*, Prescott, Arizona, July 18, 1999

[63] US Congress. House Committee on Interior and Insular Affairs. Subcommittee on Energy and the Environment. *Mill Tailings Dam Break at Church Rock, New Mexico*. 96th Cong. 1st Sess (October 22, 1979):19–24.

[64] Brugge, D.; DeLemos, J.L.; Bui, C. (2007), "The Sequoyah Corporation Fuels Release and the Church Rock Spill: Unpublicized Nuclear Releases in American Indian Communities", *American Journal of Public Health*. **97** (9): 1595–600. doi:10.2105/ajph.2006.103044. PMC 1963288. PMID 17666688

[65] Quinones, Manuel (December 13, 2011), "As Cold War abuses linger, Navajo Nation faces new mining push", *E&E News*, retrieved December 28, 2012

[66] Pasternak 2010. p. 150.

[67] Pasternak, Judy (2006-11-19). "A peril that dwelt among the Navajos". *Los Angeles Times*.

[68] U.S. EPA, Radiation Protection. "Uranium Mining Waste" 30 August 2012 Web.4 December 2012. http://www.epa.gov/radiation/tenorm/uranium.html

[69] Uranium Mining and Extraction Processes in the United States Figure 2.1. Mines and Other Locations with Uranium in the Western U.S. http://www.epa.gov/radiation/docs/tenorm/402-r-08-005-voli/402-r-08-005-v1-ch2.pdf

[70] *Laws We Use (Summaries):1978 - Uranium Mill Tailings Radiation Control Act(42 USC 2022 et seq.)*, EPA, retrieved December 16, 2012

[71] Jerome Price (1982). *The Anti-nuclear Movement*. Twayne Publishers, p. 65.

[72] Susan Brooks Thistlethwaite. Let's Take Religious Nuclear Opposition to the Next Level *Center for American Progress*, April 12, 2010.

[73] Green Party USA (undated). The Greens/Green Party USA

[74] Hevesi, Dennis (January 23, 2011). "Dagmar Wilson, Anti-Nuclear Leader, Dies at 94". *The New York Times*.

[75] Giugni, Marco (2004). *Social Protest and Policy Change: Ecology, Antinuclear, and Peace Movements* p. 45.

[76] Herman, Robin (September 24, 1979). "Nearly 200,000 Rally to Protest Nuclear Energy". *New York Times*. p. B1.

[77] Williams, Estha. Nuke Fight Nears Decisive Moment *Valley Advocate*, August 28, 2008.

[78] Colman McCarthy (February 8, 2009). "From Lafayette Square Lookout, He Made His War Protest Permanent". *The Washington Post*.

[79] "The Oracles of Pennsylvania Avenue". *Al Jazeera Documentary Channel*. April 17, 2012.

[80] Hundreds of Marchers Hit Washington in Finale of Nationwaide Peace March *Gainesville Sun*, November 16, 1986.

[81] John Byrne and Steven M. Hoffman (1996). *Governing the Atom: The Politics of Risk*. Transaction Publishers, pp. 144-145.

[82] Jerry Brown and Rinaldo Brutoco (1997). *Profiles in Power: The Anti-nuclear movement and the Dawn of the Solar Age*, pp. 195-199.

[83] Herbert P. Kitschelt. Political Opportunity and Political Protest: Anti-Nuclear Movements in Four Democracies *British Journal of Political Science*, Vol. 16, No. 1, 1986, p. 68.

[84] Lance Murdoch. Pictures: New York MayDay anti-nuke/war march Archived July 28, 2011, at the Wayback Machine. *IndyMedia*, 2 may 2005.

[85] Anti-Nuke Protests in New York *Fox News*, May 2, 2005.

[86] Lawrence S. Wittner. Nuclear Disarmament Activism in Asia and the Pacific, 1971–1996 *The Asia-Pacific Journal*, Vol. 25-5-09, June 22, 2009.

[87] Protest against nuclear reactor *Chicago Tribune*, October 16, 2008.

[88] Southeast Climate Convergence occupies nuclear facility *Indymedia UK*, August 8, 2008.

[89] Anti-Nuclear Renaissance: A Powerful but Partial and Tentative Victory Over Atomic Energy

[90] Police arrest 64 at California anti-nuclear protest *Reuters*, April 6, 2007.

[91] Anti-nuclear rally held at test site: Martin Sheen among activists cited by police *Las Vegas Review-Journal*, April 2, 2007.

[92] For decades, faith has sustained anti-nuclear movement *Seattle Times*, April 7, 2006.

[93] Bangor Protest Peaceful; 17 Anti-Nuclear Demonstrators Detained and Released *Kitsap Sun*, January 19, 2008.

[94] Twelve Arrests, But No Violence at Bangor Anti-Nuclear Protest *Kitsap Sun*, June 1, 2008.

[95] Anthony Faiola. Nuclear power regains support *The Washington Post*, November 24, 2009.

[96] Al Gore (2009). *Our Choice*, Bloomsbury, p. 152.

[97] Al Gore (2009). *Our Choice*, p. 157.

[98] Lovins, Amory (2005). Winning the Oil Endgame p. 259.

[99] Rocky Mountain Institute (1988). E88-31, Global Warming

[100] Nancy Folbre (March 28, 2011). "Renewing Support for Renewables". *New York Times*.

[101] Romm, Joe (2008). The Self-Limiting Future of Nuclear Power p. 1.

[102] Benjamin Redekop (2010). "Physicians to a Dying Planet: Helen Caldicott, Randall Forsberg, and the Anti-Nuclear Weapons Movement of the Early 1980s" (PDF). *Leadership Quarterly 21*.

[103] Worldwatch Institute (2008). Building a Low-Carbon Economy in *State of the World 2008*, p. 81.

[104] Worldwatch Institute (2008). Building a Low-Carbon Economy in *State of the World 2008*, p. 80.

[105] The Rise of the Anti-nuclear Power Movement

[106] Ancient Rockers Try to Recharge Anti-Nuclear Movement Archived November 11, 2007, at the Wayback Machine. *Business & Media Institute*, November 8, 2007.

[107] Falk, Jim (1982). *Gobal Fission:The Battle Over Nuclear Power*, p. 95.

[108] "Stewart Brand + Mark Z. Jacobson: Debate: Does the world need nuclear energy?". *TED* (published June 2010). February 2010. Retrieved 21 October 2013.

[109] Anthony Faiola. Nuclear power regains support *The Washington Post*, November 24, 2009.

[110] "Archived copy". Archived from the original on 2011-08-09. Retrieved 2010-03-27.

[111] http://www.phyast.pitt.edu/~{}blc/book/BOOK.html

[112] http://www.iea.org/Textbase/npsum/ElecCostSUM.pdf

[113] The Clean Air Task Force. 2010.

[114] Deaths per TWH by Energy Source, Next Big Future, March 2011 Quote: "The World Health Organization and other sources attribute about 1 million deaths per year to coal air pollution."

[115] Shrader-Frechette, Kristin. What Will Work: Fighting Climate Change with Renewable Energy, or Nuclear Power, Oxford University Press, 2011, pg.9, ISBN 0-19-979463-4.

[116] http://evworld.com/currents.cfm?jid=84

[117] Hvistendahl, Mara. "Coal Ash Is More Radioactive than Nuclear Waste: Scientific American". *Scientific American*, Nature America, Inc., 13 Dec. 2007. Web. 18 Mar. 2011.

[118] Monbiot, George (May 27, 2011). "Nuclear power (Environment),Renewable energy (Environment),Energy (Environment),Environment". *The Guardian*. London.

[119] Kilday, Gregg (29 May 2013). "Paul Allen Lends Support to Pro-Nuclear Doc 'Pandora's Promise'". *The Hollywood Reporter*. Retrieved 25 September 2013.

[120] O'Sullivan, Michael (13 June 2013). "'Pandora's Promise' movie review". *The Washington Post*. Retrieved 25 September 2013.

[121] Matthew Wald (April 27, 2014). "Nuclear Industry Gains Carbon-Focused Allies in Push to Save Reactors". *New York Times*.

[122] Southern Co. negotiating on nuke loan guarantees

[123] Anti-nuclear protesters reach capitol *Rutland Herald*, January 14, 2010.

[124] Wald, Matthew (February 25, 2010). "Vermont Senate Votes to Close Nuclear Plant". *The New York Times*.

[125] Rob Pavey. Groups say new Vogtle reactors need study *Augusta Chronicle*, April 21, 2010.

[126] Matthew L. Wald. Critics Challenge Safety of New Reactor Design *New York Times*, April 22, 2010.

[127] Matthew L. Wald. Sluggish Economy Curtails Prospects for Building Nuclear Reactors, *The New York Times*, October 10, 2010.

[128] Wald, Matthew L. (November 4, 2010). Vermont Nuclear Plant Up for Sale *The New York Times*.

[129] Richard Schiffman (12 March 2013). "Two years on, America hasn't learned lessons of Fukushima nuclear disaster". *The Guardian*. London.

[130] Martin Fackler (June 1, 2011). "Report Finds Japan Underestimated Tsunami Danger". *New York Times*.

[131] Leslie Kaufman (March 18, 2011). "Japan Crisis Could Rekindle U.S. Antinuclear Movement". *New York Times*.

[132] "Vermont Yankee: Countdown to closure". *WCAX*. March 21, 2011.

[133] Rochelle Becker (April 18, 2011). "Who would pay if nuclear disaster happened here?". *San Francisco Chronicle*.

[134] Martin Finucane (April 6, 2011). "Anti-nuclear sentiment regains its voice at State House rally". *Boston.com*.

[135] "Fukushima Fallout: 45 Groups and Individuals Petition NRC to Suspend All Nuclear Reactor Licensing and Conduct a "Credible" Three Mile Island-Style Review". *Nuclear Power News Today*. April 14, 2011.

[136] Renee Schoof (April 12, 2011). "Japan's nuclear crisis comes home as fuel risks get fresh look". *McClatchy*.

[137] Carly Nairn (14 April 2011). "Anti nuclear movement gears up". *San Francisco Bay Guardian*.

[138] Hsu, Jeremy (February 9, 2012). "First Next-Gen US Reactor Designed to Avoid Fukushima Repeat". Live Science (hosted on Yahoo!). Retrieved February 9, 2012.

[139] Ayesha Rascoe (Feb 9, 2012). "U.S. approves first new nuclear plant in a generation". *Reuters*.

[140] Kristi E. Swartz (February 16, 2012). "Groups sue to stop Vogtle expansion project". *The Atlanta Journal-Constitution*.

[141] Adam Piore (June 2011). "Nuclear energy: Planning for the Black Swan". *Scientific American*.

[142] Jameson Steed (March 12, 2012). "Anti nuclear groups protest San Onofre". *Daily Titan*.

[143] Media Relations (August 27, 2013O). "Entergy to Close, Decommission Vermont Yankee".

[144] Nancy Folbre (March 26, 2012). "The Nurture of Nuclear Power". *New York Times*.

[145] The Pew Research Center For The People and The Press (March 19, 2012). "As Gas Prices Pinch, Support for Oil and Gas Production Grows" (PDF).

[146] Mark Cooper (18 June 2013). "Nuclear aging: Not so graceful". *Bulletin of the Atomic Scientists*.

[147] Matthew Wald (June 14, 2013). "Nuclear Plants, Old and Uncompetitive, Are Closing Earlier Than Expected". *New York Times*.

[148] John Quiggin (8 November 2013). "Reviving nuclear power debates is a distraction. We need to use less energy". *The Guardian*.

2.9 Bibliography

See also: List of books about nuclear issues and List of films about nuclear issues

- Aron, Joan (1998). *Licensed to Kill? The Nuclear Regulatory Commission and the Shoreham Power Plant*, University of Pittsburgh Press.

- Brown, Jerry and Rinaldo Brutoco (1997). *Profiles in Power: The Anti-nuclear Movement and the Dawn of the Solar Age*, Twayne Publishers.

- Byrne, John and Steven M. Hoffman (1996). *Governing the Atom: The Politics of Risk*, Transaction Publishers.

- Clarfield, Gerald H. and William M. Wiecek (1984). *Nuclear America: Military and Civilian Nuclear Power in the United States 1940–1980*, Harper & Row.

- Cragin, Susan (2007). *Nuclear Nebraska: The Remarkable Story of the Little County That Couldn't Be Bought*, AMACOM.

- Dickerson, Carrie B. and Patricia Lemon (1995). *Black Fox: Aunt Carrie's War Against the Black Fox Nuclear Power Plant*, Council Oak Publishing Company, ISBN 1-57178-009-2

- Fradkin, Philip L. (2004). *Fallout: An American Nuclear Tragedy*, University of Arizona Press.

- Giugni, Marco (2004). *Social Protest and Policy Change: Ecology, Antinuclear, and Peace Movements in Comparative Perspective*, Rowman and Littlefield.

- Jasper, James M. (1997). *The Art of Moral Protest: Culture, Biography, and Creativity in Social Movements*, University of Chicago Press. ISBN 0-226-39481-6

- Lovins, Amory B. and Price, John H. (1975). *Non-Nuclear Futures: The Case for an Ethical Energy Strategy*, Ballinger Publishing Company, 1975, ISBN 0-88410-602-0

- McCafferty, David P. (1991). *The Politics of Nuclear Power: A History of the Shoreham Power Plant*, Kluwer.

- Miller, Byron A. (2000). *Geography and Social Movements: Comparing Anti-nuclear Activism in the Boston Area*, University of Minnesota Press.

- Natti, Susanna and Acker, Bonnie (1979). *No Nukes: Everyone's Guide to Nuclear Power*, South End Press.

- Ondaatje, Elizabeth H. (c1988). *Trends in Antinuclear Protests in the United States, 1984–1987*, Rand Corporation.

- Peterson, Christian (2003). *Ronald Reagan and Antinuclear Movements in the United States and Western Europe, 1981–1987*, Edwin Mellen Press.

- Polletta, Francesca (2002). *Freedom Is an Endless Meeting: Democracy in American Social Movements*, University of Chicago Press. ISBN 0-226-67449-5

- Pope, Daniel (2008). *Nuclear Implosions: The Rise and Fall of the Washington Public Power Supply System*, Cambridge University Press.

- Price, Jerome (1982). *The Antinuclear Movement*, Twayne Publishers.

- Smith, Jennifer (Editor). (2002). *The Antinuclear Movement*, Cengage Gale.

- Sovacool, Benjamin K. (2011). *Contesting the Future of Nuclear Power: A Critical Global Assessment of Atomic Energy*, World Scientific.

- Surbrug, Robert (2009). *Beyond Vietnam: The Politics of Protest in Massachusetts, 1974–1990*, University of Massachusetts Press.

- Walker, J. Samuel (2004). *Three Mile Island: A Nuclear Crisis in Historical Perspective*, University of California Press.

- Wellock, Thomas R. (1998). *Critical Masses: Opposition to Nuclear Power in California, 1958-1978*, The University of Wisconsin Press. ISBN 0-299-15850-0

- Wills, John (2006). *Conservation Fallout: Nuclear Protest at Diablo Canyon*, University of Nevada Press.

2.10 External links

- ALSOS Digital Library for Nuclear Issues

- Cancelled Nuclear Units Ordered in the United States

- Nuclear Reactor Shutdown List

- Public support for new nuclear power plants low, according to UN-backed poll

- Anti-nuclear renaissance: a powerful but partial and tentative victory over atomic energy

- Why Civil Resistance Works: The Strategic Logic of Nonviolent Conflict

- Nuclear Power's Global Expansion: Weighing Its Costs and Risks

- Beyond Nuclear 2013 response to the views of Hansen, Caldeira, Emanuel, and Wigley, about nuclear power.

Chapter 3

Nuclear power in the United States

regions and locations of nuclear reactors, 2008

NRC

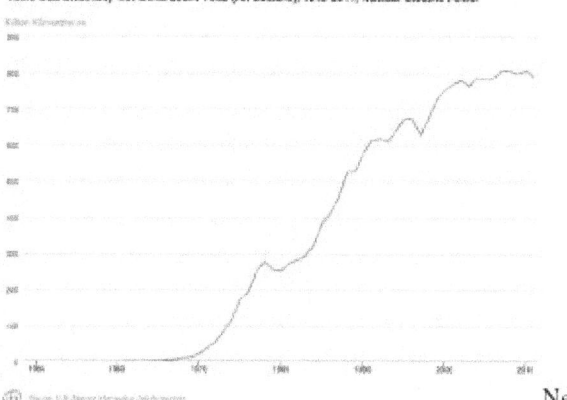

Net electrical generation from US nuclear power plants 1949-2011

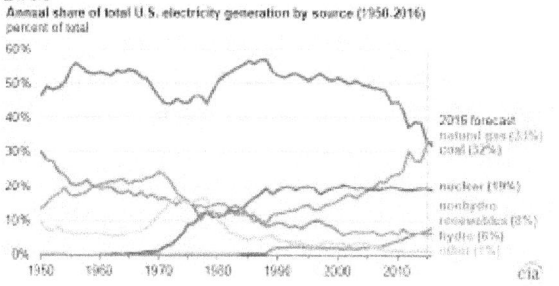

Nuclear power compared to other sources of electricity in the US, 1949-2011

Nuclear power in the United States is provided by 100 commercial reactors with a net summer capacity of 100,350 megawatts (MW), consisting of 66 pressurized water reactors and 34 boiling water reactors, producing a total of 797.2 terawatt-hours of electricity, which accounted for 19.50% of the nation's total electric energy generation in 2015. As of 2016, there are four new reactors under construction with a gross electrical capacity of 5,000 MW, while 33 reactors have been permanently shut down.[1][2] The United States is the world's largest supplier of commercial nuclear power, and in 2013 generated 33% of the world's nuclear electricity.[3]

As of October 2014, the NRC has granted license renewals providing a 20-year extension to a total of 74 reactors. In early 2014, the NRC prepared to receive the first applications of license renewal beyond 60 years of reactor life, as early as 2017, a process which by law requires public involvement.[4] Licenses for 22 reactors are due to expire before the end of the next decade if no renewals are granted.[5] The Vermont Yankee Nuclear Power Plant was the most recent nuclear power plant to be decommissioned on December 29, 2014. Another four aging reactors were permanently closed in 2013 before their licenses expired because of high maintenance and repair costs at a time when natural gas prices have fallen: San Onofre 2 and 3 in California, Crystal River 3 in Florida, and Kewaunee in Wisconsin,[6][7] and New York State is seeking to close Indian Point in Buchanan, 30 miles from New York City.[7][8]

Most reactors began construction by 1974; following the Three Mile Island accident in 1979 and changing economics, many planned projects were canceled. More than 100 orders for nuclear power reactors, many already under construction, were canceled in the 1970s and 1980s, bankrupting some companies. Up until 2013, there had also been no ground-breaking on new nuclear reactors at existing power plants since 1977. Then in 2012, the NRC approved construction of four new reactors at existing nuclear plants. Construction of the Virgil C. Summer Nuclear Generating Station Units 2 and 3 began on March 9, 2013. A few days later, on March 12, construction began on the

Vogtle Electric Generating Plant Units 3 and 4. In addition, on October 19, 2016 TVA's Unit-2 reactor at the Watts Bar Nuclear Generating Station became the first US reactor to enter commercial operation since 1996.[9]

There was a revival of interest in nuclear power in the 2000s, with talk of a "nuclear renaissance", supported particularly by the Nuclear Power 2010 Program. A number of applications were made, but facing economic challenges, and later in the wake of the Fukushima Daiichi nuclear disaster, most of these projects have been cancelled, and as of 2012, "nuclear industry officials say they expect five new reactors to enter service by 2020 – Southern's two Vogtle reactors, two at Summer in South Carolina and one at Watts Bar in Tennessee";[10] these are all at existing plants. As of August 2013, there are construction delays at Vogtle and Summer.[11]

3.1 History

3.1.1 Emergence

See also: Atomic Age
Unexpectedly high costs in the Second World War nuclear

The Shippingport reactor was the first full-scale PWR nuclear power plant in the United States.

weapons program created "...pressure on federal officials to develop a civilian nuclear power industry that could help justify the government's considerable expenditures" .[12] Research into the peaceful uses of nuclear materials began in the United States under the auspices of the Atomic Energy Commission, created by the United States Atomic Energy Act of 1946. Medical scientists were interested in the effect of radiation upon the fast-growing cells of cancer, and materials were given to them, while the military services led research into other peaceful uses.

President Jimmy Carter leaving Three Mile Island for Middletown, Pennsylvania, April 1, 1979

The Atomic Energy Act of 1954 encouraged private corporations to build nuclear reactors and a significant learning phase followed with many early partial core meltdowns and accidents at experimental reactors and research facilities.[13] This led to the introduction of the Price-Anderson Act in 1957, which was, "...an implicit admission that nuclear power provided risks that producers were unwilling to assume without federal backing." [13] The Price-Anderson Act "...shields nuclear utilities, vendors and suppliers against liability claims in the event of a catastrophic accident by imposing an upper limit on private sector liability." Without such protection, private companies were unwilling to become involved. No other technology in the history of American industry has enjoyed such continuing blanket protection.[12]

3.1.2 Power reactor research

Argonne National Laboratory was assigned by the United States Atomic Energy Commission the lead role in developing commercial nuclear energy beginning in the 1940s. Between then and the turn of the 21st century, Argonne designed, built, and operated fourteen reactors[14] at its site southwest of Chicago, and another fourteen reactors[14] at the National Reactors Testing Station in Idaho.[15] These reactors included initial experiments and test reactors that were the progenitors of today's pressurized water reactors (including naval reactors), boiling water reactors, heavy water reactors, graphite-moderated reactors, and liquid-metal cooled fast reactors, one of which[16] was the first reactor in the world to generate electricity. Argonne and a number of other AEC contractors built a total of 52 reactors at the National Reactor Testing Station. Two were never operated; except for the Neutron Radiography Facility, all the other reactors were shut down by 2000.

In the early afternoon of December 20, 1951, Argonne director Walter Zinn and fifteen other Argonne staff members witnessed a row of four light bulbs light up in a nondescript brick building in the eastern Idaho desert. Electricity from a generator connected to Experimental Breeder Reactor I (EBR-I) flowed through them. This was the first time that a usable amount of electrical power had ever been generated from nuclear fission. Only days afterward, the reactor produced all the electricity needed for the entire EBR complex.[17] One ton of natural uranium can produce more than 40 million kilowatt-hours of electricity —this is equivalent to burning 16,000 tons of coal or 80,000 barrels of oil.[18] More central to EBR-I's purpose than just generating electricity, however, was its role in proving that a reactor could create more nuclear fuel as a byproduct than it consumed during operation. In 1953, tests verified that this was the case.[19]

The US Navy took the lead, seeing the opportunity to have ships that could steam around the world at high speeds without refueling as being necessary for several decades, and the possibility of turning submarines into true full-time underwater vehicles. So, the Navy sent their "man in Engineering", then Captain Hyman Rickover, well known for his great technical talents in electrical engineering and propulsion systems in addition to his skill in project management, to the AEC to start the Naval Reactors project. Rickover's work with the AEC led to the development of the Pressurized Water Reactor (PWR), the first naval model of which was installed in the submarine USS *Nautilus*. This made the boat capable of operating under water full-time – demonstrating this ability by reaching the North Pole and surfacing through the Polar ice cap.

3.1.3   Start of commercial nuclear power

From the successful naval reactor program, plans were quickly developed for the use of reactors to generate steam to drive turbines turning generators. In April 1957, the SM-1 Nuclear Reactor in Fort Belvoir Va. was the first atomic power generator to go online and produce electrical energy to the U.S. power grid. On May 26, 1958 the first commercial nuclear power plant in the United States, Shippingport Atomic Power Station, was opened by President Dwight D. Eisenhower as part of his Atoms for Peace program. As nuclear power continued to grow throughout the 1960s, the Atomic Energy Commission anticipated that more than 1,000 reactors would be operating in the United States by 2000.[20] As the industry continued to expand, the Atomic Energy Commission's development and regulatory functions separated in 1974; the Department of Energy absorbed research and development, while the regulatory branch was spun off and turned into an independent commission known as the U.S. Nuclear Regulatory Commission (USNRC or simply NRC).

3.1.4   Opposition to nuclear power

There has been considerable opposition to the use of nuclear power in the U.S. The first U.S. reactor to face public opposition was Enrico Fermi Nuclear Generating Station in 1957. It was built approximately 30 miles from Detroit and there was opposition from the United Auto Workers Union.[21] Pacific Gas & Electric planned to build the first commercially viable nuclear power plant in the USA at Bodega Bay, north of San Francisco. The proposal was controversial and conflict with local citizens began in 1958.[22] The conflict ended in 1964, with the forced abandonment of plans for the power plant. Historian Thomas Wellock traces the birth of the anti-nuclear movement to the controversy over Bodega Bay.[22] Attempts to build a nuclear power plant in Malibu were similar to those at Bodega Bay and were also abandoned.[22]

Nuclear accidents continued into the 1960s with a small test reactor exploding at the Stationary Low-Power Reactor Number One in Idaho Falls in January 1961 and a partial meltdown at the Enrico Fermi Nuclear Generating Station in Michigan in 1966.[23] In his 1963 book *Change, Hope and the Bomb*, David Lilienthal criticized nuclear developments, particularly the nuclear industry's failure to address the nuclear waste question.[24] J. Samuel Walker, in his book *Three Mile Island: A Nuclear Crisis in Historical Perspective*, explains that the growth of the nuclear industry in the U.S. occurred in the 1970s as the environmental movement was being formed. Environmentalists saw the advantages of nuclear power in reducing air pollution, but were critical of nuclear technology on other grounds.[25] They were concerned about nuclear accidents, nuclear proliferation, high cost of nuclear power plants, nuclear terrorism and radioactive waste disposal.[26]

There were many anti-nuclear protests in the United States which captured national public attention during the 1970s and 1980s. These included the well-known Clamshell Alliance protests at Seabrook Station Nuclear Power Plant and the Abalone Alliance protests at Diablo Canyon Nuclear Power Plant, where thousands of protesters were arrested. Other large protests followed the 1979 Three Mile Island accident.[27]

In New York City on September 23, 1979, almost 200,000 people attended a protest against nuclear power.[28] Anti-nuclear power protests preceded the shutdown of the Shoreham, Yankee Rowe, Millstone I, Rancho Seco, Maine Yankee, and about a dozen other nuclear power plants.[29]

3.1.5   Over-commitment and cancellations

See also: List of canceled nuclear plants in the United States By the mid-1970s it became clear that nuclear power would

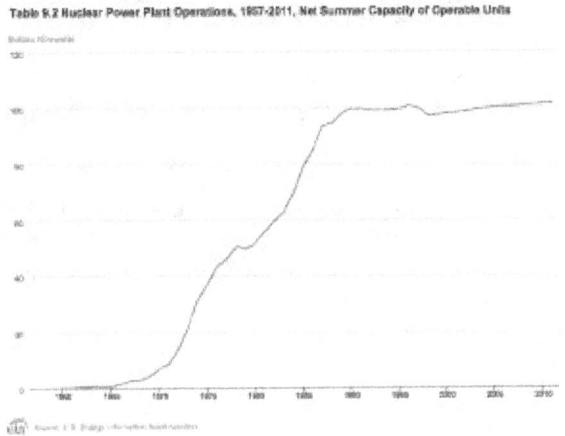

Table 9.2 Nuclear Power Plant Operations, 1957-2011, Net Summer Capacity of Operable Units

Net summer electrical generation capacity of US nuclear power plants, 1949-2011

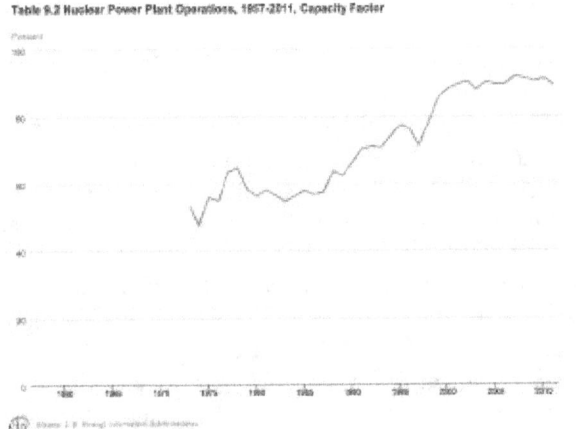

Table 9.2 Nuclear Power Plant Operations, 1957-2011, Capacity Factor

Average capacity factor of US nuclear power plants, 1957-2011

not grow nearly as quickly as once believed. Cost overruns were sometimes a factor of ten above original industry estimates, and became a major problem. For the 75 nuclear power reactors built from 1966 to 1977, cost overruns averaged 207 percent. Opposition and problems were galvanized by the Three Mile Island accident in 1979.[30]

Over-commitment to nuclear power brought about the financial collapse of the Washington Public Power Supply System, a public agency which undertook to build five large nuclear power plants in the 1970s. By 1983, cost overruns and delays, along with a slowing of electricity demand growth, led to cancellation of two WPPSS plants and a construction halt on two others. Moreover, WPPSS defaulted on $2.25 billion of municipal bonds, which is one of the largest municipal bond defaults in U.S. history. The court case that followed took nearly a decade to resolve.[31][32][33]

Eventually, more than 120 reactor orders were can-

celled,[34] and the construction of new reactors ground to a halt. Al Gore has commented on the historical record and reliability of nuclear power in the United States:

> Of the 253 nuclear power reactors originally ordered in the United States from 1953 to 2008, 48 percent were canceled, 11 percent were prematurely shut down, 14 percent experienced at least a one-year-or-more outage, and 27 percent are operating without having a year-plus outage. Thus, only about one fourth of those ordered, or about half of those completed, are still operating and have proved relatively reliable.[35]

Amory Lovins has also commented on the historical record of nuclear power in the United States:

> Of all 132 U.S. nuclear plants built (52% of the 253 originally ordered), 21% were permanently and prematurely closed due to reliability or cost problems, while another 27% have completely failed for a year or more at least once. The surviving U.S. nuclear plants produce ~90% of their full-time full-load potential, but even they are not fully dependable. Even reliably operating nuclear plants must shut down, on average, for 39 days every 17 months for refueling and maintenance, and unexpected failures do occur too.[36]

A cover story in the February 11, 1985, issue of *Forbes magazine* commented on the overall management of the nuclear power program in the United States:

> The failure of the U.S. nuclear power program ranks as the largest managerial disaster in business history, a disaster on a monumental scale ···only the blind, or the biased, can now think that the money has been well spent. It is a defeat for the U.S. consumer and for the competitiveness of U.S. industry, for the utilities that undertook the program and for the private enterprise system that made it possible.[37]

3.1.6 Three Mile Island and after

The NRC reported "(...the Three Mile Island accident...) was the most serious in U.S. commercial nuclear power plant operating history, even though it led to no deaths or injuries to plant workers or members of the nearby community."[38] The World Nuclear Association reports that

"...more than a dozen major, independent studies have assessed the radiation releases and possible effects on the people and the environment around TMI since the 1979 accident at TMI-2. The most recent was a 13-year study on 32,000 people. None has found any adverse health effects such as cancers which might be linked to the accident."[39] Other nuclear power incidents within the US (defined as safety-related events in civil nuclear power facilities between INES Levels 1 and 3[40] include those at the Davis-Besse Nuclear Power Plant, which was the source of two of the top five highest conditional core damage frequency nuclear incidents in the United States since 1979, according to the U.S. Nuclear Regulatory Commission.[41]

Despite the concerns which arose among the public after the Three Mile Island incident, the accident highlights the success of the reactor's safety systems. The radioactivity released as a result of the accident was almost entirely confined within the reinforced concrete containment structure. These containment structures, found at all nuclear power plants, were designed to successfully trap radioactive material in the event of a melt down or accident. At Three Mile Island, the containment structures operated exactly as it was designed to do, emerging successful in containing any radioactive energy. The low levels of radioactivity released post incident is considered harmless, resulting in zero injuries and deaths of residents living in proximity to the plant.

Despite many technical studies which asserted that the probability of a severe nuclear accident was low, numerous surveys showed that the public remained "very deeply distrustful and uneasy about nuclear power".[42] Some commentators have suggested that the public's consistently negative ratings of nuclear power are reflective of the industry's unique connection with nuclear weapons:[43]

> [One] reason why nuclear power is seen differently to other technologies lies in its parentage and birth. Nuclear energy was conceived in secrecy, born of war, and first revealed to the world in horror. No matter how many proponents try to separate the peaceful atom from the weapon's atom, the connection is firmly embedded in the mind of the public.[43]

Several US nuclear power plants closed well before their design lifetimes, due to successful campaigns by anti-nuclear activist groups.[44] These include Rancho Seco in 1989 in California and Trojan in 1992 in Oregon. Humboldt Bay in California closed in 1976, 13 years after geologists discovered it was built on a fault (the Little Salmon Fault). Shoreham Nuclear Power Plant was completed but never operated commercially as an authorized Emergency Evacuation Plan could not be agreed on due the political climate

after the Three Mile Island accident and Chernobyl disaster. The last permanent closure of a US nuclear power plant was in 1997.[45]

US nuclear reactors were originally licensed to operate for 40-year periods. In the 1980s, the NRC determined that there were no technical issues that would preclude longer service.[46] Over half of US nuclear reactors are over 30 years old and almost all are over twenty years old.[47] As of 2011, more than 60 reactors have received 20-year extensions to their licensed lifetimes.[48] The average capacity factor for all US reactors has improved from below 60% in the 1970s and 1980s, to 92% in 2007.[49][50]

After the Three Mile Island accident, NRC-issued reactor construction permits, which had averaged more than 12 per year from 1967 through 1978, came to an abrupt halt; no permits were issued between 1979 and 2012 (in 2012, four planned new reactors received construction permits). Many permitted reactors were never built, or the projects were abandoned. Those that were completed after Three Mile island experienced a much longer time lag from construction permit to starting of operations. The Nuclear Regulatory Commission itself described its regulatory oversight of the long-delayed Seabrook Nuclear Power Plant as "a paradigm of fragmented and uncoordinated government decision making," and "a system strangling itself and the economy in red tape."[51] The number of operating power reactors in the US peaked at 112 in 1991, far fewer than the 177 that received construction permits. By 1998 the number of working reactors declined to 104, where it remains as of 2013. The loss of electrical generation from the eight fewer reactors since 1991 has been offset by power uprates of generating capacity at existing reactors.[52]

Despite the problems following Three Mile Island, output of nuclear-generated electricity in the US grew steadily, more than tripling over the next three decades: from 255 billion kilowatt-hours in 1979 (the year of the Three Mile Island accident), to 806 billion kilowatt-hours in 2007.[53] Part of the increase was due to the greater number of operating reactors, which increased by 51%: from 69 reactors in 1979, to 104 in 2007. Another cause was a large increase in the capacity factor over that period. In 1978, nuclear power plants generated electricity at only 64% of their rated output capacity. Performance suffered even further during and after Three Mile Island, as a series of new safety regulations from 1979 through the mid-1980s forced operators to repeatedly shut down reactors for required retrofits.[54] It was not until 1990 that the average capacity factor of US nuclear plants returned to the level of 1978. The capacity factor continued to rise, until 2001. Since 2001, US nuclear power plants have consistently delivered electric power at about 90% of their rated capacity.[55] In 2016, the number of power plants was at 100 with 4 under construction.

3.1.7 Effects of Fukushima

Following the 2011 Japanese nuclear accidents. the U.S. Nuclear Regulatory Commission has announced it will launch a comprehensive safety review of the 104 nuclear power reactors across the United States, at the request of President Obama. A total of 45 groups and individuals had formally asked the NRC to suspend all licensing and other activities at 21 proposed nuclear reactor projects in 15 states until the NRC had completed a thorough post-Fukushima reactor crisis examination. The petitioners also asked the NRC to supplement its own investigation by establishing an independent commission comparable to that set up in the wake of the serious, though less severe, 1979 Three Mile Island accident.[56][57]

An industry observer noted that post-Fukushima costs were likely to go up for both current and new nuclear power plants, due to increased requirements for on-site spent fuel management and elevated design basis threats.[58][59] License extensions for existing reactors will face additional scrutiny, with outcomes depending on plants meeting new requirements, and some extensions already granted for more than 60 of the 100 operating U.S. reactors could be revisited. On-site storage, consolidated long-term storage, and geological disposal of spent fuel is "likely to be reevaluated in a new light because of the Fukushima storage pool experience".[58] Mark Cooper suggested that the cost of nuclear power, which already had risen sharply in 2010 and 2011, could "climb another 50 percent due to tighter safety oversight and regulatory delays in the wake of the reactor calamity in Japan".[60]

In 2011. London-based bank HSBC said: "With Three Mile Island and Fukushima as a backdrop, the US public may find it difficult to support major nuclear new build and we expect that no new plant extensions will be granted either. Thus we expect the clean energy standard under discussion in US legislative chambers will see a far greater emphasis on gas and renewables plus efficiency".[61]

The Obama administration "continues to support the expansion of nuclear power in the United States, despite the crisis in Japan".[62]

3.1.8 Competitiveness problems

In May 2015. a senior vice president of General Atomics stated that the U.S. nuclear industry was struggling due low U.S. fossil fuel prices, partly due to the rapid development of shale gas, and high financing costs for nuclear plants.[63]

In July 2016 Toshiba withdrew the U.S. design certification renewal for its Advanced Boiling Water Reactor because "it has become increasingly clear that energy price declines in

the US prevent Toshiba from expecting additional opportunities for ABWR construction projects".[64]

In 2016, Governor of New York Andrew Cuomo directed the New York Public Service Commission to consider ratepayer-financed subsidies similar to those for renewable sources to keep nuclear power stations profitable in the competition against natural gas..[65][66]

3.2 Safety and accidents

A clean-up crew working to remove radioactive contamination after the Three Mile Island accident.

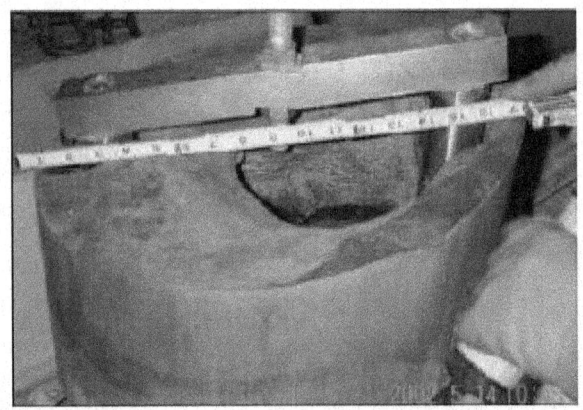

Erosion of the 6-inch-thick (150 mm) carbon steel reactor head at Davis-Besse Nuclear Power Plant in 2002, caused by a persistent leak of borated water.

Main articles: Nuclear safety in the U.S. and Nuclear

power plant accidents in the United States

Regulation of nuclear power plants in the United States is done by the Nuclear Regulatory Commission, which divides the nation into 4 administrative divisions.

On March 28, 1979, equipment failures and operator error contributed to loss of coolant and a partial core meltdown at the Three Mile Island Nuclear Power Plant in Pennsylvania. The mechanical failures were compounded by the initial failure of plant operators to recognize the situation as a loss-of-coolant accident due to inadequate training and human factors, such as human-computer interaction design oversights relating to ambiguous control room indicators in the power plant's user interface.[67] The scope and complexity of the accident became clear over the course of five days, as employees of Met Ed, Pennsylvania state officials, and members of the U.S. Nuclear Regulatory Commission (NRC) tried to understand the problem, communicate the situation to the press and local community, decide whether the accident required an emergency evacuation, and ultimately end the crisis. The NRC's authorization of the release of 40,000 gallons of radioactive waste water directly in the Susquehanna River led to a loss of credibility with the press and community.[67]

The Three Mile Island accident inspired Perrow's book *Normal Accidents*, where a nuclear accident occurs, resulting from an unanticipated interaction of multiple failures in a complex system. TMI was an example of a normal accident because it was "unexpected, incomprehensible, uncontrollable and unavoidable".[68] The World Nuclear Association has stated that cleanup of the damaged nuclear reactor system at TMI-2 took nearly 12 years and cost approximately US $973 million.[69] Benjamin K. Sovacool, in his 2007 preliminary assessment of major energy accidents, estimated that the TMI accident caused a total of $2.4 billion in property damages.[70] The health effects of the Three Mile Island accident are widely, but not universally, agreed to be very low level.[69][71] The accident triggered protests around the world.[72]

The 1979 Three Mile Island accident was a pivotal event that led to questions about U.S. nuclear safety.[73] Earlier events had a similar effect, including a 1975 fire at Browns Ferry, the 1976 testimonials of three concerned GE nuclear engineers, the GE Three. In 1981, workers inadvertently reversed pipe restraints at the Diablo Canyon Power Plant reactors, compromising seismic protection systems, which further undermined confidence in nuclear safety. All of these well-publicised events, undermined public support for the U.S. nuclear industry in the 1970s and the 1980s.[73]

On March 5, 2002, maintenance workers discovered that corrosion had eaten a football-sized hole into the reactor vessel head of the Davis-Besse plant. Although the corrosion did not lead to an accident, this was considered to be a serious nuclear safety incident.[74][75] The Nuclear Regulatory Commission kept Davis-Besse shut down until March 2004, so that FirstEnergy was able to perform all the necessary maintenance for safe operations. The NRC imposed its largest fine ever—more than $5 million—against FirstEnergy for the actions that led to the corrosion. The company paid an additional $28 million in fines under a settlement with the U.S. Department of Justice.[74]

The nuclear industry in the United States has maintained one of the best industrial safety records in the world with respect to all kinds of accidents. For 2008, the industry hit a new low of 0.13 industrial accidents per 200,000 worker-hours.[76] This is improved over 0.24 in 2005, which was still a factor of 14.6 less than the 3.5 number for all manufacturing industries.[77] However, more than a quarter of U.S. nuclear plant operators "have failed to properly tell regulators about equipment defects that could imperil reactor safety", according to a Nuclear Regulatory Commission report.[78]

As of February 2009, the NRC requires that the design of new power plants ensures that the reactor containment would remain intact, cooling systems would continue to operate, and spent fuel pools would be protected, in the event of an aircraft crash. This is an issue that has gained attention since the September 11, 2001, terrorist attacks. The regulation does not apply to the 100 commercial reactors now operating.[79] However, the containment structures of nuclear power plants are among the strongest structures ever built by mankind; independent studies have shown that existing plants would easily survive the impact of a large commercial jetliner without loss of structural integrity.[80]

Recent concerns have been expressed about safety issues affecting a large part of the nuclear fleet of reactors. In 2012, the Union of Concerned Scientists, which tracks ongoing safety issues at operating nuclear plants, found that "leakage of radioactive materials is a pervasive problem at almost 90 percent of all reactors, as are issues that pose a risk of nuclear accidents".[81] The U.S. Nuclear Regulatory Commission reports that radioactive tritium has leaked from 48 of the 65 nuclear sites in the United States.[82]

Following the Japanese Fukushima Daiichi nuclear disaster, according to Black & Veatch's annual utility survey that took place after the disaster, of the 700 executives from the US electric utility industry that were surveyed, nuclear safety was the top concern.[83] There are likely to be increased requirements for on-site spent fuel management and elevated design basis threats at nuclear power plants.[58][59] License extensions for existing reactors will face additional scrutiny, with outcomes depending on the degree to which plants can meet new requirements, and some extensions already granted for more than 60 of the

104 operating U.S. reactors could be revisited. On-site storage, consolidated long-term storage, and geological disposal of spent fuel is "likely to be reevaluated in a new light because of the Fukushima storage pool experience". [58] In March 2011, nuclear experts told Congress that spent-fuel pools at US nuclear power plants are too full. They say the entire US spent-fuel policy should be overhauled in light of the Fukushima I nuclear accidents. [84]

David Lochbaum, chief nuclear safety officer with the Union of Concerned Scientists, has repeatedly questioned the safety of the Fukushima I Plant's General Electric Mark 1 reactor design, which is used in almost a quarter of the United States' nuclear fleet. [85]

About one third of reactors in the US are boiling water reactors, the same technology which was involved in the Fukushima Daiichi nuclear disaster in Japan. There are also eight nuclear power plants located along the seismically active West coast. Twelve of the American reactors that are of the same vintage as the Fukushima Daiichi plant are in seismically active areas. [86] Earthquake risk is often measured by "Peak Ground Acceleration", or PGA, and the following nuclear power plants have a two percent or greater chance of having PGA over 0.15g in the next 50 years: Diablo Canyon, Calif.; San Onofre, Calif.; Sequoyah, Tenn.; H.B. Robinson, SC.; Watts Bar, Tenn.; Virgil C. Summer, SC.; Vogtle, GA.; Indian Point, NY.; Oconee, SC.; and Seabrook, NH. [86]

In 2013 the San Onofre Nuclear Generating Station was permanently retired when premature wear was found in the Steam Generators which had been replaced in 2010-2011.

3.3 Security and deliberate attacks

Main article: Vulnerability of nuclear plants to attack

The United States 9/11 Commission has said that nuclear power plants were potential targets originally considered for the September 11, 2001 attacks. If terrorist groups could sufficiently damage safety systems to cause a core meltdown at a nuclear power plant, and/or sufficiently damage spent fuel pools, such an attack could lead to widespread radioactive contamination. The research scientist Harold Feiveson has written that nuclear facilities should be made extremely safe from attacks that could release massive quantities of radioactivity into the community. New reactor designs have features of passive nuclear safety, which may help. In the United States, the NRC carries out "Force on Force" (FOF) exercises at all Nuclear Power Plant (NPP) sites at least once every three years. [30]

3.4 Uranium supply

Sources of uranium fuel for the US commercial nuclear power industry in 2012 (US Energy Information Administration)

A 2012 report by the International Atomic Energy Agency concluded: "The currently defined uranium resource base is more than adequate to meet high-case requirements through 2035 and well into the foreseeable future." [89]

At the start of 2013, the identified remaining worldwide uranium resources stood at 5.90 million tons, enough to supply the world's reactors at current consumption rates for more than 120 years, even if no additional uranium deposits are discovered in the meantime. Undiscovered uranium resources as of 2013 were estimated to be 7.7 million tons. Doubling the price of uranium would increase the identified reserves as of 2013 to 7.64 million tons. [90] Over the decade 2003-2013, the identified reserves of uranium (at the same price of US$130/kg) rose from 4.59 million tons in 2003 to 5.90 million tons in 2013, a increase of 28%. [91]

3.5 Fuel cycle

3.5.1 Uranium mining

Main article: Uranium mining in the United States

The United States has the 4th largest uranium reserves in the world. [92] The U.S. has its most prominent uranium reserves in New Mexico, Texas, and Wyoming. The U.S. Department of Energy has approximated there to be at least 300 million pounds of uranium in these areas. [93] Domestic production increased until 1980, after which it declined sharply due to low uranium prices. In 2012 the United States mined 17% of the uranium consumed by its nuclear power plants. The remainder was imported, principally from Canada, Russia and Australia. [92] Uranium is mined using several methods including open-pit mining, underground mining, and in-situ leaching. [94]

3.5.2   Uranium enrichment

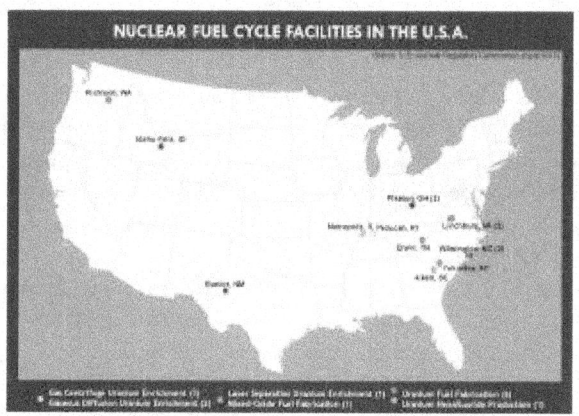

Location of nuclear reactor fuel processing facilities in the United States (US NRC)

There is one gas centrifuge enrichment plant currently in commercial operation in the US. The National Enrichment Facility, operated by URENCO east of Eunice, New Mexico, was the first uranium enrichment plant in 30 years to be built in the US. The plant started enriching uranium in 2010.[95] Two additional gas centrifuge plants have been licensed by the NRC, but are not operating. The American Centrifuge Plant in Piketown, Ohio broke ground in 2007, but stopped construction in 2009. The Eagle Rock Enrichment Facility in Bonneville County, Idaho was licensed in 2011, but construction is on hold.[96]

Currently, demonstration activities are underway in Oak Ridge, Tennessee for a future centrifugal enrichment plant. The new plant will be called the American Centrifuge Plant, which has an estimate cost of 2.3 billion USD.[97]

As of September 30, 2015, the DOE is ending its contract with the American Centrifuge Project and has stopped funding the project.[98]

3.5.3   Reprocessing

Nuclear reprocessing has been politically controversial because of the potential to contribute to nuclear proliferation, the potential vulnerability to nuclear terrorism, the political challenges of repository siting, and because of its high cost compared to the once-through fuel cycle.[99] The Obama administration has disallowed reprocessing of nuclear waste, citing nuclear proliferation concerns.[100] Critics of reprocessing worry that the recycled materials will be used for weapons. However, it is unlikely that reprocessed plutonium would be used for nuclear weapons, because it is not weapons-grade.[101] Nonetheless, it is possible that terrorists could steal these materials, because the reprocessed plutonium is less radiotoxic than spent fuel

and therefore much easier to steal. Nuclear power plants may not even notice if plutonium was stolen. It is difficult for plants to measure within even tens of kilograms, because making measurements at that accuracy is very time-consuming; consequently, it is likely that smaller amounts of plutonium could be stolen without detection.[102] Additionally, reprocessing is more expensive when compared with spent fuel storage. One study by the Boston Consulting Group estimated that reprocessing is six percent more expensive than spent fuel storage while another study by the Kennedy School of Government stated that reprocessing is 100 percent more expensive.[103]

3.5.4   Waste disposal

Recently, as plants continue to age, many on-site spent fuel pools have come near capacity, prompting creation of dry cask storage facilities as well. Several lawsuits between utilities and the government have transpired over the cost of these facilities, because by law the government is required to foot the bill for actions that go beyond the spent fuel pool.

There are some 65,000 tons of nuclear waste now in temporary storage throughout the U.S.[104] Since 1987, Yucca Mountain, in Nevada, had been the proposed site for the Yucca Mountain nuclear waste repository, but the project was shelved in 2009 following years of controversy and legal wrangling.[104][105] An alternative plan has not been proffered.[106]

At places like Maine Yankee, Connecticut Yankee and Rancho Seco, reactors no longer operate, but the spent fuel remains in small concrete-and-steel silos that require maintenance and monitoring by a guard force. Sometimes the presence of nuclear waste prevents re-use of the sites by industry.[107]

Without a long-term solution to store nuclear waste, a nuclear renaissance in the U.S. remains unlikely. Nine states have "explicit moratoria on new nuclear power until a storage solution emerges".[108][109]

Some nuclear power advocates argue that the United States should develop factories and reactors that will recycle some spent fuel. But the Blue Ribbon Commission on America's Nuclear Future said in 2012 that "no existing technology was adequate for that purpose, given cost considerations and the risk of nuclear proliferation".[109]

There is an "international consensus on the advisability of storing nuclear waste in deep underground repositories".[110] but no country in the world has yet opened such a site.[110][111][112][113][114][115] The Obama administration has disallowed reprocessing of nuclear waste, citing nuclear proliferation concerns.[100]

3.6 Water use in nuclear power production

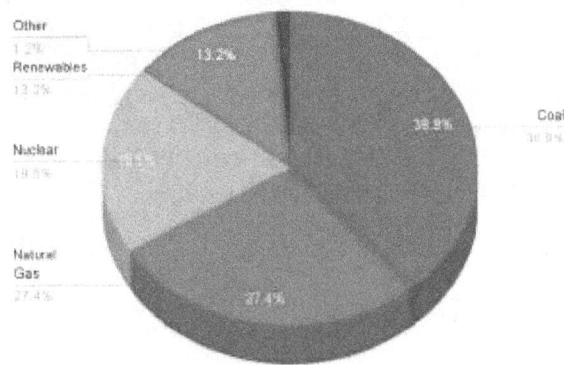

U.S. 2014 Electricity Generation By Type

U.S. 2014 Electricity Generation By Type. [116]

Once-through cooling systems, while once common, have come under attack for the possibility of damage to the environment. Wildlife can become trapped inside the cooling systems and killed, and the increased water temperature of the returning water can impact local ecosystems. US EPA regulations favors recirculating systems, even forcing some older power plants to replace existing once-through cooling systems with new recirculating systems.

A 2008 study by the Associated Press found that of the 104 nuclear reactors in the U.S., "... 24 are in areas experiencing the most severe levels of drought. All but two are built on the shores of lakes and rivers and rely on submerged intake pipes to draw billions of gallons of water for use in cooling and condensing steam after it has turned the plants' turbines." [117] much like all Rankine cycle power plants. During the 2008 southeast drought, reactor output was reduced to lower operating power or forced to shut down for safety. [117]

The Palo Verde Nuclear Generating Station is located in a desert and purchases reclaimed wastewater for cooling. [118]

3.7 Plant decommissioning

The price of energy inputs and the environmental costs of every nuclear power plant continue long after the facility has finished generating its last useful electricity. Both nuclear reactors and uranium enrichment facilities must be decommissioned, returning the facility and its parts to a safe enough level to be entrusted for other uses. After a cooling-off period that may last as long as a century, reactors must be dismantled and cut into small pieces to be packed in containers for final disposal. The process is very expensive, time-consuming, dangerous for workers, hazardous to the natural environment, and presents new opportunities for human error, accidents or sabotage. [119]

The total energy required for decommissioning can be as much as 50% more than the energy needed for the original construction. In most cases, the decommissioning process costs between US $300 million to US$5.6 billion. Decommissioning at nuclear sites which have experienced a serious accident are the most expensive and time-consuming. In the U.S. there are 13 reactors that have permanently shut down and are in some phase of decommissioning, but none of them have completed the process. [119]

New methods for decommissioning have been developed in order to minimize the usual high decommissioning costs. One of these methods is in situ decommissioning (ISD), which was implemented at the U.S. Department of Energy Savannah River Site in South Carolina for the closures of the P and R Reactors. With this tactic, the cost of decommissioning both reactors was $73 million. In comparison, the decommissioning of each reactor using traditional methods would have been an estimated $250 million. This results in a 71% decrease in cost by using ISD. [120]

3.8 Organizations

3.8.1 Fuel vendors

The following companies have active Nuclear fuel fabrication facilities in the United States. [121] These are all light water fuel fabrication facilities because only LWRs are operating in the US. The US currently has no MOX fuel fabrication facilities, though Duke Energy has expressed intent of building one of a relatively small capacity. [122]

- Areva

    Areva (formerly Areva NP) runs fabrication facilities in Lynchburg, Virginia and Richland, Washington. It also has a Generation III+ plant design, EPR (formerly the Evolutionary Power Reactor), which it plans to market in the US. [123]

- Westinghouse Electric Company

    Westinghouse operates a fuel fabrication facility in Columbia, South Carolina, [124] which processes 1,600 metric tons Uranium (MTU) per year.

It previously operated a nuclear fuel plant in Hematite, Missouri but has since closed it down.

- General Electric

  GE pioneered the BWR technology that has become widely used throughout the world. It formed the *Global Nuclear Fuel* joint venture in 1999 with Hitachi and Toshiba and later restructured into *GE-Hitachi Nuclear Energy*. It operates the fuel fabrication facility in Wilmington, North Carolina, with a capacity of 1,200 MTU per year.

- KazAtomProm

  KazAtomProm and the US company Centrus Energy have a partnership on competitive supplies of Kazakhstan's uranium to the US market.[125]

3.8.2   Industry and academic

The American Nuclear Society (ANS) scientific and educational organization has both academic and industry members. The organization publishes a large amount of literature on nuclear technology in several journals. The ANS also has some offshoot organizations such as North American Young Generation in Nuclear (NA-YGN).

The Nuclear Energy Institute (NEI) is an industry group whose activities include lobbying, experience sharing between companies and plants, and provides data on the industry to a number of outfits.

3.8.3   Anti-nuclear power groups

Some sixty anti-nuclear power groups are operating, or have operated, in the United States. These include: Abalone Alliance, Clamshell Alliance, Greenpeace USA, Institute for Energy and Environmental Research, Musicians United for Safe Energy, Nuclear Control Institute, Nuclear Information and Resource Service, Public Citizen Energy Program, Shad Alliance, and the Sierra Club.

In 1992, the chairman of the Nuclear Regulatory Commission said that "his agency had been pushed in the right direction on safety issues because of the pleas and protests of nuclear watchdog groups".[126]

Anti-nuclear protest, Boston, MA, 1977

3.9   Debate about nuclear power in the U.S.

See also: Nuclear power debate

There has been considerable public and scientific debate about the use of nuclear power in the United States, mainly from the 1960s to the late 1980s, but also since about 2001 when talk of a nuclear renaissance began. There has been debate about issues such as nuclear accidents, radioactive waste disposal, nuclear proliferation, nuclear economics, and nuclear terrorism.[26]

Some scientists and engineers have expressed reservations about nuclear power, including: Barry Commoner, S. David Freeman, John Gofman, Arnold Gundersen, Mark Z. Jacobson, Amory Lovins, Arjun Makhijani, Gregory Minor, and Joseph Romm. Mark Z. Jacobson, professor of civil and environmental engineering at Stanford University, has said: "If our nation wants to reduce global warming, air pollution and energy instability, we should invest only in the best energy options. Nuclear energy isn't one of them".[127] Arnold Gundersen, chief engineer of Fairewinds Associates and a former nuclear power industry executive, has questioned the safety of the Westinghouse AP1000, a proposed third-generation nuclear reactor.[128] John Gofman, a nuclear chemist and doctor, raised concerns about exposure to low-level radiation in the 1960s and argued against commercial nuclear power in the U.S.[129] In "Nuclear Power: Climate Fix or Folly," Amory Lovins, a physicist with the Rocky Mountain Institute, argued that expanded nuclear power "does not represent a cost-effective solution to global warming and that investors would shun it were it not for generous government subsidies lubricated by intensive lobbying efforts".[130]

Environmentalist Patrick Moore spoke out against nuclear power in 1976,[131] but today he supports it, along with renewable energy sources.[132][133][134] In Australian newspaper *The Age*, he writes "Greenpeace is wrong —we must consider nuclear power".[135] He argues that any realistic plan to reduce reliance on fossil fuels or greenhouse gas emissions need increased use of nuclear energy.[132] Phil Radford, Executive Director of Greenpeace US responded that nuclear energy is too risky, takes too long to build to address climate change, and by showing that the can U.S. shift to nearly 100% renewable energy while phasing out nuclear power by 2050.[136][137]

Environmentalist Stewart Brand wrote the book *Whole Earth Discipline*, which examines how nuclear power and some other technologies can be used as tools to address global warming.[138] Bernard Cohen, Professor Emeritus of Physics at the University of Pittsburgh, calculates that nuclear power is many times safer than other forms of power generation.[139]

President Obama early on included nuclear power as part of his "all of the above" energy strategy.[140] In a speech to the International Brotherhood of Electrical Workers in 2010, he demonstrated his commitment to nuclear power by announcing his approval of an $8 billion loan guarantee to pave the way for construction of the first new US nuclear power plant in nearly 30 years.[141][142] Then in 2012, his first post-Fukushima state-of-the-union address, Barack Obama said that America needs "an all-out, all-of-the-above strategy that develops every available source of American energy," yet pointedly omitted any mention of nuclear power.[143] But in February 2014, Energy secretary Ernest Moniz announced $6.5 billion in federal loan guarantees to enable construction of two new nuclear reactors, the first in the US since 1996.[144]

According to the Union of Concerned Scientists in March 2013 over one-third of U.S. nuclear power plants suffered safety-related incidents over the past three years, and nuclear regulators and plant operators need to improve inspections to prevent such events.[145]

Pandora's Promise is a 2013 documentary film, directed by Robert Stone. It presents an argument that nuclear energy, typically feared by environmentalists, is in fact the only feasible way of meeting humanity's growing need for energy while also addressing the serious problem of climate change. The movie features several notable individuals (some of whom were once vehemently opposed to nuclear power, but who now speak in support of it), including: Stewart Brand, Gwyneth Cravens, Mark Lynas, Richard Rhodes and Michael Shellenberger.[146] Anti-nuclear advocate Helen Caldicott appears briefly.[147]

As of 2014, the U.S. nuclear industry has begun a new lobbying effort, hiring three former senators —Evan Bayh, a Democrat; Judd Gregg, a Republican; and Spencer Abraham, a Republican —as well as William M. Daley, a for-

mer staffer to President Obama. The initiative is called Nuclear Matters, and it has begun a newspaper advertising campaign.[148]

3.9.1 Public opinion

The Gallup organization, which has periodically polled US opinion on nuclear power since 1994, found in March 2016 that, for the first time, a majority (54%) opposed nuclear power, versus 44% in favor. In polls from 2004 through 2015, a majority had supported nuclear power. support peaked at 62% in 2010, and has been in decline since.[149]

According to a CBS News poll, what had been growing acceptance of nuclear power in the United States was eroded sharply following the 2011 Japanese nuclear accidents, with support for building nuclear power plants in the U.S. dropping slightly lower than it was immediately after the Three Mile Island accident in 1979.[150] Only 43 percent of those polled after the Fukushima nuclear emergency said they would approve building new power plants in the United States.[150] A Washington Post-ABC poll conducted in April 2011 found that 64 percent of Americans opposed the construction of new nuclear reactors.[151] A survey sponsored by the Nuclear Energy Institute, conducted in September 2011, found that "62 percent of respondents said they favor the use of nuclear energy as one of the ways to provide electricity in the United States, with 35 percent opposed".[152]

According to a 2012 Pew Research Center poll, 44 percent of Americans favored and 49 percent opposed the promotion of increased use of nuclear power.[153]

A January 2014 Rasmussen poll found likely US voters split nearly evenly on whether to build more nuclear power plants, 39 percent in favor, versus 37 percent opposed, with an error margin of 3 percent.[154]

3.10 Prospects of a *nuclear renaissance*

See also: Nuclear renaissance in the United States
In the 2000s there was a renewed interest in nuclear power in the US, spurred by anticipated government curbs on carbon emissions, and a belief that fossil fuels would become more costly.[155]

3.10.1 Signs of a revival

The federal government encouraged development of nuclear power through the Nuclear Power 2010 Program,

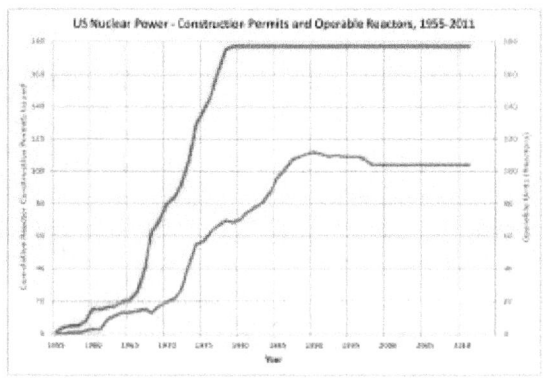

US reactor construction permits issued and operating nuclear power reactors, 1955-2011 (data from US EIA)

which coordinates efforts for building new nuclear power plants,[156] and the Energy Policy Act which has provisions favorable to nuclear.[157][158] In February 2010, President Barack Obama announced loan guarantees for two new reactors at Georgia Power's Vogtle Electric Generating Plant.[159][160] The reactors are "just the first of what we hope will be many new nuclear projects," said Carol Browner, director of the White House Office of Energy and Climate Change Policy.

In 2008, it was reported that The Shaw Group and Westinghouse would construct a factory at the Port of Lake Charles at Lake Charles, Louisiana to build components for the Westinghouse AP1000 nuclear reactor.[161] On October 23, 2008, it was reported that Northrop Grumman and Areva were planning to construct a factory in Newport News, Virginia to build nuclear reactors.[162]

As of March 2009, the U.S. Nuclear Regulatory Commission had received applications for permission to construct 26 new nuclear power reactors[163] with applications for another 7 expected.[164][165] Six of these reactors have been ordered.[166] However, not all the proposed new capacity will necessarily be built, with some applications being made to keep future options open and reserving places in a queue for government incentives available for up to the first three plants based on each innovative reactor design.[164]

In August 2011, the TVA board of directors voted to move forward with the construction of the unit one reactor at the Bellefonte Nuclear Generating Station.[167] In addition, the Tennessee Valley Authority petitioned to restart construction on the first two units at Bellefonte. But as of March 2012, many contractors have been laid off and the ultimate cost and timing for Bellefonte 1 will depend on work at another reactor TVA is completing - Watts Bar 2 in Ten-

nessee. In February 2012, TVA said the Watts Bar 2 project was running over budget and behind schedule.[168]

In 2012, The NRC approved construction permits for four new nuclear reactor units at two existing plants, the first permits in 34 years.[169] The first new permits, for two proposed reactors at the Vogtle plant, were approved in February 2012.[170] NRC Chairman Gregory Jaczko cast the lone dissenting vote, citing safety concerns stemming from Japan's 2011 Fukushima nuclear disaster: "I cannot support issuing this license as if Fukushima never happened" .[171]

The first two of the newly approved units were the Units 3 and 4 at the existing Vogtle Electric Generating Plant. As of December 2011, construction by Southern Company on the two new nuclear units had begun, and they are expected to be delivering commercial power by 2016 and 2017, respectively.[172][173] One week after Southern received the license to begin major construction on the two new reactors, a dozen environmental and anti-nuclear groups sued to stop the Plant Vogtle expansion project, saying "public safety and environmental problems since Japan's Fukushima Daiichi nuclear reactor accident have not been taken into account" .[174] The lawsuit was dismissed in July 2012.

Also in 2012, Units 2 and 3 at the SCANA Virgil C. Summer Nuclear Generating Station in South Carolina were approved, and are scheduled to come online in 2017 and 2018, respectively.[169]

A proposed nuclear power plant, the Blue Castle Project, is set to begin construction near Green River, Utah in 2023.[175] The plant will use 53,500 acre-feet of water annually from the Green River once both reactors are commissioned.[176] The first reactor is scheduled to come online in 2028, with the second reactor coming online in 2030.[175]

3.10.2   Continued problems and opposition

A number of other reactors were under consideration – a third reactor at the Calvert Cliffs Nuclear Power Plant in Maryland, a third and fourth reactor at South Texas Nuclear Generating Station, together with two other reactors in Texas, four in Florida, and one in Missouri. However, these have all been postponed or canceled.[177] But, looking ahead, experts see continuing challenges that will make it very difficult for the nuclear power industry to expand beyond a small handful of reactor projects that "government agencies decide to subsidize by forcing taxpayers to assume the risk for the reactors and mandating that ratepayers pay for construction in advance" .[60]

In May 2009, John Rowe, chairman of Exelon, which operates 17 nuclear reactors, stated that he would cancel or delay construction of two new reactors in Texas without federal loan guarantees.[111] Following the 2011 Fukushima nuclear disaster in Japan, he remarked that the nuclear renaissance was dead. Amory Lovins added that "market forces had killed it years earlier".[178]

In July 2009, the proposed Victoria County Nuclear Power Plant was delayed, as the project proved difficult to finance.[179] As of April 2009, AmerenUE has suspended plans to build its proposed plant in Missouri because the state Legislature would not allow it to charge consumers for some of the project's costs before the plant's completion. The New York Times has reported that without that "financial and regulatory certainty," the company has said it could not proceed.[180] Previously, MidAmerican Energy Company decided to "end its pursuit of a nuclear power plant in Payette County, Idaho." MidAmerican cited cost as the primary factor in their decision.[181]

In February 2010, the Vermont Senate voted 26 to 4 to block operation of the Vermont Yankee Nuclear Power Plant after 2012, citing radioactive tritium leaks, misstatements in testimony by plant officials, a cooling tower collapse in 2007, and other problems. By state law, the renewal of the operating license must be approved by both houses of the legislature for the nuclear power plant to continue operation.[182]

In 2010, demand for nuclear power softened in America, and some companies withdrew their applications for licenses to build.[183][184] In September 2010, Matthew Wald from the *New York Times* reported that "the nuclear renaissance is looking small and slow at the moment".[177]

In the first quarter of 2011, renewable energy contributed 11.7 percent of total U.S. energy production (2.245 quadrillion BTUs of energy), surpassing energy production from nuclear power (2.125 quadrillion BTUs).[185] 2011 was the first year since 1997 that renewables exceeded nuclear in US total energy production.[186]

In August 2012, the US Court of Appeals for the District of Columbia found that the NRC's rules for the temporary storage and permanent disposal of nuclear waste stood in violation of the National Environmental Policy Act, rendering the NRC legally unable to grant final licenses for any further new nuclear power plants.[187] This ruling was based on the fact that the Yucca Mountain waste repository had never received a license due to its license application being withdrawn by the DOE nor had any viable alternative waste repository been proposed. So long as this ruling stands and this impasse on waste disposal exists, no additional nuclear plants can ever be licensed for operation in the United States.

In March 2013, the concrete for the basemat of Block 2 of the Virgil C. Summer Nuclear Generating Station was poured. First concrete for Unit 3 was completed on November 4, 2013.

In March 2013, construction on unit 3 of Vogtle Electric Generating Plant started. Unit 4 was begun in November 2013.

In 2015 the Energy Information Administration estimated that nuclear power's share of U.S. generation would fall from 19% to 15% by 2040 in its central estimate (High Oil and Gas Resource case). However, as total generation increases 24% by 2040 in the central estimate, the absolute amount of nuclear generation remains fairly flat.[188]

3.11 Economics

George W. Bush signing the Energy Policy Act of 2005, which was designed to promote US nuclear reactor construction, through incentives and subsidies, including cost-overrun support up to a total of $2 billion for six new nuclear plants.[189]

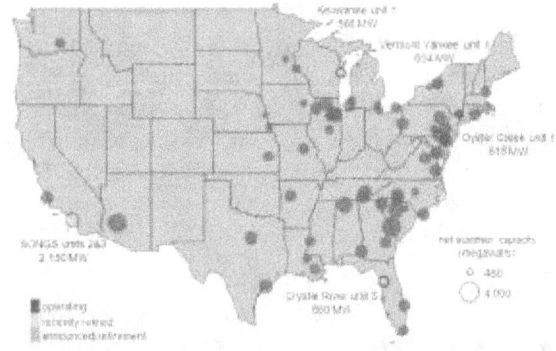

US nuclear power plants, highlighting recently and soon-to-be retired plants, as of 2013 (US EIA).

The low price of natural gas in the US since 2008 has spurred construction of gas-fired power plants as an alternative to nuclear plants. In August 2011, the head of America's largest nuclear utility said that this was not the time to build new nuclear plants, not because of political opposition or the threat of cost overruns, but because of the low price of natural gas. John Rowe, head of Exelon, said "Shale [gas] is good for the country, bad for new nuclear development" .[143]

In 2013, four older reactors were permanently closed: San Onofre 2 and 3 in California, Crystal River 3 in Florida, and Kewaunee in Wisconsin.[6][7] The state of Vermont tried to shut Vermont Yankee, in Vermont, but the plant was closed by the parent corporation for economic reasons in December 2014. New York State is seeking to close Indian Point Nuclear Power Plant, in Buchanan, 30 miles from New York City, despite this reactor being the primary contributor to Vermont's green energy fund.[7][190]

The additional cancellation of five large reactor upgrades (Prairie Island, 1 reactor, LaSalle, 2 reactors, and Limerick, 2 reactors), four by the largest nuclear company in the U.S., suggest that the nuclear industry faces "a broad range of operational and economic problems" .[191]

In July 2013, economist Mark Cooper named some nuclear power plants that face particularly intense challenges to their continued operation:[191]

- Palisades
- Ft. Calhoun
- Nine Mile Point
- Fitzpatrick
- Ginna
- Oyster Creek
- Vermont Yankee, decommissioned 2014
- Millstone
- Clinton
- Indian Point

Cooper said that the lesson for policy makers and economists is clear: "nuclear reactors are simply not competitive" .[191]

In December 2010, *The Economist* reported that the demand for nuclear power was softening in America.[184] In recent years, utilities have shown an interest in about 30 new reactors, but the number with any serious prospect of being built as of the end of 2010 was about a dozen, as some companies had withdrawn their applications for licenses to build.[183][192] Exelon has withdrawn its application for a license for a twin-unit nuclear plant in Victoria County, Texas, citing lower electricity demand projections. The decision has left the country's largest nuclear operator without

a direct role in what the nuclear industry hopes is a nuclear renaissance.[193] Ground has been broken on two new nuclear plants with a total of four reactors. The Obama administration was seeking the expansion of a loan guarantee program but as of December 2010 had been unable to commit all the loan guarantee money already approved by Congress. Since talk a few years ago of a "nuclear renaissance", gas prices have fallen and old reactors are getting license extensions. The only reactor to finish construction after 1996 was at Watts Bar, Tennessee, is an old unit, begun in 1973, whose construction was suspended in 1988, and was resumed in 2007.[194] It became operational in October 2016. Of the 100 reactors operating in the U.S., ground was broken on all of them in 1974 or earlier.[183][184]

Experts see continuing challenges that will make it very difficult for the nuclear power industry to expand beyond a small handful of reactor projects that "government agencies decide to subsidize by forcing taxpayers to assume the risk for the reactors and mandating that ratepayers pay for construction in advance" .[60]

In August 2012, Exelon stated that economic and market conditions, especially low natural gas prices, made the "construction of new merchant nuclear power plants in competitive markets uneconomical now and for the foreseeable future" .[195] In early 2013 UBS noted that some smaller reactors operating in deregulated markets may become uneconomic to operate and maintain, due to competition from generators using low priced natural gas, and may be retired early.[196] The 556 MWe Kewaunee Power Station is being closed 20 years before license expiry for these economic reasons.[190][197][198] In February 2014 the Financial Times identified Pilgrim, Indian Point, Clinton and Quad Cities power stations as potentially at risk of premature closure for economic reasons.[199]

3.12   See also

3.13   References

[1] "PRIS - Country Details" . IAEA—International Atomic Energy Agency. Retrieved October 21, 2016.

[2] "Nuclear Energy Overview" . EIA. Retrieved April 29, 2015.

[3] "International energy statistics – Nuclear Electricity Net Generation by Country" . US Energy Information Administration. Retrieved April 18, 2015.

[4] "Preparing for licensing beyond 60 years" . www.world-nuclear-news.org. 24 February 2014. Archived from the original on 4 January 2016. Retrieved 6 July 2016.

[5] "Nuclear Regulatory Commission resumes license renewals for nuclear power plants". IEA. October 29, 2014. Retrieved April 28, 2015.

[6] Mark Cooper (18 June 2013). "Nuclear aging: Not so graceful". *Bulletin of the Atomic Scientists*.

[7] Matthew Wald (June 14, 2013). "Nuclear Plants, Old and Uncompetitive, Are Closing Earlier Than Expected". *New York Times*.

[8] US Energy Information Administration. Lower power prices and high repair costs drive nuclear retirements 2 July 2013.

[9] Blau, Max (2016-10-20). "First new US nuclear reactor in 20 years goes live". *CNN.com*. Cable News Network. Turner Broadcasting System, Inc. Retrieved 2016-10-20.

[10] Ayesha Rascoe (Feb 9, 2012). "U.S. approves first new nuclear plant in a generation". *Reuters*.

[11] Nuclear Engineering International (6 August 2013). "US construction update".

[12] John Byrne and Steven M. Hoffman (1996). *Governing the Atom: The Politics of Risk*. Transaction Publishers, p. 136.

[13] Benjamin K. Sovacool. The costs of failure: A preliminary assessment of major energy accidents, 1907–2007, *Energy Policy* 36 (2008), p. 1808.

[14] "Reactors Designed by Argonne National Laboratory". Argonne National Laboratory. Retrieved 2012-05-15.

[15] "ANL-175 - Nuclear Reactors Built, Being Built, or Planned in the United States as of June 30, 1970 TID-8200 (22nd Rev.), USAEC Division of Technical Information, (1970)" (PDF). (2 MB)

[16] "Fast Reactor Technology: the EBR-I Reactor". Argonne National Laboratory. Retrieved 2012-05-15.

[17] Proving the Principle: Chapter 6

[18] "Uranium Quick Facts".

[19] "INL".

[20] Parker, Larry and Holt, Mark (March 9, 2007). Nuclear power: Outlook for new U.S. reactors CRS Report for Congress.

[21] Michael D. Mehta (2005). Risky business: nuclear power and public protest in Canada Lexington Books, p. 35.

[22] Paula Garb. Review of Critical Masses, *Journal of Political Ecology*, Vol 6, 1999.

[23] Benjamin K. Sovacool. A Critical Evaluation of Nuclear Power and Renewable Electricity in Asia, *Journal of Contemporary Asia*, Vol. 40, No. 3, August 2010, pp. 380.

[24] Wolfgang Rudig (1990). *Anti-nuclear Movements: A World Survey of Opposition to Nuclear Energy*, Longman, p. 61.

[25] Walker, J. Samuel (2004). *Three Mile Island: A Nuclear Crisis in Historical Perspective* (Berkeley: University of California Press), p. 10.

[26] Brian Martin. Opposing nuclear power: past and present, *Social Alternatives*, Vol. 26, No. 2, Second Quarter 2007, pp. 43-47.

[27] Giugni, Marco (2004). *Social Protest and Policy Change: Ecology, Antinuclear, and Peace Movements* p. 44.

[28] Herman, Robin (September 24, 1979). "Nearly 200,000 Rally to Protest Nuclear Energy". *New York Times*. p. B1.

[29] Williams, Estha. Nuke Fight Nears Decisive Moment *Valley Advocate*, August 28, 2008.

[30] Charles D. Ferguson & Frank A. Settle (2012). "The Future of Nuclear Power in the United States" (PDF). *Federation of American Scientists*.

[31] Cambridge University Press Nuclear Implosions: The Rise and Fall of the Washington Public Power Supply System Retrieved 2008-11-11

[32] "Review of 'Nuclear implosions: the rise and fall of the Washington Public Power Supply System'". *SciTech Book News*. June 2008. Retrieved 2008-11-11.

[33] Pope, Daniel (July 31, 2008). "A Northwest distaste for nuclear power". *Seattle Times*. Retrieved 2008-11-11.

[34] Nuclear Power: Outlook for New U.S. Reactors p. 3.

[35] Al Gore (2009). *Our Choice*, Bloomsbury, p. 157.

[36] Amory Lovins, Imran Sheikh, Alex Markevich (2009). Nuclear Power:Climate Fix or Folly Archived September 27, 2011, at the Wayback Machine. p. 10.

[37] "Nuclear Follies", a February 11, 1985 cover story in *Forbes magazine*.

[38] U.S. Nuclear Regulatory Commission, Federal Government of the United States (2009-08-11). "Backgrounder on the Three Mile Island Accident". Retrieved 2010-07-17.

[39] World Nuclear Association (March 2001). "Three Mile Island Accident Factsheet". Retrieved 2010-07-17. Updated version of January 2010

[40] International Atomic Energy Agency. "INES - International Nuclear and Radiological Event Scale". Retrieved 2010-07-17.)

[41] Nuclear Regulatory Commission (2004-09-16). "Davis-Besse preliminary accident sequence precursor analysis" (PDF). Retrieved 2006-06-14. and Nuclear Regulatory Commission (2004-09-20). "NRC issues preliminary risk analysis of the combined safety issues at Davis-Besse". Archived from the original on 2006-10-03. Retrieved 2006-06-14.

[42] John Byrne and Steven M. Hoffman (1996). *Governing the Atom: The Politics of Risk*. Transaction Publishers, p. 155.

[43] John Byrne and Steven M. Hoffman (1996). *Governing the Atom: The Politics of Risk*. Transaction Publishers, p. 157.

[44] "Shutting Down Rancho Seco". *Time*. 1989-06-19.

[45] Nuclear Energy Review, US Energy Information Administration, 2012.

[46] "Fact Sheet on Reactor License Renewal". *Fact Sheets*. NRC. February 16, 2011. Retrieved 2011-03-23.

[47] Half of U.S. nuclear reactors over 30 years old by Steve Hargreaves, CNNMoney.com, published March 15, 2011

[48] "Status of License Renewal Applications and Industry Activities". *Operating Reactors*. NRC. March 22, 2011. Retrieved 2011-03-23.

[49] "Nuclear power plant operations since 1957". US Energy Information Administration, 2007. File:Fig 9-2 Nuclear Power Plant Operations.jpg

[50] Findings: Energy Lessons by John Tierney, New York Times, published October 6, 2008.

[51] quoted by US EPA Commissioner Kennedy, in [Decisions of the United States Environmental Protection Agency], v.1 p.490.

[52] US Energy Information Administration, Table 9.1, Annual Review, 27 September 2012.

[53] US Energy Information Administration, Net electrical generation 1949-2011, September 2012.

[54] US Energy Information Administration, "Operating and maintenance costs for nuclear power plants in the United States". *World Energy Outlook 1994*, p.41.

[55] US Energy Information Administration, Table 9.2, Nuclear power plant operations 1957-2011, 27 Sept. 2012.

[56] "Fukushima Fallout: 45 Groups and Individuals Petition NRC to Suspend All Nuclear Reactor Licensing and Conduct a "Credible" Three Mile Island-Style Review". *Nuclear Power News Today*. April 14, 2011.

[57] Carly Nairn (14 April 2011). "Anti nuclear movement gears up". *San Francisco Bay Guardian*.

[58] Massachusetts Institute of Technology (2011). "The Future of the Nuclear Fuel Cycle" (PDF). p. xv.

[59] Mark Cooper (July 2011). "The implications of Fukushima: The US perspective". *Bulletin of the Atomic Scientists*. p. 9.

[60] "Experts: Even higher costs and more headaches for nuclear power in 2012". *MarketWatch*. 28 December 2011.

[61] HSBC (2011). Climate investment update: Japan's nuclear crisis and the case for clean energy. *HSBC Global Research*, March 18.

[62] JulieAnn McKellogg (March 18, 2011). "US Nuclear Renaissance Further Crippled by Japan Crisis". *Voice of America*.

[63] Testimony before the Committee on Science, Space and Technology

[64] "Toshiba withdraws ABWR certification application". World Nuclear News. 1 July 2016. Retrieved 5 July 2016.

[65] Yee, Vivian (July 20, 2016). "Nuclear Subsidies Are Key Part of New York's Clean-Energy Plan". *The New York Times*.

[66] "NYSDPS-DMM: Matter Master".

[67] "Minutes to Meltdown: Three Mile Island". *National Geographic*.

[68] Perrow, C. (1982). 'The President's Commission and the Normal Accident', in Sils, D., Wolf, C. and Shelanski, V. (Eds), *Accident at Three Mile Island: The Human Dimensions*, Westview, Boulder, pp.173–184.

[69] World Nuclear Association. Three Mile Island Accident January 2010.

[70] Benjamin K. Sovacool. The costs of failure: A preliminary assessment of major energy accidents, 1907–2007. *Energy Policy* 36 (2008), p. 1807.

[71] Mangano, Joseph (2004). Three Mile Island: Health study meltdown. *Bulletin of the atomic scientists*, 60(5), pp. 31–35.

[72] Mark Hertsgaard (1983). *Nuclear Inc. The Men and Money Behind Nuclear Energy*. Pantheon Books, New York, p. 95 & 97.

[73] Nathan Hultman & Jonathan Koomey (1 May 2013). "Three Mile Island: The driver of US nuclear power's decline?". *Bulletin of the Atomic Scientists*.

[74] NRC (September 2009). "Fact Sheet on Improvements Resulting From Davis-Besse Incident". *NRC Fact Sheet*.

[75] United States Government Accountability Office (2006). "Report to Congress" (PDF). p. 1.

[76] "Nuclear Industry's Safety, Operating Performance Remained Top-Notch in '08, WANO Indicators Show". *Reuters*. March 27, 2009.

[77] Fergus, Charles. "Are today's nuclear power plants safe?". *Research Penn State*.

[78] Mufson, Steven & Yang, Jia Lynn (March 24, 2011). "A quarter of U.S. nuclear plants not reporting equipment defects, report finds". *Washington Post*.

[79] "Around the Nation".

[80] "Aircraftcrashbreach - Nuclear Energy Institute".

[81] Mark Cooper (2012). "Nuclear safety and affordable reactors: Can we have both?" (PDF). *Bulletin of the Atomic Scientists*.

[82] MSNBC June 21, 2011 *Radioactive tritium leaks found at 48 US nuke sites*

[83] Wesoff, Eric (June 16, 2011). "Black & Veatch's 2011 Electric Utility Survey". *Greentechmedia*. Retrieved October 11, 2011.

[84] Clayton, Mark (March 30, 2011). "Fukushima warning: US has 'utterly failed' to address risk of spent fuel". *CS Monitor*.

[85] Northey, Hannah (March 28, 2011). "Japanese Nuclear Reactors, U.S. Safety to Take Center Stage on Capitol Hill This Week". *New York Times*.

[86] Michael D. Lemonick (24 August 2011). "What the east coast earthquake means for US nuclear plants". *The Guardian*.

[87] Sovacool, Benjamin K. (August 2010). "A Critical Evaluation of Nuclear Power and Renewable Electricity in Asia". *Journal of Contemporary Asia*. **40** (3): 393–400.

[88] Sovacool, Benjamin K. (2009). *The Accidental Century - Prominent Energy Accidents in the Last 100 Years*.

[89] Global uranium supply ensured for long term, new report shows, International Atomic Energy Agency, 26 July 2012.

[90] Uranium 2014, OECD World Nuclear Agency and International Atomic Energy Agency, 2015.

[91] Uranium 2003, OECD World Nuclear Agency and International Atomic Energy Agency, 2004.

[92] US Energy Information Administration. "The U.S. relies on foreign uranium, enrichment services to fuel its nuclear power plants." Today in Energy, 28 Aug. 2013.

[93] Union of Concerned Scientists. "How Nuclear Power Works". *Union of Concerned Scientists*. Retrieved 29 April 2014.

[94] English, Marianne. "HowStuffWorks "How Uranium Mining Works"". *HowStuffWorks*. Retrieved 29 April 2014.

[95] URENCO, URENCO USA begins enrichment of nuclear fuel, 25 June 2010.

[96] US Nuclear Regulatory Commission, Uranium enrichment, 21 Feb. 2014.

[97] "Uranium Enrichment—The American Centrifuge". USEC Inc. 2008. Retrieved 2008-05-15.

[98] "DOE Pulls Plug on Centrifuge". The Chillicothe Gazette. 2015. Retrieved 2015-09-11.

[99] Harold Feiveson; et al. (2011). "Managing nuclear spent fuel: Policy lessons from a 10-country study". *Bulletin of the Atomic Scientists*.

[100] Editorial, Nature 460, 152 (8 July 2009). "Adieu to nuclear recycling". *Nature*.

[101] Rossin, A. David. "U.S. Policy on Spent Fuel Reprocessing: The Issues". *PBS*.

[102] Union of Concerned Scientists. "Reprocessing and Nuclear Terrorism". *Union of Concerned Scientists*.

[103] Orszag, Peter R. "Costs of Reprocessing Versus Directly Disposing of Spent Nuclear Fuel". *Presentation*.

[104] Eben Harrell (August 15, 2011). "Bury Our Nuclear Waste —Before It Buries Us". *TIME*.

[105] "Nuclear industry to fight Yucca Mountain bill".

[106] "Japanese Crisis Highlights U.S. Atomic Waste Safety Problem". *Global Security Newswire*. March 24, 2011.

[107] Matthew Wald (January 24, 2012). "Wanted: Parking Space for Nuclear Waste". *New York Times*.

[108] David Biello (July 29, 2011). "Presidential Commission Seeks Volunteers to Store U.S. Nuclear Waste". *Scientific American*.

[109] Matthew Wald (January 26, 2012). "Revamped Search Urged for a Nuclear Waste Site". *New York Times*.

[110] Al Gore (2009). *Our Choice*, Bloomsbury. pp. 165-166.

[111] Suzanne Goldenberg. "US nuclear industry tries to hijack Obama's climate change bill". *the Guardian*.

[112] Motevalli, Golnar (January 22, 2008). "Nuclear power rebirth revives waste debate". *Reuters*. Retrieved 2008-05-15.

[113] "A Nuclear Power Renaissance?". *Scientific American*. April 28, 2008. Retrieved 2008-05-15.

[114] von Hippel, Frank N. (April 2008). "Nuclear Fuel Recycling: More Trouble Than It's Worth". *Scientific American*. Retrieved 2008-05-15.

[115] Kanter, James. "Is the Nuclear Renaissance Fizzling?".

[116] "EIA - Electricity Data".

[117] "Drought could close nuclear power plants - Weather - NBC News". *msnbc.com*.

[118] "Attention, Cities: You Can Sell Your Excess Wastewater to Nuclear Power Plants". *Fast Company*. 1 April 2010.

[119] Benjamin K. Sovacool (2011). *Contesting the Future of Nuclear Power: A Critical Global Assessment of Atomic Energy*, World Scientific, p. 118-119.

[120] https://www.dndkm.org/DOEKMDocuments/ BestPractices/26-EFCOG%20Best%20Practice%20-% 20SRS%20P%20and%20R%20Reactor%20Basins% 20ISD%20Final.pdf

[121] "World Nuclear Fuel Facilities". WISE Uranium Project. 14 May 2008. Retrieved 2008-05-15.

[122] "Duke Power Granted License Amendment by Nuclear Regulatory Commission To Use MOX Fuel". Duke Energy. March 3, 2005. Archived from the original on 2007-12-06. Retrieved 2008-05-15.

[123] "EPR: Generation III+ Performance" (PDF). 6 September 2007. Retrieved 2008-05-15.

[124] "Uranium Ash at Westinghouse Nuclear Fuel Plant Draws Fine". Environment News Service. 2004.

[125] "Kazakh, USA join forces in nuclear fuel supply". World Nuclear News.

[126] Matthew L. Wald. Nuclear Agency's Chief Praises Watchdog Groups. The New York Times, June 23, 1992.

[127] Mark Z. Jacobson. Nuclear power is too risky CNN, February 22, 2010.

[128] Robynne Boyd. Safety Concerns Delay Approval of the First U.S. Nuclear Reactor in Decades Scientific American, July 29, 2010.

[129] Obituary: John W. Gofman, 88, Scientist and Advocate for Nuclear Safety Dies New York Times, August 26, 2007.

[130] Nancy Folbre (March 28, 2011). "Renewing Support for Renewables". New York Times.

[131] Patrick Moore, Assault on Future Generations, Greenpeace report, p47-49, 1976 - pdf

[132] Moore, Patrick (2006-04-16). "Going Nuclear". Washington Post.

[133] Washington Post Article, Sunday, April 16, 2006 - Going Nuclear

[134] The Independent, Nuclear energy? Yes please!

[135] The Age Greenpeace is wrong —we must consider nuclear power, article by Patrick Moore, December 10, 2007

[136] Energy Revolution, Greenpeace report - pdf Archived October 7, 2013, at the Wayback Machine.

[137] "Radford, New Greenpeace Boss on Climate Change, Coal, and Nuclear Power". The Wall Street Journal. April 14, 2009.

[138] Stewart Brand (2009). Whole Earth Discipline: An Ecopragmatist Manifesto. Viking. ISBN 978-0-670-02121-5.

[139] Bernard L Cohen. "The Nuclear Energy Option".

[140] White House website. Advancing American energy, accessed 11 Oct. 2014.

[141] Henry J. Pulizzi and Christine Buurma, "Obama unveils loan guarantee for nuclear plant". Wall Street Journal, 16 Feb. 2010.

[142] Rick Jesse, Nuclear energy and an energy-independent future, White House website. 16 Feb. 2010.

[143] "America's nuclear industry struggles to get off the floor". The Economist. Feb 18, 2012.

[144] Ned Resnikoff, US to help build first new nuclear reactors in decades, MSNBC, 19 Feb. 2014.

[145] Nuclear plant inspections need to improve: report Reuters Mar 7, 2013

[146] Kilday, Gregg (29 May 2013). "Paul Allen Lends Support to Pro-Nuclear Doc 'Pandora's Promise'". The Hollywood Reporter. Retrieved 25 September 2013.

[147] O' Sullivan, Michael (13 June 2013). "'Pandora's Promise' movie review". The Washington Post. Retrieved 25 September 2013.

[148] Matthew Wald (April 27, 2014). "Nuclear Industry Gains Carbon-Focused Allies in Push to Save Reactors". New York Times.

[149] "First time majority oppose nuclear energy". Gallup. 16 March 2016.

[150] Michael Cooper (March 22, 2011). "Nuclear Power Loses Support in New Poll". The New York Times.

[151] M. V. Ramana (July 2011). "Nuclear power and the public". Bulletin of the Atomic Scientists. p. 44.

[152] "Americans' Support for Nuclear Energy Holds at Majority Level 6 Months After Japan Accident". PR Newswire. 3 October 2011.

[153] The Pew Research Center For The People and The Press (March 19, 2012). "As Gas Prices Pinch, Support for Oil and Gas Production Grows" (PDF).

[154] Rasmussen Reports, Energy Update: 39% Support Building More U.S. Nuclear Power Plants, 37% Oppose, 9 Jan. 2014.

[155] Sonja Schmid. "Nuclear Renaissance in the Age of Global Warming." Bridges. v.12. Stanford University, Dec. 2006.

[156] "The Daily Sentinel." Commission, City support NuStart. Retrieved on December 1, 2006

[157] "US energy bill favors new build reactors, new technology". Nuclear Engineering International. 12 August 2005. Retrieved 2007-12-26.

[158] Michael Grunwald & Juliet Eilperin (July 30, 2005). "Energy Bill Raises Fears About Pollution, Fraud Critics Point to Perks for Industry". Washington Post. Retrieved 2007-12-26.

[159] McCaffrey, Shannon (February 16, 2010). "Georgia Power still increasing rates". Associated Press. Retrieved 2010-02-16.

[160] A Comeback for Nuclear Power? New York Times, February 16, 2010.

[161] Louisiana goes nuclear, cnn.com, August 26, 2008

[162] Joint venture will build nuclear reactors in Newport News, The Virginian-Pilot, October 23, 2008

[163] "NRC: Combined License Applications for New Reactors".

[164] Chris Gadomski (20 February 2009). "Will nuclear rebound?". Nuclear Engineering International. Retrieved 2009-03-11.

[165] "NRC: Location of Projected New Nuclear Power Reactors".

[166] "News - The Advocate —Baton Rouge, Louisiana".

[167] "TVA board approves construction of nuclear plant". *The Tennessean*. August 18, 2011. Retrieved August 18, 2011.

[168] "TVA cuts contractors at Alabama Bellefonte nuclear site". *Reuters*. March 16, 2012.

[169] Tracy, Ryan (2012-03-30). "U.S. Approves Nuclear Plants in South Carolina". *Wall Street Journal*. Retrieved 2012-09-23.

[170] "NRC Approves Vogtle Reactor Construction". Nuclear Street. Retrieved 2012-02-09.

[171] Ayesha Rascoe (February 9, 2012). "U.S. approves first new nuclear plant in a generation". *Reuters*.

[172] "404 - - Southern Company".

[173] "404 - - Southern Company".

[174] Kristi E. Swartz (February 16, 2012). "Groups sue to stop Vogtle expansion project". *The Atlanta Journal-Constitution*.

[175] Stoddard, Patsy (January 24, 2017). "Update on the Nuclear Power Plant for Green River". Castle Dale, Utah: *Emery County Progress. Archived from the original on February 9, 2017. Retrieved February 9, 2017.*

[176] O'Donoghue, Amy Joi (October 27, 2011). "NRC holds hearing on Utah's proposed nuclear power plant". Salt Lake City, Utah: *Deseret Morning News. Archived from the original on February 7, 2017. Retrieved February 7, 2017.*

[177] Matthew L. Wald. (September 23, 2010). "Aid Sought for Nuclear Plants". *Green*. The New York Times.

[178] Amory Lovins (March–April 2012). "A Farewell to Fossil Fuels". *Foreign Affairs*.

[179] Exelon delays plan for Texas nuclear plant

[180] A key energy industry nervously awaits its 'rebirth'

[181] Reuters Editorial (29 January 2008). "MidAmerican drops Idaho nuclear project due to cost". *Reuters*.

[182] Matthew L. Wald. Vermont Senate Votes to Close Nuclear Plant *The New York Times*, February 24, 2010.

[183] Matthew L. Wald (December 7, 2010). "Nuclear 'Renaissance' Is Short on Largess". *The New York Times*.

[184] "Team France in disarray: Unhappy attempts to revive a national industry". *The Economist*. December 2, 2010.

[185] Ron Pernick and Clint Wilder (2012). "Clean Tech Nation" (PDF). p. 5.

[186] US Energy Information Administration. Total Energy.

[187] "NRC suspends final licensing decisions". 2012.

[188] "EIA predicts up to 4% fall in nuclear share of US generation by 2040". World Nuclear News. 16 April 2015. Retrieved 20 April 2015.

[189] John Quiggin (8 November 2013). "Reviving nuclear power debates is a distraction. We need to use less energy". *The Guardian*.

[190] "Vermont Green Energy Fund". Retrieved 21 February 2014.

[191] Mark Cooper (July 18, 2013). "Renaissance in reverse" (PDF). *Vermont Law School*.

[192] "Nuclear power in America: Constellation's cancellation". *The Economist*. October 16, 2010. p. 61.

[193] Matthew L. Wald (August 31, 2010). A Nuclear Giant Moves Into Wind *The New York Times*.

[194] United States of America: Nuclear Power Reactors - By Status

[195] "Exelon scraps Texas reactor project". *Nuclear Engineering International*. 29 August 2012. Retrieved 14 September 2012.

[196] "Some merchant nuclear plants could face early retirement: UBS". *Platts*. 9 January 2013. Retrieved 10 January 2013.

[197] "Dominion To Close, Decommission Kewaunee Power Station". Dominion. 22 October 2012. Retrieved 28 February 2013.

[198] Caroline Peachey (1 January 2013). "Why are North American plants dying?". Nuclear Engineering International. Retrieved 28 February 2013.

[199] Ed Crooks (19 February 2014). "Uneconomic US nuclear plants at risk of being shut down". *Financial Times*. Retrieved 25 February 2014.

3.14   External links

- GA Mansoori, N Enayati, LB Agyarko (2016). Energy: Sources, Utilization, Legislation, Sustainability, Illinois as Model State, World Sci. Pub. Co., ISBN 978-981-4704-00-7

- Comment: A US nuclear future? *Nature*, Vol. 467, 23 September 2010, pp. 391–393.

- World Nuclear Association

- US Nuclear Power Plants - General U.S. Nuclear Info

- World Nuclear Association: Nuclear energy in the world

- The Nuclear Energy Institute: The policy organization of the nuclear energy and technologies industry

- Nuclear power plant operators in the United States (SourceWatch).

- How many people live near a nuclear power plant in the United States? Data Visualization

Chapter 4

Doomsday Clock

For the Smashing Pumpkins song, see Doomsday Clock (song).

"Minutes to Midnight" redirects here. For other uses, see Minutes to Midnight (disambiguation).

The Doomsday Clock pictured at its most recent setting of "two and a half minutes to midnight".

The **Doomsday Clock** is a symbol which represents the likelihood of a global catastrophe. Maintained since 1947 by the members of *The Bulletin of the Atomic Scientists'* Science and Security Board,[1] the Clock, which hangs on a wall in *The Bulletin*'s office in the University of Chicago,[2] represents an analogy for the threat of global nuclear war. Since 2007, it has also reflected climate change[3] and new developments in the life sciences and technology that could inflict irrevocable harm to humanity.[4]

The Clock represents the hypothetical global catastrophe as "midnight", and *The Bulletin*'s opinion on how close the world is to a global catastrophe as a number of "minutes" to midnight. Its original setting in 1947 was seven minutes to midnight. It has been set backward and forward 22 times since then, the smallest ever number of minutes to midnight being two (in 1953) and the largest seventeen (in 1991). As of January 2017, the Clock is set at two and a half minutes to midnight, due to a "rise of 'strident nationalism' worldwide, United States President Donald Trump's comments over nuclear weapons, and the disbelief in the scientific consensus over climate change by the Trump Administration."[5][6] This setting is the Clock's second closest approach to midnight since its introduction, after it was set to two minutes to midnight in 1953.

4.1 History

The Doomsday Clock's origin can be traced to the international group of researchers called the Chicago Atomic Scientists, who had participated in the Manhattan Project.[2] After the atomic bombings of Hiroshima and Nagasaki, they began publishing a mimeographed newsletter and then the magazine, *Bulletin of the Atomic Scientists*, which, since its inception, has depicted the Clock on every cover. The Clock was first represented in 1947, when *The Bulletin* co-founder Hyman Goldsmith asked artist Martyl Langsdorf (wife of Manhattan Project research associate and Szilárd petition signatory Alexander Langsdorf, Jr.) to design a cover for the magazine's June 1947 issue. As Eugene Rabinowitch, another co-founder of *The Bulletin*, explained later,

> The Bulletin's clock is not a gauge to register the ups and downs of the international power struggle; it is intended to reflect basic changes in the level of continuous danger in which mankind lives in the nuclear age...[7]

In January 2007, designer Michael Bierut, who was on *The Bulletin*'s Governing Board, redesigned the Clock to give it

Cover of the 1947 Bulletin of the Atomic Scientists issue, featuring the Doomsday Clock at "seven minutes to midnight".

4.1.1 Changes

In 1947, during the Cold War, the Clock was started at seven minutes to midnight ("midnight" being a hypothetical global catastrophe) and was subsequently advanced or rewound per the state of the world and nuclear warfare prospects. The Clock's setting is decided by *The Bulletin of the Atomic Scientists*'s Science and Security Board and is completely arbitrary without a specified starting time. The Clock is not set and reset in real time as events occur; rather than respond to each and every crisis as it happens, the Science and Security Board meets twice annually to discuss global events in a deliberative manner. The closest nuclear war threat, the Cuban Missile Crisis in 1962, reached crisis, climax, and resolution before the Clock could be set to reflect that possible doomsday.[13]

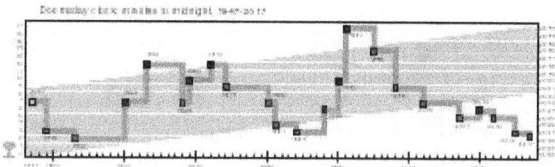

Doomsday Clock graph, 1947–2017. The lower points on the graph represent a higher probability of technologically or environmentally-induced catastrophe, and the higher points represent a lower probability.

a more modern feel. In 2009, *The Bulletin* ceased its print edition and became one of the first print publications in the U.S. to become entirely digital; the Clock is now found as part of the logo on *The Bulletin*'s website. Information about the Doomsday Clock Symposium,[8] a timeline of the Clock's settings,[5] and multimedia shows about the Clock's history and culture[9] can also be found on *The Bulletin*'s website.

The 5th Doomsday Clock Symposium[8] was held on November 14, 2013, in Washington, D.C.; it was a daylong event that was open to the public and featured panelists discussing various issues on the topic "Communicating Catastrophe". There was also an evening event at the Hirshhorn Museum and Sculpture Garden in conjunction with the Hirshhorn's current exhibit, "Damage Control: Art and Destruction Since 1950".[10] The panel discussions, held at the American Association for the Advancement of Science, were streamed live from *The Bulletin*'s website and can still be viewed there.[11] Reflecting international events dangerous to humankind, the Clock has been adjusted 22 times since its inception in 1947,[12] when it was set to "seven minutes to midnight".

4.2 In popular culture

The Clock is featured in the 1986–87 comic book series Watchmen and the 2009 film *Watchmen*, where it is set at five minutes to midnight.[25] It is also featured in an *xkcd* comic, where the main character pulls the hour hand forward an hour in accordance with the start of Daylight Saving Time.[26] The title of Iron Maiden's song "2 Minutes to Midnight" is a reference to the Clock.[27]

4.3 See also

- Apocalypticism
- Doomsday device
- Eschatology
- Mutual assured destruction
- Risks to civilization, humans, and planet Earth
- Svalbard Global Seed Vault
- *The Bomb* (film)

4.4 References

[1] "Science and Security Board". *The Bulletin of the Atomic Scientists*.

[2] "Doomsday Clock moving closer to midnight?". *The Spokesman-Review*. October 16, 2006.

[3] Stover, Dawn (September 26, 2013). "How Many Hiroshimas Does it Take to Describe Climate Change?". *The Bulletin of the Atomic Scientists*.

[4] "'Doomsday Clock' Moves Two Minutes Closer To Midnight". *The Bulletin of the Atomic Scientists*. Retrieved June 29, 2013.

[5] "Timeline". *The Bulletin of the Atomic Scientists*. January 2015.

[6] Science and Security Board Bulletin of the Atomic Scientists. "It is two and a half minutes to midnight" (PDF). Bulletin of the Atomic Scientists. Retrieved January 26, 2017.

[7] "The Doomsday Clock". *The Southeast Missourian*. February 22, 1984.

[8] "Doomsday Clock Symposium". *The Bulletin of the Atomic Scientists*. Retrieved September 10, 2013.

[9] "A Timeline of Conflict, Culture, and Change". *The Bulletin of the Atomic Scientists*. Retrieved June 20, 2013.

[10] "Damage Control: Art and Destruction Since 1950". *Hirshhorn Museum and Sculpture Garden*. 2013.

[11] "5th Doomsday Clock Symposium". *The Bulletin of the Atomic Scientists*. Retrieved September 14, 2013.

[12] "Doomsday Clock ticks closer to midnight". *Washington Post*. January 10, 2012. Retrieved January 10, 2012.

[13] "Remembering the Cuban Missile Crisis". *Bulletin of the Atomic Scientists*. Retrieved 6 October 2016.

[14] "Doomsday Clock at 3'til midnight". *The Daily News*. December 21, 1983.

[15] "Hands of the 'Doomsday Clock' turned back three minutes". *The Reading Eagle*. December 17, 1987.

[16] "The North Korean nuclear test". *The Bulletin of the Atomic Scientists*. 2009. Retrieved 2009-08-04.

[17] "'Doomsday Clock' Moves Two Minutes Closer To Midnight". *The Bulletin of Atomic Scientists*. 17 January 2007. Retrieved 6 April 2015.

[18] "Nukes, climate push 'Doomsday Clock' forward". *MSNBC*. 2012-01-15. Retrieved 2012-01-15.

[19] "Doomsday Clock moves to five minutes to midnight". *Bulletin of the Atomic Scientists*. Retrieved 2013-06-29.

[20] Casey, Michael (22 January 2015). "Doomsday Clock moves two minutes closer to midnight". *CBS News*. Retrieved 23 January 2015.

[21] "Board moves the Clock ahead". *Bulletin of the Atomic Scientists*. 26 January 2017. Retrieved January 26, 2017.

[22] Holley, Peter; Ohlheiser, Abby; Wang, Amy B. "The Doomsday Clock just advanced, 'thanks to Trump': It's now just 2½ minutes to 'midnight.'". *Washington Post*. Retrieved 26 January 2017.

[23] Bromwich, Jonah Engel (26 January 2017). "Doomsday Clock Moves Closer to Midnight, Signaling Concern Among Scientists". *The New York Times*. Retrieved 26 January 2017.

[24] Chappell, Bill. "The Doomsday Clock Is Reset: Closest To Midnight Since The 1950s". *NPR.org*. Retrieved 26 January 2017.

[25] Couch, Aaron. "Real-Life Doomsday Clock Is Closer to Midnight Than in 'Watchmen'". *The Hollywood Reporter*. Retrieved January 28, 2017.

[26] xkcd, "Doomsday Clock"

[27] Bowen, LB (January 24, 2017). "Doomsday Clock: Iron Maiden – Two Minutes to Midnight". *OnStage Magazine*. Retrieved February 11, 2017.

4.5 External links

- *Bulletin of the Atomic Scientists*
- Timeline of the Doomsday Clock

Chapter 5

Nuclear safety in the United States

A clean-up crew working to remove radioactive contamination after the Three Mile Island accident.

Nuclear safety in the U.S. is governed by federal regulations issued by the Nuclear Regulatory Commission (NRC). The NRC regulates all nuclear plants and materials in the U.S. except for of nuclear plants and materials controlled by the U.S. government, as well those powering naval vessels.[1][2]

The 1979 Three Mile Island accident was a pivotal event that led to questions about U.S. nuclear safety.[3] Earlier events had a similar effect, including a 1975 fire at Browns Ferry, the 1976 testimonials of three concerned GE nuclear engineers, the GE Three. In 1981, workers inadvertently reversed pipe restraints at the Diablo Canyon Power Plant reactors, compromising seismic protection systems, which further undermined confidence in nuclear safety. All of these well-publicised events, undermined public support for the U.S. nuclear industry in the 1970s and the 1980s.[3] In 2002, the USA had what former NRC Commissioner

Victor Gilinsky termed "its closest brush with disaster" since Three Mile Island's 1979 meltdown; a workman at the Davis-Besse reactor found a large rust hole in the top of the reactor pressure vessel.[4]

Recent concerns have been expressed about safety issues affecting a large part of the nuclear fleet of reactors. In 2012, the Union of Concerned Scientists, which tracks ongoing safety issues at operating nuclear plants, found that "leakage of radioactive materials is a pervasive problem at almost 90 percent of all reactors, as are issues that pose a risk of nuclear accidents".[5]

Following the Japanese Fukushima Daiichi nuclear disaster, according to Black & Veatch's annual utility survey that took place after the disaster, of the 700 executives from the US electric utility industry that were surveyed, nuclear safety was the top concern.[6] There are likely to be increased requirements for on-site spent fuel management and elevated design basis threats at nuclear power plants.[7][8] License extensions for existing reactors will face additional scrutiny, with outcomes depending on the degree to which plants can meet new requirements, and some of the extensions already granted for more than 60 of the 104 operating U.S. reactors could be revisited. On-site storage, consolidated long-term storage, and geological disposal of spent fuel is "likely to be reevaluated in a new light because of the Fukushima storage pool experience".[7]

In October 2011, the Nuclear Regulatory Commission instructed agency staff to move forward with seven of the 12 safety recommendations put forward by the federal task force in July. The recommendations include "new standards aimed at strengthening operators' ability to deal with a complete loss of power, ensuring plants can withstand floods and earthquakes and improving emergency response capabilities". The new safety standards will take up to five years to fully implement.[9]

5.1 Scope

The topic of **nuclear safety** covers:

- The research and analysis of possible or potential incidents or events at nuclear facilities,

- The equipment and procedures designed to prevent those incidents or events from having serious consequences,

- The actions to reduces the consequences of those incidents or events,

- The calculation of the probabilities, and the seriousness, of equipment, procedures or actions failing,

- The evaluation of the possible timing and scope of those consequences,

- The actions taken to protect the public during a release of radioactivity,

- The training and rehearsals performed to ensure readiness in case an incident/event occurs.

This article will also consider accidents that have occurred.

In the following, the names of federal regulations will be abbreviated in the standard way. For example, "Code of Federal Regulations, Title 10, Part 100, Section 23" will be given as "10CFR100.23".

5.2 Issues

More than a quarter of U.S. nuclear plant operators "have failed to properly tell regulators about equipment defects that could imperil reactor safety", according to a Nuclear Regulatory Commission report.[10]

In February 2011, a major manufacturer in the nuclear industry reported a potential "substantial safety hazard" with control rods at more than two dozen reactors around the USA. GE Hitachi Nuclear Energy said it had discovered extensive cracking and "material distortion," and recommended that the boiling water reactors using its Marathon control rod blades replace them more frequently than previously told. If the design life is not revised, it "could result in significant control blade cracking and could, if not corrected, create a substantial safety hazard and is considered a reportable condition," the company said in its report to the NRC.[11]

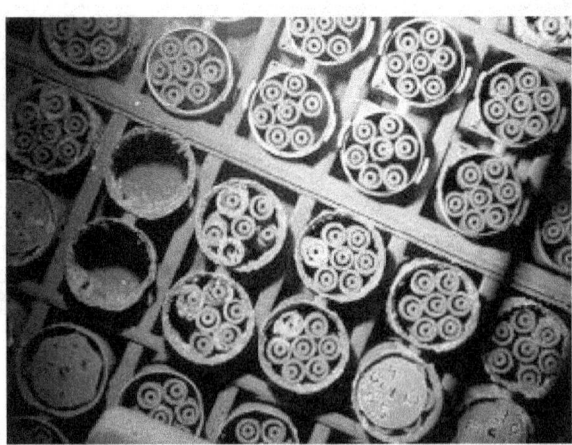

Spent nuclear fuel stored underwater and uncapped at the Hanford site in Washington, USA.

5.2.1   Radioactive waste storage

The Fukushima nuclear disaster has reopened questions about the risks of U.S. nuclear reactors, and especially the pools that store spent fuel. In March 2011, nuclear experts told Congress that spent-fuel pools at US nuclear power plants are too full. A fire at a spent-fuel pool could release cesium-137. Experts say the entire US spent-fuel policy should be overhauled in light of Fukushima I.[12][13]

With the cancellation of the Yucca Mountain nuclear waste repository in Nevada, more nuclear waste is being loaded into sealed metal casks filled with inert gas. Many of these casks will be stored in coastal or lakeside regions where a salt air environment exists, and the Massachusetts Institute of Technology is studying how such dry casks perform in salt environments. Some hope that the casks can be used for 100 years but cracking related to corrosion could occur in 30 years or less.[14] Robert Alvarez, a former Department of Energy official who oversaw nuclear issues, said dry casks would provide safer storage until a permanent nuclear repository was built and loaded, a process that would take decades.[15]

At places like Maine Yankee, Connecticut Yankee and Rancho Seco, reactors no longer operate, but the spent fuel remains in small concrete-and-steel silos that require maintenance and monitoring by a guard force. Sometimes the presence of nuclear waste prevents re-use of the sites by industry.[16]

Without a long-term solution to store nuclear waste, a nuclear renaissance in the U.S. remains unlikely. Nine states have "explicit moratoria on new nuclear power until a storage solution emerges".[17]

Some nuclear power advocates argue that the United States should develop factories and reactors that will recycle some of the spent fuel. (It is not now the policy of the United States to recycle its spent nuclear fuel.) But the Blue Ribbon Commission on America's Nuclear Future said in 2012 that "no existing technology was adequate for that purpose, given cost considerations and the risk of nuclear proliferation".[18]

5.2.2   Earthquake risk

About one third of reactors in the US are boiling water reactors, the same technology which was involved in the Fukushima Daiichi nuclear disaster in Japan. There are also eight nuclear power plants located along the seismically active West coast. Twelve of the American reactors that are of the same vintage as the Fukushima Daiichi plant are in seismically active areas.[19] Earthquake risk is often measured by "Peak Ground Acceleration", or PGA. The following nuclear power plants have a two percent or greater chance of having PGA over 0.15g in the next 50 years: Diablo Canyon, Calif.; San Onofre, Calif.; Sequoyah, Tenn.; H.B. Robinson, SC.; Watts Bar, Tenn.; Virgil C. Summer, SC.; Vogtle, GA.; Indian Point, NY.; Oconee, SC.; and Seabrook, NH.[19]

5.2.3   GE Mark 1 reactor containment design

Experts have long criticized General Electric's Mark I reactor containment design, because it offered a relatively weak containment vessel.[20] Three GE scientists resigned 35 years ago in protest of the design of the Mark I containment system.[21] David Lochbaum, chief nuclear safety officer with the Union of Concerned Scientists, has repeatedly questioned the safety of the Fukushima I Plant's GE Mark 1 reactor containment design.[22] In a 2012 nuclear power safety report, David Lochbaum and Edwin Lyman said:

> The designs of the Fukushima reactors closely resemble those of many U.S. reactors, and the respective emergency response procedures are comparable as well. But while most U.S. reactors may not be vulnerable to that site's specific earthquake/ tsunami sequence, they are vulnerable to other severe natural disasters. Moreover, similarly serious conditions could be created by a terrorist attack.[23]

5.2.4   Aging of nuclear reactors

An important concern in the nuclear safety field is the aging of nuclear reactors. Quality Assurance Technicians, weld inspectors and radiographers use ultrasonic waves to look for cracks and other defects in hot metal parts, in order to identify "microscale" defects that lead to big cracks.[14]

5.2.5   Population considerations

111 million people live within 50 miles of a U.S. nuclear power plant.[24]

5.2.6   Terrorist attack

In February 1993, a man drove his car past a check point the Three Mile Island Nuclear plant, then broke through an entry gate. He eventually crashed the car through a secure door and entered the Unit 1 reactor turbine building. The intruder, who had a history of mental illness, hid in a building and was not apprehended for four hours. Stephanie Cooke asks: "What if he'd been a terrorist armed with a ticking bomb?"[25]

After 9/11, it would seem prudent for nuclear plants to be prepared for an attack by a large, well-armed terrorist group. But the Nuclear Regulatory Commission, in revising its security rules, decided not to require that plants be able to defend themselves against groups carrying sophisticated weapons. According to a study by the Government Accountability Office, the N.R.C. appeared to have based its revised rules "on what the industry considered reasonable and feasible to defend against rather than on an assessment of the terrorist threat itself".[26][27]

The Protected Area encloses the Exclusion Zone (as defined in 10CFR100.3 [28]). It also serves as a security zone, within which only trusted, FBI background-checked and badged individuals are allowed to walk unescorted. The Protected Area is surrounded by a number of closely monitored, motion-detection protected fences, and the gap in between the fences is electronically monitored. There are many layers of gates, and those are well guarded. Numerous other security measures are in effect.[29]

The missile shield protecting the containment structure is intended to protect not only from natural forces, such as tornadoes, but is designed to be strong enough to withstand a direct hit from a 747 jetliner. One plant, Florida's Turkey Point NGS, survived a direct hit by Category 5 Hurricane Andrew in 1992, with no damage to the containment. No actual missile shield has been subjected to an aircraft impact test. However, a highly similar test was done at Sandia National Laboratories and filmed (see Containment building), and the target was essentially undamaged (reinforced concrete is strongly resistant both to impact and to fire). The NRC's Chairman has said "Nuclear power plants are inherently robust structures that our studies provide adequate protection in a hypothetical attack by an airplane. The NRC has also taken actions that require nuclear power plant operators to be able to manage large fires or explosions - no matter what has caused them." [30]

5.2.7   Flood risks

Fort Calhoun Nuclear Generating Station surrounded by the 2011 Missouri River Floods on June 16, 2011

In 2012, Larry Criscione and Richard H. Perkins publicly accused the US Nuclear Regulatory Commission of downplaying flood risks for nuclear plants which are sited on waterways downstream from large reservoirs and dams. They are engineers with over 20 years of combined government and military service who work for the NRC. Other nuclear safety advocates have supported their complaints. [31]

5.2.8   Procedures

In the U.S., the Operating License is granted by the government and carries the force of law. The Final Safety Analysis Report (FSAR) is part of the Operating License, and the plant's Technical Specifications (which contain the restrictions the operators consult during operation) are a chapter of the FSAR. All procedures are checked against the Technical Specifications and also by a Transient Analysis engineer, and each copy of an approved procedure is numbered and the copies controlled (so that updating all copies at once can be assured). In a U.S. nuclear power plant, unlike in most other industries, approved procedures carry the force of law and to deliberately violate one is a criminal act.

5.2.9   Reactor Protective System (RPS)

Main article: Reactor Protective System

5.2.10   Design Basis Events

"Design Basis Events [DBE] are defined as conditions of normal operation, including anticipated operational occurrences, design basis accidents, external events, and natural phenomena for which the plant must be designed to ensure functions (b)(1)(i) (A) through (C)" of 10CFR50-49. [32] These include (A) maintaining the integrity of the reactor coolant pressure boundary; (B) maintaining the capability to shut down the reactor and maintain it in a safe shutdown condition; *OR* (C) maintaining the capability to prevent or mitigate the consequences of accidents that could result in potential offsite exposures. The normal DBE evaluated is loss-of-coolant accident (LOCA).

The Fukushima I nuclear accident was caused by a "beyond design basis event," the tsunami and associated earthquakes were more powerful than the plant was designed to accommodate, and the accident is directly due to the tsunami overflowing the too-low seawall. [33] Since then, the possibility of unforeseen beyond design basis events has been a major concern for plant operators. [34]

5.2.11   Whistleblowers

Main article: Nuclear whistleblowers

There have been a number of nuclear whistleblowers, often nuclear engineers, who have identified safety concerns at nuclear power plants in the United States. In 1976 Gregory Minor, Richard Hubbard, and Dale Bridenbaugh "blew the whistle" on safety problems at nuclear power plants in the United States. The three nuclear engineers gained the attention of journalists and their disclosures about the threats of nuclear power had a significant impact. George Galatis was a senior nuclear engineer who reported safety problems at the Millstone 1 Nuclear Power Plant, relating to reactor refueling procedures, in 1995. [35] [36] Other nuclear whistleblowers include Arnold Gundersen and David Lochbaum.

5.2.12   Assessments of risks

The NRC (and its predecessors) have over the decades produced three major analyses of the risks of nuclear power: a fourth, all-encompassing one (the State-of-the-Art Reactor Consequence Analyses, or *SOARCA*, study) is in generation now. The new study will be based on actual test results, on probabilistic risk assessment (PRA) methodology, and on the evaluated actions of government agencies.

The existing studies are:

- NUREG-1150 (1991)

- CRAC-II (1982) (based on WASH-1400 results)

- WASH-1400 (1975)

- WASH-740 (1957) (not PRA-based)

Reactor vendors now routinely calculate probabilistic risk assessments of their nuclear power plant designs. General Electric has recalculated maximum core damage frequencies per year per plant for its nuclear power plant designs:[37]

BWR/4 —1×10^{-5} (a typical plant)

BWR/6 —1×10^{-6} (a typical plant)

ABWR —2×10^{-7} (now operating in Japan)

ESBWR —3×10^{-8} (submitted for Final Design Approval by NRC)

The proposed AP1000 has a maximum core damage frequency of 5.09×10^{-7} per plant per year. The European Pressurized Reactor (EPR) has a maximum core damage frequency of 4×10^{-7} per plant per year.[38]

According to the Nuclear Regulatory Commission, 20 states in the USA have requested stocks of potassium iodide which the NRC suggests should be available for those living within 10 miles (16 km) of a nuclear power plant in the unlikely event of a severe accident.[39] Radioactive iodine (radioiodine) is one of the products that can be released in a serious nuclear power plant accident. Potassium iodide (KI) is a non radioactive form of iodine that may be taken to reduce the amount of radioactive iodine absorbed by the body's thyroid gland. When taken before or shortly after a radiological exposure, potassium iodide blocks the thyroid glands ability to absorb radioactive iodine. Potassium iodide should be taken by the public during an emergency only when directed by public health officials.

5.3   Accidents

Main articles: Nuclear accidents and List of canceled nuclear plants in the United States

5.3.1   Emergency Classifications

The NRC established a classification scale for nuclear power plant events to ensure consistency in the communications and emergency response.

- Unusual Event—This is the lowest of the four emergency classifications. This classification indicates that a small problem has occurred. No release of radioactive material is expected and federal, state and county officials are notified.

- Alert—Events are in process or have occurred which involve an actual or potential substantial degradation in the level of safety of the plant. Any releases of radioactive material from the plant are expected to be limited to a small fraction of the Environmental Protection Agency (EPA) Protective Action Guide for Nuclear Incidents (PAGs)

- Site Area Emergency—Involves events in process or which have occurred that result in actual or likely major failures of plant functions needed for protection of the public. Any releases of radioactive material are not expected to exceed the levels established by the EPA PAGs except near the site boundary.

- General Emergency—The most serious emergency classification and indicates a serious problem. A general emergency involves actual or imminent substantial core damage or melting of reactor fuel with the potential for loss of containment integrity. Emergency sirens will be sounded and federal, state and county officials will act to ensure public safety. Radioactive releases during a general emergency can reasonably be expected to exceed the EPA PAGs for more than the immediate site area.

5.3.2   Rocky Flats Plant

The Rocky Flats Plant, a former U.S. nuclear weapons production facility in the state of Colorado, caused radioactive contamination within and outside its boundaries and also produced "area-wide contamination of the Denver area."[40][41] The contamination resulted from decades of emissions, leaks and fires that released radioactive isotopes, largely plutonium (Pu-239), into the environment. The plant was located about 15 miles upwind from Denver and has since been shut down and its buildings demolished and completely removed from the site. Public protests and a combined Federal Bureau of Investigation (FBI) and United States Environmental Protection Agency (EPA) raid in 1989 stopped production at the Rocky Flats Plant.[42]

As noted in a scientific journal, "Exposures of a large population in the Denver area to plutonium and other radionuclides in the exhaust plumes from the plant date back to 1953."[43] Moreover, in 1957 there was a major Pu-239 fire at the plant, followed by another major fire in 1969. Both of these fires resulted in this radioactive material being released into the atmosphere, with the then-secret 1957

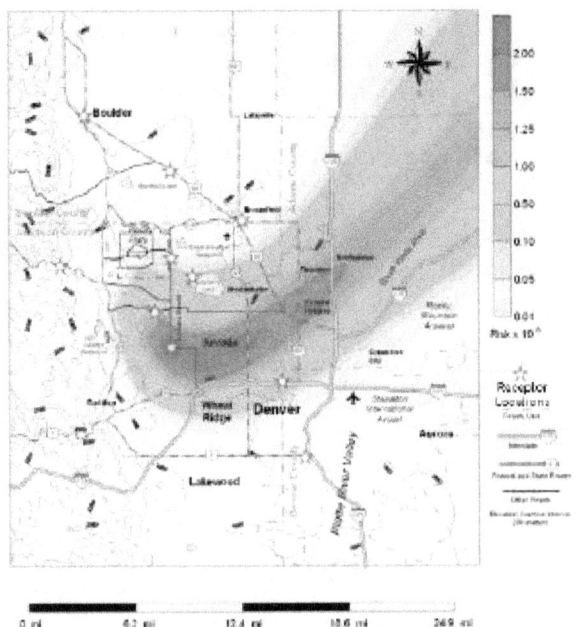

The Hanford site represents two-thirds of America's high-level radioactive waste by volume. Nuclear reactors line the riverbank at the Hanford Site along the Columbia River in January 1960.

One of four example estimates of the plutonium (Pu-239) plume from the 1957 fire at the Rocky Flats nuclear weapons plant. More info.

fire being the more serious of the two. The contamination of the Denver area by plutonium from these fires and other sources was not reported until the 1970s, and as of 2011 the U.S. Government continues to withhold data on post-Superfund cleanup contamination levels. Elevated levels of plutonium have been found in the remains of cancer victims living near the Rocky Flats site, and breathable plutonium outside the former boundaries of the plant was found in August 2010.[42][44][45][46]

5.3.3 Hanford Site

The Hanford Site is a mostly decommissioned nuclear production complex on the Columbia River in the U.S. state of Washington, operated by the United States federal government. Plutonium manufactured at the site was used in the first nuclear bomb, tested at the Trinity site, and in Fat Man, the bomb detonated over Nagasaki, Japan. During the Cold War, the project was expanded to include nine nuclear reactors and five large plutonium processing complexes, which produced plutonium for most of the 60,000 weapons in the U.S. nuclear arsenal.[47][48] Many of the early safety procedures and waste disposal practices were inadequate, and government documents have since confirmed that Hanford's operations released significant amounts of radioactive materials into the air and the Columbia River, which still threatens the health of residents and ecosystems.[49] The weapons production reac-

tors were decommissioned at the end of the Cold War, but the decades of manufacturing left behind 53 million US gallons (200,000 m^3) of high-level radioactive waste,[50] an additional 25 million cubic feet (710,000 m^3) of solid radioactive waste, 200 square miles (520 km^2) of contaminated groundwater beneath the site[51] and occasional discoveries of undocumented contaminations that slow the pace and raise the cost of cleanup.[52] The Hanford site represents two-thirds of the nation's high-level radioactive waste by volume.[53] Today, Hanford is the most contaminated nuclear site in the United States[54][55] and is the focus of the nation's largest environmental cleanup.[47]

5.3.4 SL-1 meltdown

The SL-1, or Stationary Low-Power Reactor Number One, was a United States Army experimental nuclear power reactor which underwent a steam explosion and meltdown on January 3, 1961, killing its three operators. The direct cause was the improper withdrawal of the central control rod, responsible for absorbing neutrons in the reactor core. The event is the only known fatal reactor accident in the United States.[56][57] The accident released about 80 curies (3.0 TBq) of iodine-131,[58] which was not considered significant due to its location in a remote desert of Idaho. About 1,100 curies (41 TBq) of fission products were released into the atmosphere.[59]

5.3.5 Three Mile Island

Main article: Three Mile Island accident

On March 28, 1979, equipment failures and operator error contributed to loss of coolant and a partial core meltdown

This image of the SL-1 core served as a sober reminder of the damage that a nuclear meltdown can cause.

President Jimmy Carter leaving Three Mile Island for Middletown, Pennsylvania, April 1, 1979.

at the Three Mile Island Nuclear Power Plant in Pennsylvania. The mechanical failures were compounded by the initial failure of plant operators to recognize the situation as a loss-of-coolant accident due to inadequate training and human factors, such as human-computer interaction design oversights relating to ambiguous control room indicators in the power plant's user interface. In particular, a hidden indicator light led to an operator manually overriding the automatic emergency cooling system of the reactor because the operator mistakenly believed that there was too much

coolant water present in the reactor and causing the steam pressure release.[60] The scope and complexity of the accident became clear over the course of five days, as employees of Met Ed, Pennsylvania state officials, and members of the U.S. Nuclear Regulatory Commission (NRC) tried to understand the problem, communicate the situation to the press and local community, decide whether the accident required an emergency evacuation, and ultimately end the crisis. The NRC's authorization of the release of 40,000 gallons of radioactive waste water directly in the Susquehanna River led to a loss of credibility with the press and community.[60]

The 1979 Three Mile Island accident inspired Perrow's book *Normal Accidents*, where a nuclear accident occurs, resulting from an unanticipated interaction of multiple failures in a complex system. TMI was an example of a normal accident because it was "unexpected, incomprehensible, uncontrollable and unavoidable" .[61]

> Perrow concluded that the failure at Three Mile Island was a consequence of the system's immense complexity. Such modern high-risk systems, he realized, were prone to failures however well they were managed. It was inevitable that they would eventually suffer what he termed a 'normal accident'. Therefore, he suggested, we might do better to contemplate a radical redesign, or if that was not possible, to abandon such technology entirely.[62]

A fundamental issue contributing to a nuclear power system's complexity is its extremely long lifetime. The timeframe from the start of construction of a commercial nuclear power station through the safe disposal of its last radioactive waste, may be 100 to 150 years.[63]

The World Nuclear Association has stated that cleanup of the damaged nuclear reactor system at TMI-2 took nearly 12 years and cost approximately US $973 million.[64] Benjamin K. Sovacool, in his 2007 preliminary assessment of major energy accidents, estimated that the TMI accident caused a total of $2.4 billion in property damages.[65] The health effects of the Three Mile Island accident are widely, but not universally, agreed to be very low level.[64][66] The accident triggered protests around the world.[67]

5.3.6   List of accidents

See also: Nuclear power plant accidents in the United States

The United States Government Accountability Office reported more than 150 incidents from 2001 to 2006 alone of nuclear plants not performing within acceptable safety

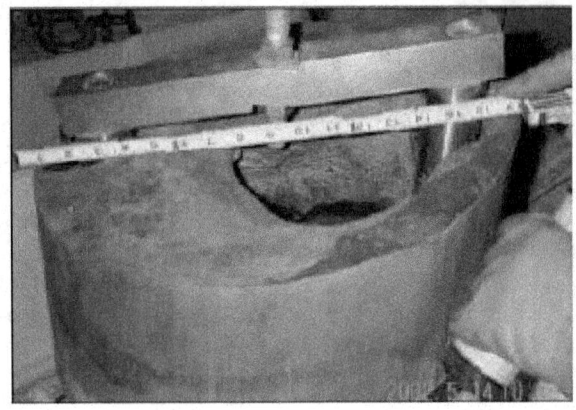

Erosion of the 6-inch-thick (150 mm) carbon steel reactor head at Davis-Besse Nuclear Power Plant in 2002, caused by a persistent leak of borated water.

Following the 2011 Japanese Fukushima nuclear disaster, authorities shut down the nation's 54 nuclear power plants. As of 2013, the Fukushima site remains highly radioactive, with some 160,000 evacuees still living in temporary housing, and some land will be unfarmable for centuries. The difficult cleanup job will take 40 or more years, and cost tens of billions of dollars. [24] [73]

guidelines. In 2006, it said: "Since 2001, the ROP has resulted in more than 4,000 inspection findings concerning nuclear power plant licensees' failure to fully comply with NRC regulations and industry standards for safe plant operation, and NRC has subjected more than 7.5 percent (79) of the 103 operating plants to increased oversight for varying periods".[68] Seventy-one percent of all recorded major nuclear accidents, including meltdowns, explosions, fires, and loss of coolants, occurred in the United States, and they happened during both normal operations as well as emergency situations such as floods, droughts, and earthquakes.[69]

5.3.7 Chernobyl

Experts have disagreed about whether an accident as serious as the Chernobyl disaster could occur in the USA.[72] In 1986, Commissioner Asselstine testified before Congress that:

> While we hope that their occurrence is unlikely, there are accident sequences for U.S. plants that can lead to rupture or by-passing of the containment in U.S. reactors which would result in the off-site release of fission products comparable or worse than the releases estimated by the NRC to have taken place during the Chernobyl accident.[72]

5.3.8 Fukushima implications

Following the Fukushima Daiichi nuclear disaster, according to Black & Veatch's annual utility survey that took place after the disaster, of the 700 executives from the US electric utility industry that were surveyed, nuclear safety

was the top concern.[6] There are likely to be increased requirements for on-site spent fuel management and elevated design basis threats at nuclear power plants.[7][8] License extensions for existing reactors will face additional scrutiny, with outcomes depending on the degree to which plants can meet new requirements, and some of the extensions already granted for more than 60 of the 104 operating U.S. reactors could be revisited. On-site storage, consolidated long-term storage, and geological disposal of spent fuel is "likely to be reevaluated in a new light because of the Fukushima storage pool experience".[7]

In October 2011, the Nuclear Regulatory Commission instructed agency staff to move forward with seven of the 12 safety recommendations put forward by the federal task force in July. The recommendations include "new standards aimed at strengthening operators' ability to deal with a complete loss of power, ensuring plants can withstand floods and earthquakes and improving emergency response capabilities". The new safety standards will take up to five years to fully implement.[9]

On February 9, 2012 Jaczko cast the lone dissenting vote on plans to build the first new nuclear power plant in more than 30 years when the NRC voted 4-1 to allow Atlanta-based Southern Co to build and operate two new nuclear power reactors at its existing Vogtle nuclear power plant in Georgia. He cited safety concerns stemming from Japan's 2011 Fukushima nuclear disaster, saying "I cannot support issuing this license as if Fukushima never happened".[74]

5.4 Recent developments

According to senior scientist Edwin Lyman from the UCS, despite the events of September 11, the Nuclear Regulatory

Commission (NRC) has voted to delay implementation of safety and security upgrades in ways that will weaken protection of nuclear power plants."[75]

Experience has shown that having a good security plan on paper is no guarantee that it could be implemented in practice. Yet, upgraded NRC-run "force-on-force" security exercises (using a team of mock nuclear terrorists) have been delayed. Also, the schedule for developing new requirements for protecting dry cask spent fuel storage from sabotage has been put back by five years, to the end of 2023."[75]

Lyman says that these new moves illustrate an "ominous trend". Pressure from the nuclear industry to delay tighter security arrangements has succeeded, with the full support of the NRC' s commissioners. Commissioners backing of these retrograde measures could be seen as providing industry protection rather than defending the safety of the public."[75]

5.5   See also

- Nuclear safety
- Nuclear power
- Nuclear power in the United States
- List of nuclear reactors
- Institute of Nuclear Power Operations
- Nuclear Power 2010 Program
- List of books about nuclear issues
- Lists of nuclear disasters and radioactive incidents
- Nuclear whistleblowers
- Radiological protection

5.6   References

[1] About NRC, U.S. Nuclear Regulatory Commission. Retrieved 2007-6-1.

[2] Our Governing Legislation. U.S. Nuclear Regulatory Commission. Retrieved 2007-6-1.

[3] Nathan Hultman & Jonathan Koomey (1 May 2013). "Three Mile Island: The driver of US nuclear power's decline?". Bulletin of the Atomic Scientists.

[4] Stephanie Cooke (March 19, 2011). "Nuclear power is on trial". CNN. Retrieved April 29, 2011.

[5] Mark Cooper (2012-68-61). "Nuclear safety and affordable reactors: Can we have both?" (PDF). Bulletin of the Atomic Scientists. Check date values in: |date= (help)

[6] Eric Wesoff, Greentechmedia. "Black & Veatch' s 2011 Electric Utility Survey." June 16, 2011. Retrieved October 11, 2011.

[7] Massachusetts Institute of Technology (2011). "The Future of the Nuclear Fuel Cycle" (PDF). p. xv.

[8] Mark Cooper (July 2011). "The implications of Fukushima: The US perspective". Bulletin of the Atomic Scientists. p. 9.

[9] Andrew Restuccia (2011-10-20). "Nuke regulators toughen safety rules". The Hill.

[10] Steven Mufson & Jia Lynn Yang (March 24, 2011). "A quarter of U.S. nuclear plants not reporting equipment defects, report finds". Washington Post.

[11] Dave Gram (February 17, 2011). "Possible fuel rod hazard seen at some nuke plants". Bloomberg.

[12] Mark Clayton (March 30, 2011). "Fukushima warning: US has 'utterly failed' to address risk of spent fuel". CS Monitor.

[13] "Nuclear fuel disposal now in spotlight". UPI. March 31, 2011.

[14] Matthew Wald (August 9, 2011). "Researching Safer Nuclear Energy". New York Times.

[15] Renee Schoof (April 12, 2011). "Japan's nuclear crisis comes home as fuel risks get fresh look". McClatchy.

[16] Matthew Wald (January 24, 2012). "Wanted: Parking Space for Nuclear Waste". New York Times.

[17] David Biello (July 29, 2011). "Presidential Commission Seeks Volunteers to Store U.S. Nuclear Waste". Scientific American.

[18] Matthew Wald (January 26, 2012). "Revamped Search Urged for a Nuclear Waste Site". New York Times.

[19] Michael D. Lemonick (24 August 2011). "What the east coast earthquake means for US nuclear plants". The Guardian. London.

[20] John Byrne and Steven M. Hoffman (1996). Governing the Atom: The Politics of Risk, Transaction Publishers. p. 132.

[21] Anupam Chander (April 1, 2011). "Who's to blame for Fukushima?". LA Times.

[22] Hannah Northey (March 28, 2011). "Japanese Nuclear Reactors, U.S. Safety to Take Center Stage on Capitol Hill This Week". New York Times.

[23] David Lochbaum & Edwin Lyman (March 2012). "U.S. NUCLEAR POWER SAFETY ONE YEAR AFTER FUKUSHIMA" (PDF). Union of Concerned Scientists.

[24] Richard Schiffman (12 March 2013). "Two years on, America hasn't learned lessons of Fukushima nuclear disaster". *The Guardian*. London.

[25] Stephanie Cooke (March 19, 2011). "Nuclear power is on trial". *CNN*.

[26] Elizabeth Kolbert (28 March 2011). "The Nuclear Risk". *The New Yorker*.

[27] Daniel Hirsch et al. The NRC's Dirty Little Secret, *Bulletin of the Atomic Scientists*, May 1, 2003, vol. 59 no. 3, pp. 44-51.

[28] 10CFR100

[29] Nuclear Power Plants Are Most Secure Industrial Facilities in U.S., NEI Tells Congress

[30] "Statement from Chairman Dale Klein on Commission's Affirmation of the Final DBT Rule". Nuclear Regulatory Commission. Retrieved 2007-04-07.

[31] Tom Zeller Jr. (December 4, 2012). "Nuclear Power Whistleblowers Charge Federal Regulators With Favoring Secrecy Over Safety". *Huff Post Green*.

[32] 10CFR50.49

[33] "Genesis of a disaster: Moment tsunami swamps Japan's doomed Fukushima nuclear plant". *Daily Mail*. London.

[34] Declan Butler (21 April 2011). "Reactors, residents and risk". *Nature*.

[35] Eric Pooley. Nuclear Warriors *Time Magazine*, March 4, 1996.

[36] NRC Failure to Adequately Regulate - Millsone Unit 1, 1995

[37] Hinds, David; Chris Maslak (January 2006). "Next-generation nuclear energy: The ESBWR" (PDF). Nuclear News. Retrieved 2008-05-13.

[38] (PDF) Archived March 8, 2007, at the Wayback Machine.

[39] "Consideration of Potassium Iodide in Emergency Planning". U.S. Nuclear Regulatory Commission. Retrieved 2006-11-10.

[40] Moore 2007

[41] Iversen, Kristen (2012-03-10). "Fallout at a Former Nuclear Weapon Plant". *The New York Times*.

[42] "The September 1957 Rocky Flats fire: A guide to records series of the Department of Energy". United States Department of Energy. Retrieved September 3, 2011.

[43] http://www.jstor.org/pss/4312671

[44] "Rocky Flats Nuclear Site Too Hot for Public Access, Citizens Warn". Environment News Service. August 5, 2010. Retrieved September 17, 2011.

[45] Hooper, Troy (August 4, 2011). "Invasive weeds raise nuclear concerns at Rocky Flats". The Colorado Independent. Retrieved September 17, 2011.

[46] "1969 Fire Page 7". Colorado.edu. Retrieved 2011-10-27.

[47] "Hanford Site: Hanford Overview". United States Department of Energy. Archived from the original on June 5, 2012. Retrieved February 13, 2012.

[48] "Science Watch: Growing Nuclear Arsenal". *The New York Times*. April 28, 1987. Retrieved January 29, 2007.

[49] "An Overview of Hanford and Radiation Health Effects". Hanford Health Information Network. Archived from the original on 2010-01-06. Retrieved January 29, 2007.

[50] "Hanford Quick Facts". Washington Department of Ecology. Archived from the original on June 24, 2008. Retrieved January 19, 2010.

[51] Hanford Facts

[52] Stang, John (December 21, 2010). "Spike in radioactivity a setback for Hanford cleanup". *Seattle Post-Intelligencer*.

[53] Harden, Blaine; Dan Morgan (June 2, 2007). "Debate Intensifies on Nuclear Waste". *Washington Post*. p. A02. Retrieved January 29, 2007.

[54] Dininny, Shannon (April 3, 2007). "U.S. to Assess the Harm from Hanford". *Seattle Post-Intelligencer*. Associated Press. Retrieved January 29, 2007.

[55] Schneider, Keith (February 28, 1989). "Agreement for a Cleanup at Nuclear Site". *The New York Times*. Retrieved January 30, 2008.

[56] Stacy, Susan M. (2000). *Proving the Principle: A History of The Idaho National Engineering and Environmental Laboratory, 1949-1999* (PDF). U.S. Department of Energy, Idaho Operations Office. ISBN 0-16-059185-6. Chapter 16.

[57] "The SL-1 Reactor Accident".

[58] The Nuclear Power Deception Table 7: Some Reactor Accidents

[59] Horan, J. R., and J. B. Braun, 1993, *Occupational Radiation Exposure History of Idaho Field Office Operations at the INEL*, EGG-CS-11143, EG&G Idaho, Inc., October, Idaho Falls, Idaho.

[60] Minutes to Meltdown: Three Mile Island Archived April 29, 2011, at the Wayback Machine. - National Geographic

[61] Perrow, C. (1982), 'The President's Commission and the Normal Accident', in Sils, D., Wolf, C. and Shelanski, V. (Eds), *Accident at Three Mile Island: The Human Dimensions*, Westview, Boulder, pp.173–184.

[62] Nick Pidgeon (22 September 2011 Vol 477). "In retrospect: Normal accidents". *Nature*. Check date values in: |date= (help);

[63] Storm van Leeuwen, Jan (2008). Nuclear power – the energy balance

[64] World Nuclear Association. Three Mile Island Accident January 2010.

[65] Benjamin K. Sovacool. The costs of failure: A preliminary assessment of major energy accidents, 1907–2007. *Energy Policy* 36 (2008), p. 1807.

[66] Mangano, Joseph (2004). Three Mile Island: Health study meltdown. *Bulletin of the atomic scientists*, 60(5), pp. 31 – 35.

[67] Mark Hertsgaard (1983). *Nuclear Inc. The Men and Money Behind Nuclear Energy*, Pantheon Books, New York, p. 95 & 97.

[68] United States Government Accountability Office (2006). "Report to Congress" (PDF). p. 4.

[69] Alexander Ochs (2012-03-16). "The End of the Atomic Dream: One Year After Fukushima, the Shortfalls of Nuclear Energy Are Clearer Than Ever" . *Worldwatch*.

[70] Benjamin K. Sovacool. A Critical Evaluation of Nuclear Power and Renewable Electricity in Asia, *Journal of Contemporary Asia*, Vol. 40, No. 3, August 2010, pp. 393–400.

[71] Benjamin K. Sovacool (2009). The Accidental Century - Prominent Energy Accidents in the Last 100 Years

[72] John Byrne and Steven M. Hoffman (1996). *Governing the Atom: The Politics of Risk*, Transaction Publishers, p. 152.

[73] Martin Fackler (June 1, 2011). "Report Finds Japan Underestimated Tsunami Danger" . *New York Times*.

[74] Ayesha Rascoe (Feb 9, 2012). "U.S. approves first new nuclear plant in a generation" . *Reuters*.

[75] Edwin Lyman, Ominous Votes by the NRC, *All things nuclear*, October 23, 2015.

5.7   External links

- The US Nuclear Regulatory Commission supervises the US Nuclear industry

- Nuclear Power Safety and Security Information

Chapter 6

Nuclear weapons debate

Since the atomic bombings of Hiroshima and Nagasaki, nuclear weapons have remained highly controversial and contentious objects in the forum of public debate.

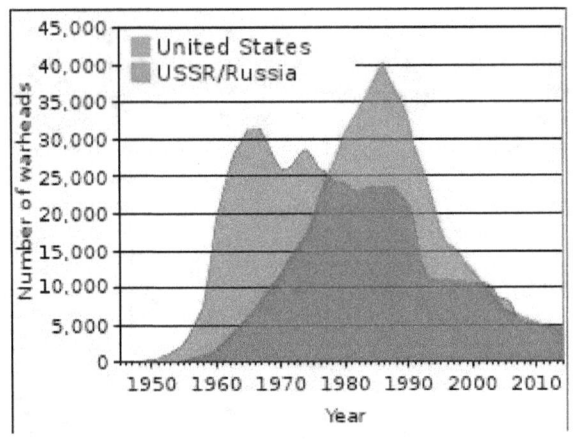

U.S. and USSR/Russian nuclear weapons stockpiles, 1945–2005.

The **nuclear weapons debate** refers to the controversies surrounding the threat, use and stockpiling of nuclear weapons. Even before the first nuclear weapons had been developed, scientists involved with the Manhattan Project were divided over the use of the weapon. The only time nuclear weapons have been used in warfare was during the final stages of World War II when United States Army Air Forces B-29 Superfortress bombers dropped atomic bombs on the Japanese cities of Hiroshima and Nagasaki in early August 1945. The role of the bombings in Japan's surrender and the U.S.'s ethical justification for them have been the subject of scholarly and popular debate for decades.

Nuclear disarmament refers both to the act of reducing or eliminating nuclear weapons and to the end state of a nuclear-free world. Proponents of disarmament typically condemn a priori the threat or use of nuclear weapons as immoral and argue that only total disarmament can eliminate the possibility of nuclear war. Critics of nuclear disarmament say that it would undermine deterrence and make conventional wars more likely, more destructive, or both. The debate becomes considerably complex when considering various scenarios for example, total vs partial or unilateral vs multilateral disarmament.

6.1 History

Even before the first nuclear weapons had been developed, scientists involved with the Manhattan Project were divided over the use of the weapon. Some—notably a number at the University of Chicago Metallurgical Laboratory, represented in part by Leó Szilárd—lobbied early on that the atomic bomb should only be built as a deterrent against Nazi Germany getting a bomb, and should not be used against populated cities. The Franck Report argued in June 1945 that instead of being used against a city, the first atomic bomb should be "demonstrated" to the Japanese on an uninhabited area.*[1] This recommendation was not agreed with by the military commanders, the Los Alamos Target Committee (made up of other scientists), or the politicians who had input into the use of the weapon. Because the Manhattan Project was considered to be "top secret", there was no public discussion of the use of nuclear arms, and even within the U.S. government, knowledge of the bomb was extremely limited.

The Little Boy atomic bomb was detonated over the

The Fat Man mushroom cloud resulting from the nuclear explosion over Nagasaki.

Japanese city of Hiroshima on 6 August 1945. Exploding with a yield equivalent to 12,500 tonnes of TNT, the blast and thermal wave of the bomb destroyed nearly 50,000 buildings (including the headquarters of the 2nd General Army and Fifth Division) and killed approximately 75,000 people, among them 20,000 Japanese soldiers and 20,000 Koreans.[2] Detonation of the "Fat Man" atomic bomb exploded over the Japanese city of Nagasaki three days later on 9 August 1945, destroying 60% of the city and killing approximately 35,000 people, among them 23,200-28,200 Japanese civilian munitions workers and 150 Japanese soldiers.[3] The role of the bombings in Japan's surrender and the U.S.'s ethical justification for them has been the subject of scholarly and popular debate for decades. J. Samuel Walker suggests that "the controversy over the use of the bomb seems certain to continue".[4]

After the bombings of Hiroshima and Nagasaki, the world's nuclear weapons stockpiles grew,[5] and nuclear weapons have been detonated on over two thousand occasions for testing and demonstration purposes. Countries known to have detonated nuclear weapons—and that acknowledge possessing such weapons—are (chronologically) the United States, the Soviet Union (succeeded as a nuclear power by Russia), the United Kingdom, France, the People's Republic of China, India, Pakistan, and North Korea.[6]

In the early 1980s, following a revival of the nuclear arms race, a popular nuclear disarmament movement emerged. In October 1981 half a million people took to the streets in several cities in Italy, more than 250,000 people protested in Bonn, 250,000 demonstrated in London, and 100,000 marched in Brussels.[7] The largest antinuclear protest was held on June 12, 1982, when one million people demonstrated in New York City against nuclear weapons.[8][9][10] In October 1983, nearly 3 million people across western Europe protested nuclear missile deployments and demanded an end to the arms race.[11]

6.2   Arguments

Under the scenario of total multilateral disarmament, there is no possibility of nuclear war. Under scenarios of partial disarmament there is disagreement as to how the probability of nuclear war would change. Critics of nuclear disarmament say that it would undermine the ability of governments to threaten sufficient retaliation upon attack to deter aggression against them. Application of game theory to questions of strategic nuclear warfare during the Cold War resulted in the doctrine of mutually assured destruction (MAD), a concept developed, by Robert S. McNamara, among others, in the mid-1960s.[12] The success of MAD in averting nuclear war was theorized to depend upon the "readiness at any time before, during, or after an attack to destroy the adversary as a functioning society." [13] Those who believe governments should develop or maintain nuclear-strike capability, usually justify their position with reference to MAD and the Cold War, claiming that a "nuclear peace" was the result of both the U.S. and the U.S.S.R. possessing mutual second-strike retaliation capability. Since the end of the cold war, theories of deterrence in international relations have been further developed and generalized in the concept of the stability–instability paradox[14][15] Proponents of disarmament call into question the assumption that political leaders are rational actors who place the protection of their citizens above other considerations, and highlight, as McNamara himself later acknowledged with the benefit of hindsight, the non-rational choices, chance and contingency which played a significant role in averting nuclear war, for example during the Cuban Missile Crisis of 1962 and the Able Archer 83 crisis of 1983,[16] thus, they argue, evidence trumps theory and deterrence theories cannot be reconciled with the historical record.

Kenneth Waltz argues in favor of the continued proliferation of nuclear weapons[17] In the July 2012 issue of *Foreign Affairs* Waltz took issue with the view of most U.S., European, and Israeli, commentators and policymakers that a nuclear-armed Iran would be unacceptable. Instead Waltz argues that it would probably be the best possible outcome, as it would restore stability to the Middle East by balancing Israel's regional monopoly on nuclear weapons.[18]

Professor John Mueller of Ohio State University, author of *Atomic Obsession*[19] has also dismissed the need to interfere with Iran's nuclear program and expressed that arms control measures are counterproductive.[20] During a 2010 lecture at the University of Missouri, which was broadcast by C-Span, Dr. Mueller has also argued that the threat from nuclear weapons, including that from terrorists, has been exaggerated, both in the popular media, and by officials.[21]

In contrast, various American government officials, including Henry Kissinger, George Shultz, Sam Nunn, and William Perry.[22][23][24] who were in office during the Cold War period, are now advocating the elimination of nuclear weapons in the belief that the doctrine of mutual Soviet-American deterrence is obsolete, and that reliance on nuclear weapons for deterrence is becoming increasingly hazardous and decreasingly effective in the post cold war era[22] A 2011 article in *The Economist* argues along similar lines, that risks are more acute in rivalries between relatively new nuclear states that lack the "security safeguards" developed by America and the Soviet Union and that additional risks are posed by the emergence of pariah states, such as North Korea (possibly soon to be joined by Iran), armed with nuclear weapons as well as the declared ambition of terrorists to steal, buy or build a nuclear device.[25]

6.3   See also

- Agency for the Prohibition of Nuclear Weapons in Latin America and the Caribbean
- Anti-nuclear protests in the United States
- Comprehensive Test Ban Treaty
- Debate over the atomic bombings of Hiroshima and Nagasaki
- Effects of nuclear explosions
- Effects of nuclear explosions on human health
- History of the anti-nuclear movement
- International Court of Justice advisory opinion on legality of nuclear weapons
- Lists of nuclear disasters and radioactive incidents
- List of states with nuclear weapons
- List of nuclear weapons
- List of nuclear close calls
- Nth Country Experiment

- Nuclear disarmament
- Nuclear Non-Proliferation Treaty
- Nuclear peace
- Nuclear power debate
- Nuclear proliferation
- Nuclear Tipping Point
- Nuclear weapons and the United Kingdom
- Nuclear weapons and the United States
- Strategic Arms Limitation Talks
- Three Non-Nuclear Principles, of Japan
- United Nations Security Council Resolution 1194
- Uranium mining debate

6.4   References

[1] Schollmeyer, Josh (January–February 2005). *"Minority Report"*. *"Bulletin of the Atomic Scientists"*. Retrieved 2009-08-04. External link in |publisher= (help)

[2] Emsley, John (2001). "Uranium". *Nature's Building Blocks: An A to Z Guide to the Elements*. Oxford: Oxford University Press. p. 478. ISBN 0-19-850340-7.

[3] *Nuke-Rebuke: Writers & Artists Against Nuclear Energy & Weapons (The Contemporary anthology series)*. The Spirit That Moves Us Press. May 1, 1984. pp. 22–29.

[4] Walker, J. Samuel (April 2005). "Recent Literature on Truman's Atomic Bomb Decision: A Search for Middle Ground". *Diplomatic History*. **29** (2): 334. doi:10.1111/j.1467-7709.2005.00476.x.

[5] Mary Palevsky, Robert Futrell, and Andrew Kirk. Recollections of Nevada's Nuclear Past *UNLV FUSION*, 2005, p. 20.

[6] "Federation of American Scientists: Status of World Nuclear Forces". Fas.org. Retrieved 2010-01-12.

[7] David Cortright (2008). *Peace: A History of Movements and Ideas*, Cambridge University Press, p. 147.

[8] Jonathan Schell. The Spirit of June 12 *The Nation*, July 2, 2007.

[9] David Cortright (2008). *Peace: A History of Movements and Ideas*, Cambridge University Press, p. 145.

[10] 1982 - a million people march in New York City

[11] David Cortright (2008). *Peace: A History of Movements and Ideas*, Cambridge University Press, p. 148.

[12] Elliot, Jeffrey M. and Robert Reginald. (1989). *The Arms Control, Disarmament, and Military Security Dictionary*. Santa Barbara: ABC-CLIO, Inc.

[13] Gertcher, Frank L., and William J. Weida. (1990). *Beyond Deterrence*, Boulder: Westview Press, Inc.

[14] http://www.stimson.org/images/uploads/research-pdfs/ESCCONTROLCHAPTER1.pdf

[15] Krepon, Michael (November 2, 2010). "The Stability-Instability Paradox". *Arms Control Wonk*. Retrieved 2016-10-04.

[16] James G Blight, Janet M. Lang. The Fog of War: Lessons from the Life of Robert S. McNamara, page 60.

[17] Waltz, Kenneth (1981). "The Spread of Nuclear Weapons: More May Be Better". *Adelphi Papers*. London: International Institute for Strategic Studies (171).

[18] Waltz, Kenneth (July–August 2012). "Why Iran Should Get the Bomb: Nuclear Balancing Would Mean Stability". *Foreign Affairs*.

[19] "Atomic Obsession - Hardback - John Mueller - Oxford University Press".

[20] Bloggingheads.tv from 19:00 to 26:00 minutes

[21] "[Atomic Obsession]". *C-SPAN.org*. Retrieved 2016-10-04.

[22] George P. Shultz, William J. Perry, Henry A. Kissinger and Sam Nunn. A World Free of Nuclear Weapons *Wall Street Journal*, January 4, 2007, page A15.

[23] Hugh Gusterson (30 March 2012). "The new abolitionists". *Bulletin of the Atomic Scientists*.

[24] "A World Free of Nuclear Weapons - - Publications - Nuclear Security Project". *www.nuclearsecurityproject.org*. Retrieved 2017-02-05.

[25] "Nuclear endgame: The growing appeal of zero". *The Economist*. June 16, 2011.

[ ] Murphy, Arthur W. (1976). *The Nuclear Power Controversy*, Prentice-Hall.

[ ] Malheiros, Tania. *Brasiliens geheime Bombe: Das brasilianische Atomprogramm*. Tradução: Maria Conceição da Costa e Paulo Carvalho da Silva Filho. Frankfurt am Main: Report-Verlag, 1995.

[ ] Malheiros, Tania. *Brasil, a bomba oculta: O programa nuclear brasileiro*. Rio de Janeiro: Gryphus, 1993. (Portuguese)

[ ] Malheiros, Tania. *Histórias Secretas do Brasil Nuclear*. (WVA Editora; ISBN 85-85644087) (Portuguese)

[ ] Walker, J. Samuel (2004). *Three Mile Island: A Nuclear Crisis in Historical Perspective*, University of California Press.

[ ] Williams, Phil (Ed.) (1984). *The Nuclear Debate: Issues and Politics*, Routledge & Keagan Paul, London.

[ ] Wittner, Lawrence S. (2009). *Confronting the Bomb: A Short History of the World Nuclear Disarmament Movement*, Stanford University Press.

6.5   Further reading

See also: List of books about nuclear issues

- M. Clarke and M. Mowlam (Eds) (1982). *Debate on Disarmament*, Routledge and Kegan Paul.

- Cooke, Stephanie (2009). *In Mortal Hands: A Cautionary History of the Nuclear Age*, Black Inc.

- Falk, Jim (1982). *Global Fission: The Battle Over Nuclear Power*, Oxford University Press.

Chapter 7

List of nuclear whistleblowers

2000 candles in memory of the Chernobyl disaster in 1986, at a commemoration 25 years after the nuclear accident, as well as for the Fukushima nuclear disaster of 2011.

There have been a number of **nuclear whistleblowers**, often nuclear engineers, who have identified safety concerns about nuclear power and nuclear weapons production. In 1976 Gregory Minor, Richard Hubbard and Dale Bridenbaugh "blew the whistle" on safety problems at nuclear power plants in the United States. George Galatis was a senior nuclear engineer who reported safety problems at the Millstone 1 Nuclear Power Plant, relating to reactor refueling procedures, in 1996.[1] Other nuclear power whistleblowers include Arnold Gundersen and David Lochbaum.

7.1 Karen Silkwood

Main article: Karen Silkwood

The first prominent nuclear whistleblower was Karen Silkwood, who worked as a chemical technician at a Kerr-McGee nuclear fuel plant. Silkwood became an activist in the Oil, Chemical and Atomic Workers International Union in order to protest health and safety issues. In 1974, she testified to the United States Atomic Energy Commission about her concerns.[2] A few months later she died in a car crash under mysterious conditions on the way to a meet-

ing with a *New York Times* reporter and a national union leader.[3] The 1983 film *Silkwood* is an account of this story.

7.2 The "GE Three"

Main article: GE Three

On February 2, 1976, Gregory C. Minor, Richard B. Hubbard, and Dale G. Bridenbaugh (known as the GE Three) "blew the whistle" on safety problems at nuclear power plants, and their action has been called "an exemplary instance of whistleblowing".[4]

The three engineers gained the attention of journalists and their disclosures about the threats of nuclear power had a significant impact. They timed their statements to coincide with their resignations from responsible positions in General Electric's nuclear energy division, and later established themselves as consultants on the nuclear power industry for state governments, federal agencies, and overseas governments. The consulting firm they formed, MHB Technical Associates, was technical advisor for the movie, "The China Syndrome." The three engineers participated in Congressional hearings which their disclosures precipitated.[5][6]

Browns Ferry nuclear power plant construction went underway in 1966. It was located in Alabama and in 1967 it earned a Federal construction permit. The plant received new design standards which call for "physical separation of electrical cables."[7] There was an issue with the instructions on how to accomplish this so the AEC inspector F.U. Bower requested that the AEC elaborate; however, there was no response from the organization and installation went on. Still, no instructions were issued after five failed inspections in 1970. The lack of cable separation instructions led to the sacrifice of safety coolant systems in two of the units in order to improve one with severe safety violation. The ignorance of the AEC led to the fire that occurred

Browns Ferry Unit 1 under construction

on March 22, 1975, that almost led to a radiation leak. The substance separating the wires caught fire when tests to find air leaks with a candle ignited it thus resulting in damage to the control systems. With damage to the control systems, the cooling system that keeps the units from leaking radiation did not work properly. Somehow the situation was avoided and the units were put out of service. Throughout the occurrence of these events Bridenbaugh had been discussing his reservations on the safety at the plant in vain and in 1976 a year later Bridenbaugh, Hubbard and Minor resigned.

7.3   Crystal River 3 and Lou Putney

Main article: Crystal River 3 Nuclear Power Plant

Lou Putney came on the scene of the Crystal River 3 plant after receiving a call from a plant engineer. The engineer claimed that the managers hired engineers based on "good ol' boy mentality." [8] The plant had experience numerous shut downs since 1978. Along with this concern, the engineer was not confident that the manager possessed the qualifications to be a manager. Although the engineer pursued nothing further with his complaint, it prompted Putney to purchase shares of stock in the company that would allow him to file "shareholder resolutions." Putney had looked into

the nuclear reactors that were built of an unsafe material for emergency cooling procedures. The NRC had placed Crystal River on the top 14 worst reactors list because of this. So, the shares were purchased in 1981, which is when Putney filed his first shareholder resolution requesting the plant be shut down. This tradition was upheld by Putney for seven years until he was required to purchase more stock in order to continue filing resolutions. Over the course of sixteen years, Putney filed a total of fourteen shareholder resolutions. All of these resolutions were ignored and were met with offers to buy out his shares so he could no longer file the resolutions. [8] The plant was officially decommissioned in September 2009.

7.4   Ronald Goldstein

Ronald J. Goldstein was a supervisor employed by EBASCO, which was a major contractor for the construction of the South Texas plants. In the summer of 1985, Goldstein identified safety problems to SAFETEAM, an internal compliance program established by EBASCO and Houston Lighting, including noncompliance with safety procedures, the failure to issue safety compliance reports, and quality control violations affecting the safety of the plant.

SAFETEAM was promoted as an independent safe haven for employees to voice their safety concerns. The two companies did not inform their employees that they did not believe complaints reported to SAFETEAM had any legal protection. After he filed his report to SAFETEAM, Goldstein was fired. Subsequently, Golstein filed suit under federal nuclear whistleblower statutes. [9]

The U.S. Department of Labor ruled that his submissions to SAFETEAM were protected and his dismissal was invalid, a finding upheld by Labor Secretary Lynn Martin. The ruling was appealed and overturned by the Fifth Circuit Court of Appeals, which ruled that private programs offered no protection to whistleblowers. After Goldstein lost his case, Congress amended the federal nuclear whistleblower law to provide protection reports made to internal systems and prevent retaliation against whistleblowers. [10]

7.5   Fernald Nuclear Incidents

Main article: Fernald Feed Materials Production Center

The Fernald Feed Materials Production Center was built in Crosby Township, Ohio in 1951, and decommissioned in 1989. Fernald processed uranium trioxide and uranium tetrafluoride, among other radioactive materials, to produce

Fernald Production-era Aerial

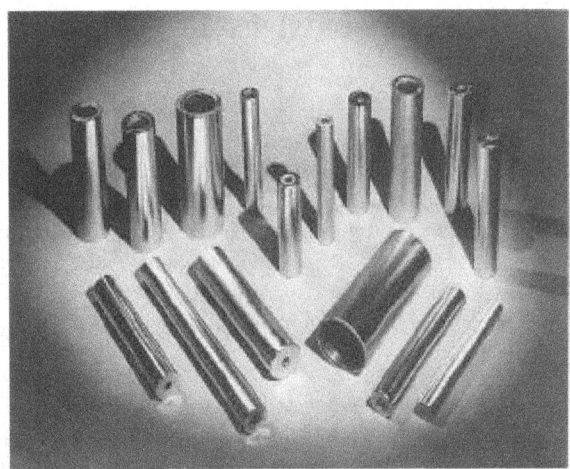

Uranium components fabricated at Fernald

the uranium fuel cores for nuclear weapons. It was shrouded in suspicion with many manager changes and the people of the town ill-informed of the purpose of the plant.[11] The Fernald Feed Materials Production Center also conducted an evaluation of how much material was contaminated by Radium. Using 138 pieces of the CR-39 film assays, they were able to determine that people working in the area where K-65 silos (Underground chamber used to store missiles) had lower levels of exposure of materials contaminated by Radon than the Q-11 silos between the period of 1952-1988 Journal of Exposure Science and Environmental Epidemiology.Throughout 1951-1995 the plant had numerous scandals including faking numbers for contamination and disregarding evidence of ground water pollution. Among the citizens affected by the pollution was Mrs. Lisa Crawford who took action. Crawford and other residents filed a lawsuit in 1985 and became president of the organization FRESH (Fernald Residents for Environmental Safety and Health).[12] A lawsuit was then filed once again against Fernald by former employees several years later in 1990. After several years of being heavily

advised not to blow the whistle, the workers earned themselves a $15 million settlement and lifelong medical monitoring.[12] In 1992, FERMCO was hired to construct a cleanup plan for the plant and in 1996, around accusations of wasteful spending, the cleanup of ground water and soil was completed.[12]

7.6 Mordechai Vanunu

Mordechai Vanunu 2009

Main article: Mordechai Vanunu

Mordechai Vanunu blew the whistle on the nuclear plant in Dimona, Israel in an interview with The Sunday Times that was published on the 5th of October, 1986. According to Vanunu, this plant had been producing nuclear weapons for 10 to 20 years.[13] It is estimated that there may be around 200 nuclear weapons in possession of Israel's nuclear weapons program.[14] Vanunu demonstrated his knowledge to Frank Barnaby and John Steinbach and they confirmed the credibility of his story. Frank Barnaby wrote in his *Declaration of Frank Barnaby in the Matter of Mordechai Vanunu* that Vanunu had the bare minimum knowledge of nuclear physics that a technician should have and accurately described the makeup of the nuclear plant in Dimona.[15] Vanunu has been in and out of jail after serving his 18 years issued by a closed door trial. Despite the whistle blown towards the operation of the nuclear weapons program in Israel, the Israeli government denied the existence of all allegations. [16]

7.7   Arnold Gundersen

Main article: Arnold Gundersen

In 1990 Arnold Gundersen discovered radioactive material in an accounting safe at Nuclear Energy Services in Danbury, Connecticut, the consulting firm where he held a $120,000-a-year job as senior vice-president.[17] Three weeks after he notified the company president of what he believed to be radiation safety violations, Gundersen was fired. According to the *New York Times*, for three years, Gundersen "was awakened by harassing phone calls in the middle of the night" and he "became concerned about his family's safety". Gundersen believes he was blacklisted, harassed and fired for doing what he thought was right.[17]

The New York Times reports that Gundersen's case is not uncommon, especially in the nuclear industry. Even though nuclear workers are encouraged to report potential safety hazards, those who do risk demotion and dismissal. Instead of correcting the problems, whistleblowers say, industry management and government agencies attack them as the cause of the problem. Driven out of their jobs and shunned by neighbors and co-workers, whistleblowers often turn to each other for support.[17]

The Whistleblower Support Fund is an organization that has compiled resources for whistleblowers to access if they are considering whistleblowing. It was founded by Donald Ray Soeken, who has counseled whistleblowers for 35 years. In addition, a social network to connect whistleblowers to other whistleblowers will be implemented. It will be a private discussion where whistleblowers can safely seek support.[18]

7.8   David Lochbaum

Main article: David Lochbaum

In the early 1990s, nuclear engineer David Lochbaum and a colleague, Don Prevatte, identified a safety problem in a plant where they were working, but were ignored when they raised the issue with the plant manager, the utility and the U.S. Nuclear Regulatory Commission (NRC).[19] After bringing their concerns to Congress, the problem was corrected not just at the original nuclear plant but at plants

across the country.[20]

7.9   Gerald W. Brown

Main article: Gerald W. Brown
Gerald W. Brown was the whistleblower on the Thermo-

Gerald W. Brown

Lag scandal, as well as on silicone foam firestop issues in the US and Canada, exposing the fact that fireproofing of wiring between control rooms and reactors did not function as intended and exposing bounding and combustibility issues with organic firestops.

7.10   George Galatis

Main article: George Galatis

George Galatis was a senior nuclear engineer and whistleblower who reported safety problems at the Millstone 1 Nuclear Power Plant, relating to reactor refueling procedures, in 1996.[1][21] The unsafe procedures meant that spent fuel rod pools at Unit 1 had the potential to boil, possibly releasing radioactive steam throughout the plant.[22] Galatis eventually took his concerns to the Nuclear Regulatory Commission, to find that they had "known about the unsafe procedures for years". As a result of going to the NRC, Galatis experienced "subtle forms of harassment, retaliation, and intimidation".[21]

7.11   Rainer Moormann

Rainer Moormann is a German chemist and nuclear power whistleblower. Since 1976 he has been working at the

Moormann rainer 04

Forschungszentrum Jülich, doing research on safety problems with pebble bed reactors, fusion power and spallation neutron sources. In 2008 Moormann published a critical paper on the safety of pebble bed reactors,[23][24] which raised attention among specialists in the field, and managed to distribute it via the media, facing considerable opposition. For doing this despite the occupational disadvantages he had to accept as a consequence, Moormann was awarded the whistleblower award of the Federation of German Scientists (VDW)[25] and of the German section of the International Association of Lawyers Against Nuclear Arms (IALANA).

7.12 Setsuo Fujiwara

Setsuo Fujiwara, who used to design reactors, said he clashed with supervisors over an inspection audit he conducted in March 2009 at the Tomari nuclear plant in Japan. Fujiwara refused to approve a routine test by the plant's operator, Hokkaido Electric Power, saying the test was flawed. A week later, he was summoned by his supervisor, who ordered him to *correct* his written report to indicate that the test had been done properly. After Fujiwara refused, his employment contract was not renewed. "They told me my job was just to approve reactors, not to raise doubts about them", said Fujiwara, 62, who is now suing the nuclear safety organization to get rehired. In a written response to questions from *The New York Times*, the agency

said it could not comment while the court case was under way.[26] Along with the lawsuit Mr. Fujiwara filed against the agency he used to work for, he had gone to the Tokyo District Court to further write several complaints about how the JNES (Japan Nuclear Energy Safety Organization) failed to follow the UN laws concerning how to properly inspect nuclear energy reactors. Mr. Fujiwara also submitted several documents and emails that dealt with how the reactor inspections were improperly handled by JNES even though JNES denies all allegations. [42]

7.13 Walter Tamosaitis

The Hanford site resulted in a number of whistleblowers during the efforts to clean the site up. Walter Tamosaitis blew the whistle on the Energy Department's plan for waste treatment at the Hanford site in 2011. Tamosaitis's concern was the possibility of explosive hydrogen gas being built up inside tanks that the company was to store the harmful chemical sludge they were trying to put into hibernation for its chemical life. Shortly after this Tamosaitis was demoted and two years later, fired which triggered his lawsuit for wrongful termination. A $4.1 million settlement was offered to Tamosaitis from AECOM on the 12th of August 2015.[27] Tamosaitis has since been reinstated.

Donna Busche blew the whistle resulting in her 2013 lawsuit with claims that the URS "retaliated against her.[27] She was head of nuclear safety and a URS employee around the time when she expressed her concerns.[28][29]

Gary Brunson reported 34 safety and engineering violations after quitting in 2012. Brunson was federal engineering chief before he quit.[30]

Shelly Doss earned "$20,000 in emotional distress and $10,000 in callous disregard of her rights" as well as reinstatement in 2014.[27] Doss was an environmental specialist at the time of her firing in 2011 working for Washington River Protection Solutions.

7.14 Larry Criscione and Richard H. Perkins

In 2012, Larry Criscione and Richard H. Perkins publicly accused the US Nuclear Regulatory Commission of downplaying flood risks for nuclear plants which are sited on waterways downstream from large reservoirs and dams. They are engineers with over 20 years of combined government and military service who work for the NRC. Other nuclear safety advocates have supported their complaints.[31]

7.15   John Pace and Dan Parks

In 1959, John Pace was employed at the Santa Susana Field Lab, 36 miles from Los Angeles, at the time of the partial meltdown during the Sodium Reactor Experiment. According to Pace, the reactor building doors and reactor exhaust stack were opened, releasing radiation into the atmosphere. Dan Parks also worked at the Santa Susana Field Lab, as a health physicist in the 1960s, where he witnessed the burning of radioactive waste in burn pits, and frequently saw workers illegally vent radiation into the atmosphere. In 2015 they spoke out about the incident in an investigative report by NBC News.*[32]*[33]

7.16   Other nuclear whistleblowers

- Chuck Atkinson*[34]

- Dale G. Bridenbaugh

- Joe Carson *[35]

- Larry Criscione*[36]

- Mark Gillespie*[37]

- Lars-Olov Höglund*[38]

- Carl Hocevar*[39]

- David Hoffman*[40]

- Avon Hudson

- Rainer Moormann

- Carl Patrickson*[41]

- Richard H. Perkins*[36]

- Robert Pollard*[39]

- John P. Shannon*[42]

- Don Ranft*[43]

- Zhores Medvedev*[44]*[45]

- Ronald A. Sorri*[46]

7.17   See also

- Nuclear accidents in the United States

- Nuclear safety

- Anti-nuclear movement in the United States

7.18   References

[1] Eric Pooley. Nuclear Warriors *Time Magazine*, March 4, 1996.

[2] Rashke, Richard L. (1981). *The Killing of Karen Silkwood: The Story Behind the Kerr-McGee Plutonium Case*. Houghton Mifflin Company. ISBN 978-0801486678.

[3] Tolley, Laura (April 24, 1994). "Karen Silkwood Case Returns to Haunt Parents : Whistle-blower: The disclosure that Los Alamos lab has had her bone fragments since 1974 outrages family. A worker in a plutonium-processing plant, she died in a car crash while on her way to meet a reporter. She had promised to bring proof the plant was unsafe; no documents were in the car.". Los Angeles Times. Associated Press. Retrieved 18 September 2016.

[4] Whistleblower on Nuclear Plant Safety

[5] "Environment: The San Jose Three". 16 February 1976. Retrieved 15 June 2015.

[6] "Environment: The Struggle over Nuclear Power". *TIME.com*. 8 March 1976. Retrieved 15 June 2015.

[7] Weil, Vivian. "Moral Responsibility and Whistleblowing in the Nuclear Industry: Browns Ferry and Three Mile Island." (1983).

[8] Riggs, Stephanie. "FLORIDA POWER AND THE CRYSTAL RIVER NUCLEAR POWER PLANT." (1997).

[9] Zuniga, Sonia (April 26, 2016). "20 memorable whistle-blowers in the U.S." . Houston Chronicle. Retrieved 18 September 2016.

[10] Kohn, Stephen Martin (2011). *The Whistleblower's Handbook: A Step-by-Step Guide to Doing What's Right and Protecting Yourself*. Guilford, CT: Globe Pequot Press. pp. 116–18. ISBN 9780762774791.

[11] Bonfield, Tim (February 11, 1996). "History repeats itself". The Cincinnati Enquirer. Retrieved 18 September 2016.

[12] "The Enquirer's Fernald Investigation". *enquirer.com*. Retrieved 2016-06-27.

[13] Barnaby, Frank (1 January 1987). "The Nuclear Arsenal in the Middle East". *Journal of Palestine Studies*. **17** (1): 97–106. doi:10.2307/2536653. Retrieved 7 December 2016 – via JSTOR.

[14] The Emirates Center for Strategic Studies and Research (2009). *Nuclear Energy in the Gulf*. I. B.Tauris & Company, Limited. p. 336. ISBN 9789948141174.

[15] Barnaby, Frank (14 June 2004). "Expert Opinion Of Charles Frank Barnaby in The Matter of Mordechai Vanunu" (PDF). *fas*. Retrieved 21 June 2016.

[16] http://search.proquest.com/docview/1714013298

[17] Julie Miller (February 12, 1995). "Paying The Price For Blowing The Whistle". *The New York Times*.

[18] "Whistleblower Support Fund". *Whistleblower Support Fund*. Retrieved 2016-06-13.

[19] Heidorn Jr., Rich (January 31, 1999). "Whistle-blower Tells Of Nuclear-plant Fears He Disagreed With Bosses Over Safety Issues In Allentown And Was Fired. Osha Sided With Him.". The Philadelphia Inquirer. Retrieved 18 September 2016.

[20] Kyle Rabin (2011-06-30). "Our Hero: David Lochbaum of the Union of Concerned Scientists". *Ecocentric*.

[21] William H. Shaw. *Business Ethics* 2004, pp. 267-268.

[22] Adam Bowles. A Cry in the Nuclear Wilderness *Christianity Today*, October 2, 2000.

[23] "JuSER" (PDF). Retrieved 15 June 2015.

[24] "JuSER" (PDF). Retrieved 15 June 2015.

[25] Press statement of the VWD (in German), official short version in English: http://ialana.net/uploads/media/Program_Whistleblower_Award_2011.pdf, inofficial English translation of the press release: https://euzicasa.wordpress.com/2011/06/10/presentation-of-whistleblower-award-2011_via-hintergrund/

[26] "Warnings on Fukushima ignored, insiders say: They attribute failure to cozy ties between government and industry". *Power Engineering*. March 11, 2012.

[27] Times, Los Angeles. "Hanford nuclear weapons site whistleblower wins $4.1-million settlement". *latimes.com*. Retrieved 2016-06-28.

[28] CBS News (February 19, 2014). "Second whistleblower Donna Busche fired at troubled Wash. State Hanford nuke plant". CBS This Morning. Retrieved 18 September 2016.

[29] Winograd, David (February 18, 2014). "Whistleblower Fired After Raising Safety Concerns at Federal Nuclear Site". TIME. Retrieved 18 September 2016.

[30] LaFlure, Rebecca (November 19, 2013). "Hanford nuclear site clean-up: The mess gets worse". NBC News. Retrieved 18 September 2016.

[31] Tom Zeller Jr. (April 4, 2012). "Nuclear Power Whistleblowers Charge Federal Regulators With Favoring Secrecy Over Safety". *Huff Post Green*.

[32] Grover, Joel; Glasser, Matthew. "L.A.'s Nuclear Secret". NBC. National Broadcasting Network. Retrieved 31 October 2016.

[33] Grover, Joel; Glasser, Matthew (2015). "L.A.'s Nuclear Secret: Part 1". NBC News - Investigative Team. National Broadcasting Company. Retrieved 31 October 2016.

[34] "Whistleblowing and nonviolence, by Brian Martin". Retrieved 7 December 2016.

[35] "DOE from the Perspective of a DOE Safety Employee (Whistle-blower Joe Carson addresses a meeting of sick USEC Paducah workers)". October 31, 2011. Retrieved 22 September 2013.

[36] Tom Zeller Jr. (12/04/2012). "Nuclear Power Whistleblowers Charge Federal Regulators With Favoring Secrecy Over Safety". *Huff Post Green*. Check date values in: |date= (help)

[37] "Well-Liked Leaders Know The Secret: Make Us Laugh". Retrieved 7 December 2016.

[38] "Former Safety Chief Blows Whistle on Poor Safety Culture - Interview with Lars-Olov Höglund on the Forsmark incident - nonuclear.se". Retrieved 7 December 2016.

[39] Falk, Jim (1982). *Gobal Fission:The Battle Over Nuclear Power*, p. 95.

[40] "Court papers: Nuclear feud at Fla. plant". Retrieved 7 December 2016.

[41] "Nuke Watchdog Urges New Look at Whistleblower Case - The NewStandard". Retrieved 7 December 2016.

[42] "Corruption at the Knolls Atomic Power Laboratory...". December 21, 2001. Retrieved 14 September 2013.

[43] Heidorn, Rich. "Whistle-blower Tells Of Nuclear-plant Fears He Disagreed With Bosses Over Safety Issues In Allentown And Was Fired. Osha Sided With Him". *philly.com*. Philadelphia Inquirer: Collections • OSHA. Retrieved 18 September 2016.

[44] Medvedev, Zhores (1979). *The Nuclear Disaster in the Urals*. New York: W.W. Norton.

[45] Rabi, Thomas (2012). "The Nuclear Disaster of Kyshtym 1957 and the Politics of the Cold War". *Environment & Society Portal, Arcadia 2012, no. 20. Rachel Carson Center for Environment and Society*. **20**. Retrieved 27 December 2016.

[46] "Ronald A. Sorri, Complainant v. L&M Technologies Inc. and Sandia National Laboratories; OHA Case No. LWA-0001" (PDF). Department of Energy Whistleblower Cases. Retrieved 1 January 2017.

*[1] *[2]

7.19 External links

- Watching the Watchdogs

- A Nuclear Plant Gets New Equipment and a New Attitude

- Government Accountability Project Whistleblower protection Org.

- Ed Yong (28 November 2013). "3 ways to blow the whistle" (PDF). *Nature (journal) Vol 503.*

- National Whistleblowers Center - Nuclear Whistleblowers

[1] "Estimation of radon exposures to workers at the Fernald Feed Materials Production Center 1952-1988" . *Journal of Exposure Science and Environmental Epidemiology.* **18**.

[2] Hayashi, Yuka. 2011. " World News: Japan: Ex-Inspector Complains of Regulators' Practices." *The Wall Street Journal Asia,* Jun 16

Chapter 8

Nuclear power phase-out

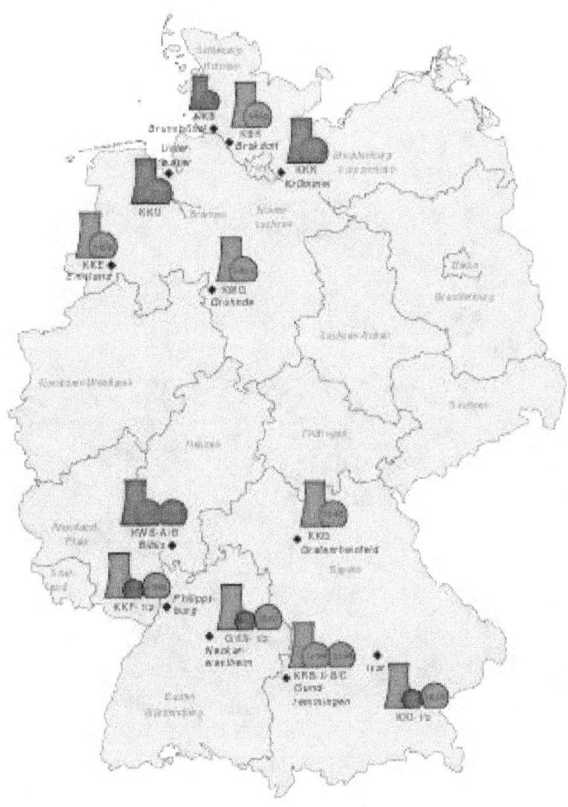

Eight German nuclear power reactors (Biblis A and B, Brunsbuettel, Isar 1, Kruemmel, Neckarwestheim 1, Philippsburg 1 and Unterweser) were permanently shut down on 6 August 2011, following the Japanese Fukushima nuclear disaster.[1]

A **nuclear power phase-out** is the discontinuation of usage of nuclear power for energy production. Often initiated because of concerns about nuclear power, phase-outs usually include shutting down nuclear power plants and looking towards fossil fuels and renewable energy.

Three nuclear accidents have influenced the discontinuation of nuclear power: the 1979 Three Mile Island partial nuclear meltdown in the United States, the 1986 Chernobyl disaster in the USSR, and the 2011 Fukushima nuclear disaster in Japan.

Following the March 2011 Fukushima nuclear disaster, Germany has permanently shut down eight of its 17 reactors and pledged to close the rest by the end of 2022.[2] Italy voted overwhelmingly to keep their country non-nuclear.[3] Switzerland and Spain have banned the construction of new reactors.[4] Japan's prime minister has called for a dramatic reduction in Japan's reliance on nuclear power.[5] Taiwan's president did the same. Shinzō Abe, the new prime minister of Japan since December 2012, announced a plan to re-start some of the 54 Japanese nuclear power plants (NPPs) and to continue some NPP sites under construction.[6]

As of 2016, countries including Australia, Austria, Denmark, Greece, Ireland, Italy, Latvia, Liechtenstein, Luxembourg, Malaysia, Malta, New Zealand, Norway, Philippines, and Portugal have no nuclear power stations and remain opposed to nuclear power.[7][8] Belgium, Germany, Spain and Switzerland are **phasing-out nuclear power**.[8][9][10][11] Globally, more nuclear power reactors have closed than opened in recent years but overall capacity has increased.[10]

Italy is the only country that has closed all of functioning nuclear plants. Lithuania, Kazakhstan have shut down their only nuclear plants, but plan to built new ones to replace them. Armenia shut down its only nuclear plant but subsequently restarted it. Austria never used its first nuclear plant that was completely built. Due to financial, politic and technical reasons Cuba, Libya, North Korea and Poland never completed the construction of their first nuclear plants (although North Korea and Poland plan to). Azerbaijan, Georgia, Ghana, Ireland, Kuwait, Oman, Peru, Singapore, Venezuela have planned, but not constructed their first nuclear plants.

8.1 Overview

A popular movement against nuclear power has gained strength in the Western world, based on concerns about

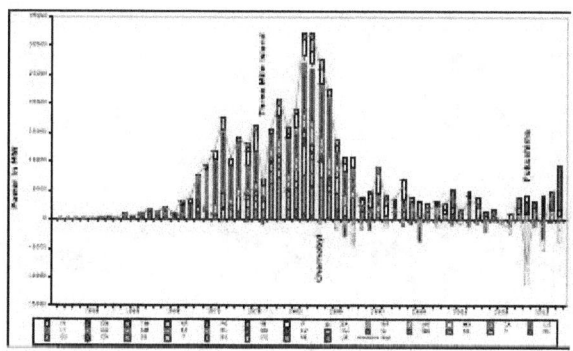

Timeline of commissioned and decommissioned nuclear capacity since the 1950s[12]

120,000 people attended an anti-nuclear protest in Bonn, Germany, on October 14, 1979, following the Three Mile Island accident.[13]

more nuclear accidents and concerns about nuclear waste. Anti-nuclear critics see nuclear power as a dangerous, expensive way to boil water to generate electricity.[14] The 1979 Three Mile Island accident and the 1986 Chernobyl disaster played a key role in stopping new plant construction in many countries. Major anti-nuclear power groups include Friends of the Earth, Greenpeace, Institute for Energy and Environmental Research, Nuclear Information and Resource Service, and Sortir du nucléaire (France).

Several countries, especially European countries, have abandoned the construction of new of nuclear power plants.[15] Austria (1978), Sweden (1980) and Italy (1987) voted in referendums to oppose or phase out nuclear power, while opposition in Ireland prevented a nuclear program there. Countries that have no nuclear plants and have restricted new plant constructions comprise Australia, Austria, Denmark, Greece, Italy, Ireland and Norway.[16][17] Poland stopped the construction of a plant.[16][18] Belgium, Germany, Spain, and Sweden decided not to build new plants or intend to phase out nuclear power, although still mostly relying on nuclear energy.[16][19]

New reactors under construction in Finland and France, which were meant to lead a nuclear new build, have been substantially delayed and are running over-budget.[20][21][22] However, China has 27 new reactors under construction,[23] and there are also new reactors being built in Belarus, Brazil, India, Japan, Pakistan, Russia, Slovakia, South Korea, Turkey, United Arab Emirates ,United Kingdom and the United States of America. At least 100 older and smaller reactors will "most probably be closed over the next 10-15 years" .[24]

Countries that wish to shut down nuclear power plants must find alternatives for electricity generation; otherwise, they are forced to become dependent on imports. Therefore, the discussion of a future for nuclear energy is intertwined with discussions about fossil fuels or an energy transition to renewable energy.

8.2   Countries that have decided on a phase-out

Main article: Nuclear energy policy

8.2.1   Austria

See also: Anti-nuclear movement in Austria

A nuclear power station was built during the 1970s at Zwentendorf, Austria, but its start-up was prevented by a popular vote in 1978. On July 9, 1997, the Austrian Parliament voted unanimously to maintain the country's anti-nuclear policy.[25]

8.2.2   Belgium

Belgium's nuclear phase-out legislation was agreed in July 1999 by the Liberals (VLD and MR), the Socialists (SP.A and PS) and the Greens party (Groen! and Ecolo). The phase-out law calls for each of Belgium's seven reactors to close after 40 years of operation with no new reactors built subsequently. When the law was being passed, it was speculated it would be overturned again as soon as an administration without the Greens was in power.[26]

In 2003, a new government was elected without the Greens. In September 2005, the government decided to partially overturn the previous decision, extending the phase-out period for another 20 years, with possible further extensions. It remains unknown if additional nuclear plants will be built.

In July 2005, the Federal Planning Bureau published a new

report, which states that oil and other fossil fuels generate 90% of Belgian energy use, while nuclear power accounts for 9% and renewable energy for 1%. Electricity only amounts to 16% of total energy use, and while nuclear-powered electricity amounts to 9% of use in Belgium, in many parts of Belgium, especially in Flanders, it makes up more than 50% of the electricity provided to households and businesses.[27] This was one of the major reasons to revert the earlier phase-out, since it was impossible to provide more than 50% of the electricity by 'alternative' energy-production, and a revert to the classical coal-driven electricity would mean inability to adhere to the Kyoto Protocol.

It is projected that within 25 years renewable energy will increase to at most 5% of the energy use, because of high costs. The current plan of the government is for all nuclear power stations to shut down by 2025. The report raises concerns about greenhouse gases and sustainability.[28]

In August 2005, French SUEZ offered to buy the Belgian Electrabel, which runs nuclear power stations.[29] At the end of 2005, Suez had some 98.5% of all Electrabel shares. Beginning 2006, Suez and Gaz de France announced a merger.

In the 2010–2011 Belgian government formation negotiations, the phase-out was emphasized again, with concrete plans to shut off three of the country's seven reactors by 2015.[30]

8.2.3 Germany

See also: Anti-nuclear movement in Germany and Nuclear power in Germany § Closures and phase-out
More than 80% of Germany's shut down nuclear power has been replaced with coal power plants,[31] which release 100 times as much radiation as a nuclear power plant of the same wattage.[32]

In 2000, the German government, consisting of the SPD and Alliance '90/The Greens, officially announced its intention to phase out the use of nuclear energy. The power plants in Stade and Obrigheim were turned off on 14 November 2003, and 11 May 2005, respectively. The plants' dismantling was scheduled to begin in 2007.[33]

The year 2000 Renewable Energy Sources Act provided for a feed-in tariff in support of renewable energy. The German government, declaring climate protection as a key policy issue, announced a carbon dioxide reduction target by the year 2005 compared to 1990 by 25%.[34] In 1998, the use of renewables in Germany reached 284 PJ of primary energy demand, which corresponds to 5% of the total electricity demand. By 2010, the German government wanted to reach 10%.[26]

A nuclear power plant at Grafenrheinfeld, Germany. Chancellor Angela Merkel's coalition announced on May 30, 2011, that Germany's 17 nuclear power stations will be shut down by 2022, in a policy reversal following Japan's Fukushima Daiichi nuclear disaster.[2]

Anti-nuclear activists have argued the German government had been supportive of nuclear power by providing financial guarantees for energy providers. Also it has been pointed out, there were, as yet, no plans for the final storage of nuclear waste. By tightening safety regulations and increasing taxation, a faster end to nuclear power could have been forced. A gradual closing down of nuclear power plants had come along with concessions in questions of safety for the population with transport of nuclear waste throughout Germany.[35] This latter point has been disagreed with by the Minister of Environment, Nature Conservation and Nuclear Safety.[36]

Critics of a phase-out in Germany argue that the power output from the nuclear power stations will not be adequately compensated and predict an energy crisis. They also argue that only coal-powered plants could compensate for nuclear power and CO_2 emissions will increase tremendously (with the use of oil and fossils). Energy may have to be imported from France's nuclear power facilities or Russian natural gas, despite Russia's not being perceived as a safe partner in much of Western Europe.[37]

In 2011, Deutsche Bank analysts concluded that "the global impact of the Fukushima accident is a fundamental shift in public perception with regard to how a nation prioritizes and values its populations health, safety, security, and natural environment when determining its current and future energy pathways". There were many anti-nuclear protests and, on 29 May 2011, Merkel's government announced that it would close all of its nuclear power plants by 2022.[38][39] Following the March 2011 Fukushima nuclear disaster, Germany has permanently shut down eight of its 17 reactors. Between 2011 and 2014 Germany burned more coal, an additional 9.5 million tonnes of oil equivalent.[40] Galvanised by the Fukushima nuclear disaster, first anniversary anti-nuclear demonstrations were held in Germany in March 2012. Organisers say more than 50,000 people in six regions took part.[41]

The German *Energiewende* designates a significant change in energy policy from 2010. The term encompasses a transition by Germany to a low carbon, environmentally sound, reliable, and affordable energy supply.[42] On 6 June 2011, following Fukushima, the government removed the use of nuclear power as a bridging technology as part of their policy.[43]

In September 2011, German engineering giant Siemens announced it will withdraw entirely from the nuclear industry, as a response to the Fukushima nuclear disaster in Japan, and said that it would no longer build nuclear power plants anywhere in the world. The company's chairman, Peter Löscher, said that "Siemens was ending plans to cooperate with Rosatom, the Russian state-controlled nuclear power company, in the construction of dozens of nuclear plants throughout Russia over the coming two decades".[44][45] Also in September 2011, IAEA Director General Yukiya Amano said the Japanese nuclear disaster "caused deep public anxiety throughout the world and damaged confidence in nuclear power".[46]

A 2016 study shows that the security of electricity supply in Germany has improved in 2014 despite dropping after the nuclear phaseout. The study was conducted near the halfway point of the phaseout, 9 plants having been shut and a further 8 still in operation. Between 2011 and 2015 Germany's use of fossil fuels for electricity increased.[47][48]

In early-October 2016 Vattenfall began litigation against the German government for its 2011 decision to accelerate the phase-out of nuclear power. Hearing are taking place at the World Bank's International Centre for Settlement of Investment Disputes (ICSID) in Washington DC and Vattenfall is claiming almost €4.7 billion in damages. The German government regards the action as "inadmissible and unfounded".[49] These proceedings were ongoing in December 2016, despite Vattenfall commencing civil litigation within Germany.[50]

On 5 December 2016, the Federal Constitutional Court (*Bundesverfassungsgericht*) ruled that the nuclear plant operators affected by the accelerated phase-out of nuclear power following the Fukushima disaster are eligible for "adequate" compensation. The court found that the nuclear exit was essentially constitutional but that the utilities are entitled to damages for the "good faith" investments they made in 2010. The utilities can now sue the German government under civil law. E.ON, RWE, and Vattenfall are expected to seek a total of €19 billion under separate suits.[51][52][53] Six cases were registered with courts

in Germany, as of 7 December 2016.[50][54]

8.2.4 Italy

Nuclear power phase-out commenced in Italy in 1987, one year after the Chernobyl accident. Following a referendum in that year, Italy's four nuclear power plants were closed down, the last in 1990. A moratorium on the construction of new plants, originally in effect from 1987 until 1993, has since been extended indefinitely.[55]

In recent years, Italy has been an importer of nuclear-generated electricity, and its largest electricity utility Enel S.p.A. has been investing in reactors in both France and Slovakia to provide this electricity in the future, and also in the development of the EPR technology.

In October 2005, there was a seminar sponsored by the government about the possibility of reviving Italian nuclear power.[56] The fourth cabinet led by Silvio Berlusconi tried to implement a new nuclear plan but a referendum held in June 2011 stopped any project.

8.2.5 Philippines

See also: Anti-nuclear movement in the Philippines

In the Philippines, in 2004, President Gloria Macapagal-Arroyo outlined her energy policy. She wants to increase indigenous oil and gas reserves through exploration, develop alternative energy resources, enforce the development of natural gas as a fuel and coco diesel as alternative fuel, and build partnerships with Saudi Arabia, Asian countries, China and Russia. She also made public plans to convert the Bataan Nuclear Power Plant into a gas-powered facility.[57]

8.2.6 Sweden

Main article: Nuclear power phase-out in Sweden

After the Three Mile Island accident in 1979, there was a referendum in 1980 about nuclear power. As a result of this, the Swedish parliament decided that no further nuclear power plants should be built, and that a nuclear power phase-out should be completed by 2010.

After the 1986 Chernobyl disaster in Ukraine, the question of security of nuclear energy was again raised. In 1997, the Riksdag, the Swedish parliament, decided to shut down one of the reactors at Barsebäck by July 1, 1998, and the second before July 1, 2001, although under the condition that their energy production would be compensated. At Barsebäck, block 1 was shut down on November 30, 1999 and block 2 on June 1, 2005.

On 5 February 2009, the Government of Sweden effectively ended the phase-out policy.[58] In 2010, Parliament approved for new reactors to replace existing ones.[59]

An opinion poll in April 2016 showed that about half of Swedes want to phase out nuclear power, 30 percent want its use continued, and 20 percent are undecided.[60] Prior to the Fukushima Daiichi nuclear disaster in 2011, "a clear majority of Swedes" had been in favour of nuclear power.[60]

In October 2015, decisions were take to phase out two reactors at Oskarshamn[61] and two at Ringhals,[62] reducing the number of remaining reactors from 12 in 1999 to 6 in 2020. They were shuttered for economical reasons, one of which was high costs for safety upgrades and another lower electricity prices than expected.

The competitive disadvantage for nuclear was partly a result of policy. The renewable portfolio standard (Elcertifikat) system was introduced in 2003 by a parliamentary majority including three anti-nuclear parties and the governing Social Democrats which was in the process of phasing out the Barsebäck power station. The Elcertifikat forced in successively more renewables in the electricity mix. As electricity consumption in Sweden dropped considerably from 2001 on, from about 150 to 140 TWh/year, nuclear power had to compete with subsidised wind power and for a shrinking market.

In June 2016 the opposition parties and the government reached an agreement on Swedish nuclear power.[63] The agreement is to phase out the output tax on nuclear power, and allow ten new replacement reactors to be built at current nuclear plants.[64]

8.2.7 Switzerland

●Gösgen
●Leibstadt
●Mühleberg
●*Lucens*
Switzerland Nuclear power plants (view)
● Active plants

● *Closed plants*

See also: Nuclear power in Switzerland and Anti-nuclear movement in Switzerland

As of 2013, the five operational Swiss nuclear reactors were

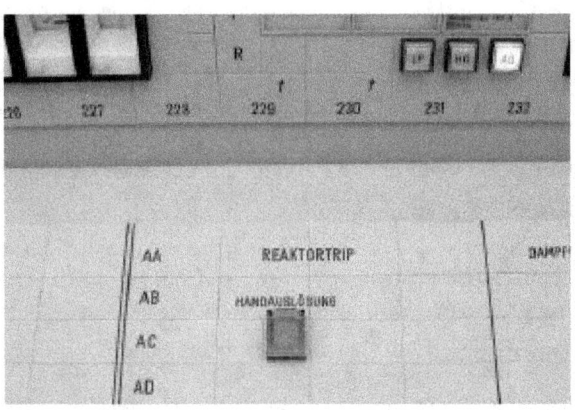

An emergency switch-off button of the Beznau Nuclear Power Plant. In 2011, the federal authorities decided to gradually phase out nuclear power in Switzerland.

Beznau 1 and 2, Gösgen, Leibstadt, and Mühleberg—all located in the German speaking part of the country. Nuclear power accounted for 36.4% of the national electricity generation, while 57.9% came from hydroelectricity. The remaining 5.7% was generated by other conventional and non-hydro renewable power stations.[65]

On 25 May 2011, the Federal Council decided on a slow phase-out by not extending running times or building new power plants.[66] The first power plant, Mühleberg, will stop running in 2019, the last in 2034.[67]

There have been many Swiss referenda on the topic of nuclear energy, beginning in 1979 with a citizens' initiative for nuclear safety, which was rejected. In 1984, there was a vote on an initiative "for a future without further nuclear power stations" with the result being a 55 to 45% vote against. On September 23, 1990, Switzerland had two more referenda about nuclear power. The initiative "stop the construction of nuclear power stations", which proposed a ten-year moratorium on the construction of new nuclear power plants, was passed with 54.5% to 45.5%. The initiative for a phase-out was rejected with by 53% to 47.1%. In 2000, there was a vote on a green tax for support of solar energy. It was rejected by 67–31%. On May 18, 2003, there were two referenda: "Electricity without Nuclear", asking for a decision on a nuclear power phase-out, and "Moratorium Plus", for an extension of the earlier-decided moratorium on the construction of new nuclear power plants. Both were turned down. The results were: Moratorium Plus: 41.6% Yes, 58.4% No; Electricity without Nuclear: 33.7% Yes, 66.3% No.[68]

The program of the "Electricity without Nuclear" petition was to shut down all nuclear power stations by 2033, starting with Unit 1 and 2 of Beznau nuclear power stations, Mühleberg in 2005, Gösgen in 2009, and Leibstadt in 2014. "Moratorium Plus" was for an extension of the moratorium for another ten years, and additionally a condition to stop the present reactors after 40 years of operation. In order to extend the 40 years by ten more years, another referendum would have to be held (at high administrative costs). The rejection of the Moratorium Plus had come as a surprise to many, as opinion polls before the referendum had showed acceptance. Reasons for the rejections in both cases were seen as the worsened economic situation.[69]

8.3 Other significant places

8.3.1 Europe

See also: Anti-nuclear movement in Spain

In Spain a moratorium was enacted by the socialist government in 1983[70][71] and in 2006 plans for a phase-out of seven reactors were being discussed anew.[72]

In Ireland, a nuclear power plant was first proposed in 1968. It was to be built during the 1970s at Carnsore Point in County Wexford. The plan called for first one, then ultimately four plants to be built at the site, but it was dropped after strong opposition from environmental groups, and Ireland has remained without nuclear power since. Despite opposing nuclear power (and nuclear fuel reprocessing at Sellafield), Ireland is to open an interconnector to the mainland UK to buy electricity, which is, in some part, the product of nuclear power.

Slovenian nuclear plant in Krško (co-owned with Croatia) is scheduled to be closed by 2023, and there are no plans to build further nuclear plants. The debate on whether and when to close the Krško plant was somewhat intensified after the 2005/06 winter energy crisis. In May 2006 the Ljubljana-based daily *Dnevnik* claimed Slovenian government officials internally proposed adding a new 1000 MW block into Krško after the year 2020.

Greece operates only a single small nuclear reactor in the Greek National Physics Research Laboratory in Demokritus Laboratories for research purposes.

The future of nuclear power in the United Kingdom is currently under review. The country has a number of reactors which are currently reaching the end of their working life, and it is currently undecided how they will be replaced. The UK is also currently failing to reach its targets for reduction on CO_2 emissions, which situation may be made worse if new nuclear power stations are not built. The UK also uses a large proportion of gas-fired power stations, which produce half the CO_2 emissions as coal, but there have been recent difficulties in obtaining adequate gas supplies. The UK government has just appointed a new pro-nuclear energy minister.

8.3.2 The Netherlands

In the Netherlands, in 1994, the Dutch parliament voted to phase out after a discussion of nuclear waste management. The power station at Dodewaard was shut down in 1997. In 1997 the government decided to end Borssele's operating license, at the end of 2003. In 2003 the shut-down was postponed by the government to 2013.[73][74] In 2005 the decision was reversed and research in expanding nuclear power has been initiated. Reversal was preceded by the publication of the Christian Democratic Appeal's report on sustainable energy.[75] Other coalition parties then conceded. In 2006 the government decided that Borssele will remain open until 2033, if it can comply with the highest safety standards. The owners, Essent and DELTA will invest 500 million euro in sustainable energy, together with the government, money which the government claims otherwise should have been paid to the plants owners as compensation.

8.3.3 Australia

See also: Anti-nuclear movement in Australia

New Zealand enacted the *New Zealand Nuclear Free Zone, Disarmament, and Arms Control Act 1987* which prohibits the stationing of nuclear weapons on the territory of New Zealand and the entry into New Zealand waters of nuclear armed or propelled ships. This Act of Parliament, however, does not prevent the construction of nuclear power plants.

In Australia there are no nuclear power plants. Australia has very extensive, low-cost coal reserves and substantial natural gas and majority political opinion is still opposed to domestic nuclear power on both environmental and economic grounds.

8.3.4 Asia

Renewable energy, mainly hydropower, is gaining share.[76][77]

For North Korea, two PWRs at Kumho were under construction until that was suspended in November 2003. On September 19, 2005 North Korea pledged to stop building nuclear weapons and agreed to international inspections in return for energy aid, which may include one or more light water reactors – the agreement said "The other parties expressed their respect and agreed to discuss at an appropriate time the subject of the provision of light-water reactor" [sic]."[78]

In July 2000, the Turkish government decided not to build four reactors at the controversial Akkuyu Nuclear Power Plant, but later changed its mind. The official launch ceremony took place in April 2015, and the first unit is expected to be completed in 2020."[79]

Taiwan has 3 active plants and 6 reactors. Active seismic faults run across the island, and some environmentalists argue Taiwan is unsuited for nuclear plants."[80] Construction of the Lungmen Nuclear Power Plant using the ABWR design has encountered public opposition and a host of delays, and in April 2014 the government decided to halt construction."[81] Construction will be halted from July 2015 to 2017 in order to allow time for a referendum to be held."[82] The 2016 election was won by a government with stated policies that included phasing out nuclear power generation."[83]

India has 20 reactors operating, 6 reactors under construction, and is planning an additional 24."[84]

Vietnam had developed detailed plans for 2 nuclear power plants with 8 reactors, but in November 2016 decided to abandon nuclear power plans as they were "not economically viable because of other cheaper sources of power." "[85]

8.3.5 Japan

See also: Anti-nuclear power movement in Japan
Once a nuclear proponent, Prime Minister Naoto Kan be-

Three of the reactors at Fukushima I overheated, causing meltdowns that eventually led to hydrogen explosions, which released large amounts of radioactive gases into the air."[86]

came increasingly anti-nuclear following the Fukushima nuclear disaster. In May 2011, he closed the aging Hamaoka

Anti-Nuclear Power Plant Rally on 19 September 2011 at Meiji Shrine complex in Tokyo. Sixty thousand people marched chanting "Sayonara nuclear power" and waving banners, calling on Japan's government to abandon nuclear power, following the Fukushima disaster."[87]"[88]

Nuclear Power Plant over earthquake and tsunami fears, and said he would freeze plans to build new reactors. In July 2011, Kan said that "Japan should reduce and eventually eliminate its dependence on nuclear energy ... saying that the Fukushima accident had demonstrated the dangers of the technology" ."[89] In August 2011, the Japanese government passed a bill to subsidize electricity from renewable energy sources."[90] A 2011 Japanese Cabinet energy white paper says "public confidence in safety of nuclear power was greatly damaged" by the Fukushima disaster, and calls for a reduction in the nation's reliance on nuclear power."[91] As of August 2011, the crippled Fukushima nuclear plant is still leaking low levels of radioactivity and areas surrounding it could remain uninhabitable for decades."[92]

By March 2012, one year after the disaster, all but two of Japan's nuclear reactors were shut down; some were damaged by the quake and tsunami. The following year, the last two were taken off-line. Authority to restart the others after scheduled maintenance throughout the year was given to local governments, and in all cases local opposition prevented restarting.

Prime Minister Shinzo Abe's government, reelected on a platform of restarting nuclear power, plans to have nuclear power account for 20 to 22 percent of the country's total electricity supply by 2030, compared with roughly 30 percent before the disaster at the Fukushima complex.

So far two reactors at Sendai nuclear power plant have been restarted."[93]

8.3.6 United States

See also: Anti-nuclear movement in the United States

The United States is, as of 2013, undergoing a practical phase-out independent of stated goals and continued official support. This is not due to concerns about the source or anti-nuclear groups, but due to the rapidly falling prices of natural gas and the reluctance of investors to provide funding for long-term projects when short term profitability of turbine power is available.

Through the 2000s a number of factors led to greatly increased interest in new nuclear reactors, including rising demand, new lower-cost reactor designs, and concerns about global climate change. By 2009, about 30 new reactors were planned, and a large number of existing reactors had applied for upgrades to increase their output. In total, 39 reactors have had their licences renewed, three Early Site Permits have been applied for, and three consortiums have applied for Combined Construction-Operating Licences under the *Nuclear Power 2010 Program*. In addition, the Energy Policy Act of 2005 contains incentives to further expand nuclear power.[94]

However, by 2012 the vast majority of these plans were cancelled, and several additional cancellations followed in 2013. Currently only five new reactors are under construction, and one, at Watts Bar, was originally planned in the 1970s and only under construction now. Construction of the new AP1000 design is underway at two locations in the United States in Georgia and South Carolina. Plans for additional reactors in Florida were cancelled in 2013.

Some smaller reactors operating in deregulated markets have become uneconomic to operate and maintain, due to competition from generators using low priced natural gas, and may be retired early.[95] The 556 MWe Kewaunee Power Station is being closed 20 years before license expiry for these economic reasons.[96][97] Duke Energy's Crystal River 3 Nuclear Power Plant in Florida closed, as it could not recover the costs needed to fix its containment building.[98]

As a result of these changes, after reaching peak production in 2007, US nuclear capacity has been undergoing constant reduction every year.

8.3.7 South America

In Brazil, nuclear energy, produced by two reactors at Angra, accounts for about 4% of the country's electricity – about 13 TWh per year.[99] Angra III is under construction and due to come online in 2018. Brazil plans to build seven more reactors by 2025.[100]

In Argentina, about 6% of the electricity comes from 3 operational reactors: The *Embalse Río Tercero* plant, a CANDU6 reactor, the *Atucha 1* plant, a PHWR German design, and the *Atucha 2* plant, also a PHWR German design. Argentina also has some other research reactors, and exports nuclear technology.

8.4 Pros and cons of nuclear power

8.4.1 The nuclear debate

Main article: Nuclear power debate

The **nuclear power debate** is about the controversy[101][102][103][104][105] which has surrounded the deployment and use of nuclear fission reactors to generate electricity from nuclear fuel for civilian purposes. The debate about nuclear power peaked during the 1970s and 1980s, when it "reached an intensity unprecedented in the history of technology controversies", in some countries.[106][107]

Proponents of nuclear energy argue that nuclear power is a sustainable energy source which reduces carbon emissions and can increase energy security if its use supplants a dependence on imported fuels.[108] Proponents advance the notion that nuclear power produces virtually no air pollution, in contrast to the chief viable alternative of fossil fuel. Proponents also believe that nuclear power is the only viable course to achieve energy independence for most Western countries. They emphasize that the risks of storing waste are small and can be further reduced by using the latest technology in newer reactors, and the operational safety record in the Western world is excellent when compared to the other major kinds of power plants.[109]

Opponents say that nuclear power poses many threats to people and the environment. These threats include health risks and environmental damage from uranium mining, processing and transport, the risk of nuclear weapons proliferation or sabotage, and the unsolved problem of radioactive nuclear waste.[110][111][112] They also contend that reactors themselves are enormously complex machines where many things can and do go wrong, and there have been many serious nuclear accidents.[113][114] Critics do not believe that these risks can be reduced through new technology.[115] They argue that when all the energy-intensive stages of the nuclear fuel chain are considered, from uranium mining to nuclear decommissioning, nuclear power is not a low-carbon electricity source.[116][117][118]

8.4.2 Economics

Main article: Economics of nuclear power plants

The **economics of new nuclear power plants** is a controversial subject, since there are diverging views on this topic, and multi-billion dollar investments ride on the choice of an energy source. Nuclear power plants typically have high capital costs for building the plant, but low direct fuel costs (with much of the costs of fuel extraction, processing, use and long term storage externalized). Therefore, comparison with other power generation methods is strongly dependent on assumptions about construction timescales and capital financing for nuclear plants. Cost estimates also need to take into account plant decommissioning and nuclear waste storage costs. On the other hand measures to mitigate global warming, such as a carbon tax or carbon emissions trading, may favor the economics of nuclear power versus fossil fuels.

In recent years there has been a slowdown of electricity demand growth and financing has become more difficult, which affects large projects such as nuclear reactors, with very large upfront costs and long project cycles which carry a large variety of risks.[119] In Eastern Europe, a number of long-established projects are struggling to find finance, notably Belene in Bulgaria and the additional reactors at Cernavoda in Romania, and some potential backers have pulled out.[119] Where cheap gas is available and its future supply relatively secure, this also poses a major problem for nuclear projects.[119]

Analysis of the economics of nuclear power must take into account who bears the risks of future uncertainties. To date all operating nuclear power plants were developed by state-owned or regulated utility monopolies[120] where many of the risks associated with construction costs, operating performance, fuel price, and other factors were borne by consumers rather than suppliers. Many countries have now liberalized the electricity market where these risks, and the risk of cheaper competitors emerging before capital costs are recovered, are borne by plant suppliers and operators rather than consumers, which leads to a significantly different evaluation of the economics of new nuclear power plants.[121]

Following the 2011 Fukushima Daiichi nuclear disaster, costs are likely to go up for currently operating and new nuclear power plants, due to increased requirements for on-site spent fuel management and elevated design basis threats.[122]

8.4.3 Environment

Main article: Environmental impact of nuclear power

The **environmental impact of nuclear power** results from

Nuclear power activities involving the environment: mining, enrichment, generation and geological disposal.

the nuclear fuel cycle, operation, and the effects of nuclear accidents.

The greenhouse gas emissions from nuclear fission power are small relative to those associated with coal, oil, gas, solar and biomass. They are about equal to those associated with wind and hydroelectric.[123]

The routine health risks from nuclear fission power are very small relative to those associated with coal, oil, gas, solar, biomass, wind and hydroelectric.[124]

However, there is a "catastrophic risk" potential if containment fails,[125] which in nuclear reactors can be brought about by over-heated fuels melting and releasing large quantities of fission products into the environment. The public is sensitive to these risks and there has been considerable public opposition to nuclear power. Even so, in comparing the fatalities for major accidents alone in the energy sector it is still found that the risks associated with nuclear power are extremely small relative to those associated with coal, oil, gas and hydroelectric.[124]

The 1979 Three Mile Island accident and 1986 Chernobyl disaster, along with high construction costs, ended the rapid growth of global nuclear power capacity.[125] A further disastrous release of radioactive materials followed the 2011 Japanese tsunami which damaged the Fukushima I Nuclear Power Plant, resulting in hydrogen gas explosions and partial meltdowns classified as a Level 7 event. The large-scale release of radioactivity resulted in people being evacuated from a 20 km exclusion zone set up around the power plant, similar to the 30 km radius Chernobyl Exclusion Zone still in effect.

8.4.4   Accidents

Main article: Nuclear and radiation accidents

The effect of nuclear accidents has been a topic of debate

The abandoned city of Pripyat with Chernobyl plant in the distance

practically since the first nuclear reactors were constructed. It has also been a key factor in public concern about nuclear facilities.[126] Some technical measures to reduce the risk of accidents or to minimize the amount of radioactivity released to the environment have been adopted. Despite the use of such measures, human error remains, and "there have been many accidents with varying effects as well near misses and incidents" .[126][127]

Benjamin K. Sovacool has reported that worldwide there have been 99 accidents at nuclear power plants.[128] Fifty-seven accidents have occurred since the Chernobyl disaster, and 57% (56 out of 99) of all nuclear-related accidents have occurred in the USA.[128] Serious nuclear power plant accidents include the Fukushima Daiichi nuclear disaster (2011), Chernobyl disaster (1986), Three Mile Island accident (1979), and the SL-1 accident (1961).[129] Stuart Arm states, "apart from Chernobyl, no nuclear workers or members of the public have ever died as a result of exposure to radiation due to a commercial nuclear reactor incident." [130]

The International Atomic Energy Agency maintains a website reporting recent accidents.[131]

8.4.5   Safety

Main article: Nuclear safety and security

Nuclear safety and security covers the actions taken to prevent nuclear and radiation accidents or to limit their consequences. This covers nuclear power plants as well as all other nuclear facilities, the transportation of nuclear materials, and the use and storage of nuclear materials for medical, power, industry, and military uses.

Although there is no way to guarantee that a reactor will always be designed, built and operated safely, the nuclear power industry has improved the safety and performance of reactors, and has proposed safer reactor designs, though many of these designs have yet to be tested at industrial or commercial scales.[132] Mistakes do occur and the designers of reactors at Fukushima in Japan did not anticipate that a tsunami generated by an earthquake would disable the backup systems that were supposed to stabilize the reactor after the earthquake.[133][134] According to UBS AG, the Fukushima I nuclear accidents have cast doubt on whether even an advanced economy like Japan can master nuclear safety.[135] Catastrophic scenarios involving terrorist attacks are also conceivable.[132]

An interdisciplinary team from MIT have estimated that given the expected growth of nuclear power from 2005 – 2055, at least four serious nuclear accidents would be expected in that period.[136][137] To date, there have been five serious accidents (core damage) in the world since 1970 (one at Three Mile Island in 1979; one at Chernobyl in 1986; and three at Fukushima-Daiichi in 2011), corresponding to the beginning of the operation of generation II reactors. This leads to on average one serious accident happening every eight years worldwide.[134] Despite these accidents, the safety record of nuclear power, in terms of lives lost (ignoring nonfatal illnesses) per unit of electricity delivered, is better than every other major source of power in the world, and on par with solar and wind.[124][138][139]

8.5   Energy transition

See also: energy transition, 100% renewable energy, nuclear power debate, and green movement

The Energy transition is the shift by several countries to sustainable economies by means of renewable energy, energy efficiency and sustainable development. The final goal is the abolishment of coal and other non-renewable energy sources.[141]

Renewable energy encompasses wind, biomass (such as landfill gas and sewage gas), hydropower, solar power (thermal and photovoltaic), geothermal, and ocean power. These renewable sources serve as alternatives to conventional power generation such as coal power, oil power, and natural gas power. Piecemeal measures often have only limited potential, so a timely implementation for the energy transition requires multiple approaches in parallel. Energy conser-

Photovoltaic array and wind turbines at the Schneebergerhof wind farm in the German state of Rheinland-Pfalz

Parabolic trough power plant for electricity production, near the town of Kramer Junction in California's San Joaquin Valley

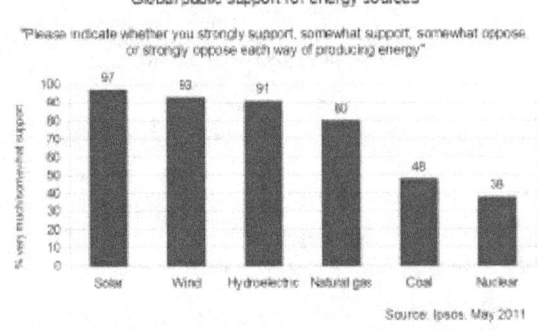

Global public support for energy sources, based on a survey by Ipsos (2011).[140]

vation and improvements in energy efficiency thus play a major role. An example of an effective energy conserva-

tion measure is improved insulation for buildings; an example of improved energy efficiency is cogeneration of heat and power. Smart electric meters can schedule energy consumption for times when electricity is available inexpensively.

Issues exist that currently prevent a shift over to a 100% renewable technologies. There is debate over the environmental impact of solar power, and the environmental impact of wind power. Some argue that the pollution produced and requirement of rare earth elements offsets many of the benefits compared to other alternative power sources such as hydroelectric, geothermal, and nuclear power.[142] According to the 2013 *Post Carbon Pathways* report, which reviewed many international studies, the key roadblocks are: climate change denial, the fossil fuels lobby, political inaction, unsustainable energy consumption, outdated energy infrastructure, and financial constraints.[143] However, according to a research paper published in 2014, renewable energy by itself, will not be able to stop climate change.[144]

Google spent $30 million on their RE<C project to develop renewable energy and stave off catastrophic climate change. The project was cancelled after concluding that a best-case scenario for rapid advances in renewable energy could only result in emissions 55 percent below the fossil fuel projections for 2050.[145] Current developments towards a 100% renewable energy policy require solutions to low storage capacity, low energy density, and high cost.[146]

8.6 See also

- Nuclear renaissance
- Anti-nuclear movement
- Energy conservation
- Energy development
- Fossil fuel phase-out
- List of energy topics
- Nuclear Non-Proliferation Treaty
- Nuclear energy policy
- Nuclear power controversy
- Oil phase-out in Sweden
- Nuclear power in France
- Renewable energy commercialization
- Wind power

8.7   Notes and references

[1] IAEA (2011 Highlights). "Power Reactor Information System". Check date values in: ldate= (help)

[2] Annika Breidthardt (May 30, 2011). "German government wants nuclear exit by 2022 at latest". Reuters.

[3] "Italy Nuclear Referendum Results". 13 June 2011. Archived from the original on 25 March 2012.

[4] Henry Sokolski (November 28, 2011) "Nuclear Power Goes Rogue". Newsweek.

[5] Tsuyoshi Inajima & Yuji Okada (October 28, 2011). "Nuclear Promotion Dropped in Japan Energy Policy After Fukushima". Bloomberg.

[6] He is fighting a continuing economic crisis with Abenomics

[7] "Nuclear power: When the steam clears". The Economist. March 24, 2011.

[8] Duroyan Fertl (June 5, 2011). "Germany: Nuclear power to be phased out by 2022". Green Left.

[9] Erika Simpson and Ian Fairlie, Dealing with nuclear waste is so difficult that phasing out nuclear power would be the best option, Lfpress, February 26, 2016.

[10] "Difference Engine: The nuke that might have been". The Economist. Nov 11, 2013.

[11] James Kanter (May 25, 2011). "Switzerland Decides on Nuclear Phase-Out". New York Times.

[12] "The Database on Nuclear Power Reactors". IAEA.

[13] Herbert P. Kitschelt. Political Opportunity and Political Protest: Anti-Nuclear Movements in Four Democracies British Journal of Political Science. Vol. 16, No. 1, 1986, p. 71.

[14] Helen Caldicott (2006). Nuclear Power is Not the Answer to Global Warming or Anything Else, Melbourne University Press. ISBN 0-522-85251-3, p. xvii

[15] Netherlands: Court case on closure date Borssele NPP, article from anti-nuclear organization (WISE), dated June 29, 2001.

[16] Nuclear Power in the World Energy Outlook, by the Uranium Institute, 1999.

[17] Anti-nuclear resolution of the Austrian Parliament Archived 23 February 2006 at the Wayback Machine., as summarised by an anti-nuclear organisation (WISE).

[18] Nuclear news from Poland, article from the Web site of the European Nuclear Society, April 2005.

[19] Germany Starts Nuclear Energy Phase-Out, article from Deutsche Welle, November 14, 2003.

[20] James Kanter. In Finland, Nuclear Renaissance Runs Into Trouble New York Times, May 28, 2009.

[21] James Kanter. Is the Nuclear Renaissance Fizzling? Green, 29 May 2009.

[22] Rob Broomby. Nuclear dawn delayed in Finland BBC News, 8 July 2009.

[23] Nuclear Power in China

[24] Michael Dittmar. Taking stock of nuclear renaissance that never was Sydney Morning Herald, August 18, 2010.

[25] "Coalition of Nuclear-Free Countries". WISE News Communique. 26 September 1997. Archived from the original on 23 February 2006. Retrieved 2006-05-19.

[26] Ruffles, Philip; Michael Burdekin; Charles Curtis; Brian Eyre; Geoff Hewitt; William Wilkinson (July 2003). "An Essential Programme to Underpin Government Policy on Nuclear Power" (PDF). Nuclear Task Force. Retrieved 2012-09-11.

[27] Henry, Alain (July 12, 2005). Quelle énergie pour un développement durable ?. Working Paper 14-05 (in French). Federal Planning Bureau.

[28] Addicted to nuclear energy? < Belgian news | Expatica Belgium. Expatica.com. Retrieved on 2011-06-04.

[29] Kanter, James (2005-08-10). "Big French Utility Offers a Full Buyout in Belgium". The New York Times.

[30] "Belgium plans to phase out nuclear power". BBC News. 2011-10-31.

[31] Things worse than nuclear power

[32] Coal Combustion: Nuclear resource or danger

[33] German nuclear energy phase-out begins with first plant closure. Terradaily.com (2003-11-14). Retrieved on 2011-06-04.

[34] http://www.agores.org/Publications/EnR/GermanyREPolicy2000.pdf

[35] Kommunikation Wissenschaft

[36] 'Nuclear phase-out in Germany and the Challenges for Nuclear Regulation'. Bmu.de. Retrieved on 2011-06-04.

[37] "Germany split over green energy". BBC News. 2005-02-25.

[38] Caroline Jorant (July 2011). "The implications of Fukushima: The European perspective". Bulletin of the Atomic Scientists. p. 15.

[39] Knight, Ben (15 March 2011). "Merkel shuts down seven nuclear reactors". Deutsche Welle. Retrieved 15 March 2011.

[40] http://www.bp.com/content/dam/bp/pdf/ Energy-economics/statistical-review-2015/ bp-statistical-review-of-world-energy-2015-coal-section. pdf pg5

[41] "Anti-nuclear demos across Europe on Fukushima anniversary". *Euronews*. 11 March 2011.

[42] Federal Ministry of Economics and Technology (BMWi); Federal Ministry for the Environment, Nature Conservation and Nuclear Safety (BMU) (28 September 2010). *Energy concept for an environmentally sound, reliable and affordable energy supply* (PDF). Berlin, Germany: Federal Ministry of Economics and Technology (BMWi). Retrieved 2016-05-01.

[43] *The Federal Government's energy concept of 2010 and the transformation of the energy system of 2011* (PDF). Bonn, Germany: Federal Ministry for the Environment, Nature Conservation, and Nuclear Safety (BMU). October 2011. Retrieved 2016-06-16.

[44] John Broder (October 10, 2011). "The Year of Peril and Promise in Energy Production". *New York Times*.

[45] "Siemens to quit nuclear industry". *BBC News*. 18 September 2011.

[46] "IAEA sees slow nuclear growth post Japan". *UPI*. September 23, 2011.

[47] "Supply security is even more stable despite nuclear phase-out —fossil reserve power is replaceable" (PDF) (Press release). Hamburg, Germany: Greenpeace Energy. 5 September 2016. Retrieved 2016-09-08.

[48] Huneke, Fabian; Lizzi, Philipp; Lenck, Thorsten (August 2016). *The consequences so far of Germany's nuclear phase-out on the security of energy supply —A brief analysis commissioned by Greenpeace Energy eG in Germany* (PDF). Berlin, Germany: Energy Brainpool. Retrieved 2016-09-08. This reference provides a good overview of the phase-out.

[49] "Showdown in Germany's nuclear phase-out". *Clean Energy Wire (CLEW)*. Berlin, Germany. 10 October 2016. Retrieved 2016-10-24.

[50] "Nuclear plant operators continue lawsuits". *Clean Energy Wire (CLEW)*. Berlin, German. 8 December 2016. Retrieved 2016-12-08.

[51] "German utilities eligible for "adequate" nuclear exit compensation". *Clean Energy Wire (CLEW)*. Berlin, Germany. 6 December 2016. Retrieved 2016-12-06.

[52] "The thirteenth amendment to the Atomic Energy Act is for the most part compatible with the Basic Law" (Press release). Karlsruhe, Germany: Bundesverfassungsgericht. 6 December 2016. Retrieved 2016-12-06.

[53] "German utilities win compensation for nuclear phaseout". *Deutsche Welle (DW)*. Bonn, Germany. 5 December 2016. Retrieved 2016-12-06. Provides a history of the nuclear exit.

[54] "Atomausstieg: Konzerne klagen weiter – auf Auskunft" [Nuclear exit: corporations sue further - for information]. *Der Tagesspiegel* (in German). Berlin, German. Retrieved 2016-12-08.

[55] http://energytrends.pnl.gov/italy/it004.htm

[56] "Prospettive dell'energia nucleare in Italia". Retrieved 5 May 2012.

[57] Archived 23 February 2006 at the Wayback Machine.

[58] Borgenäs, Johan (November 11, 2009). "Sweden Reverses Nuclear Phase-out Policy". Nuclear Threat Initiative.

[59] "Sweden to replace existing nuclear plants with new ones". *BBC News Online*. 18 June 2010.

[60] "30 years after Chernobyl: Half of Swedes oppose nuclear power". *Sveriges Radio*. 26 April 2016.

[61] http://okg.se/sv/Press/2015/ Beslut-fattat-om-fortida-stangning-av-O1-och-O2/

[62] https://corporate.vattenfall.se/press-och-media/engelska/ r1-and-r2-in-operation-until-2020-and-2019/

[63] "Sweden strikes deal to continue nuclear power". *The Local*. 10 June 2016.

[64] Juhlin, Johan. "Klart i dag: så blir den svenska energipolitiken". *svt.se* (in Swedish). Sveriges Television. Retrieved 2016-06-13.

[65] Swiss Federal Office of Energy (SFOE) Electricity statistics 2013 (in French and German), 23 June 2014

[66] «Mutiger Entscheid» bis «Kurzschlusshandlung» (Politik, Schweiz, NZZ Online). Nzz.ch. Retrieved on 2011-06-04.

[67] Schweiz plant Atomausstieg – Schweiz – derStandard.at › International. Derstandard.at. Retrieved on 2011-06-04.

[68] Bundesamt für Energie BFE – Startseite. Energieschweiz.ch. Retrieved on 2011-06-04.

[69] Archived 13 December 2004 at the Wayback Machine.

[70] "Spain halts nuclear power". WISE News Communique. May 24, 1991. Retrieved 2006-05-19.

[71] "Nuclear Power in Spain". World Nuclear Association. May 2006. Archived from the original on 22 February 2006. Retrieved 2006-05-19.

[72] "404 error". Archived from the original on 18 February 2005. Retrieved 2006-05-19.

[73] Archived 23 February 2005 at the Wayback Machine.

[74] Archived 12 June 2007 at the Wayback Machine.

[75] Archived 23 September 2006 at the Wayback Machine.

[76] EIA – 1000 Independence Avenue, SW, Washington, DC 20585. Eia.doe.gov. Retrieved on 2011-06-04. Archived 1 March 2008 at the Wayback Machine.

[77] Archived 28 September 2006 at the Wayback Machine.

[78] Archived 20 September 2005 at the Wayback Machine.

[79] "Ground broken for Turkey's first nuclear power plant". World Nuclear News. 15 April 2015. Retrieved 19 April 2015.

[80] Andrew Jacobs (12 January 2012). "Vote Holds Fate of Nuclear Power in Taiwan". New York Times. Retrieved 13 January 2012.

[81] "Taiwan to halt construction of fourth nuclear power plant". Reuters. 28 April 2014. Retrieved 28 April 2014.

[82] Lin, Sean (4 February 2015). "AEC approves plan to shutter fourth nuclear facility". Taipei Times. Retrieved 5 March 2015.

[83] "EDITORIAL: Taiwan bows to public opinion in pulling plug on nuclear power". The Asahi Shimbun. 31 October 2016. Retrieved 31 October 2016.

[84]

[85] "Vietnam ditches nuclear power plans". Deutsche Welle. Associated Press. 10 November 2016. Retrieved 11 November 2016.

[86] Martin Fackler (June 1, 2011). "Report Finds Japan Underestimated Tsunami Danger". New York Times.

[87] "Thousands march against nuclear power in Tokyo". USA Today. September 2011.

[88] David H. Slater (9 November 2011). "Fukushima women against nuclear power: finding a voice from Tohoku". The Asia-Pacific Journal. Archived from the original on 14 February 2014.

[89] Hiroko Tabuchi (July 13, 2011). "Japan Premier Wants Shift Away From Nuclear Power". New York Times.

[90] Chisaki Watanabe (August 26, 2011). "Japan Spurs Solar, Wind Energy With Subsidies, in Shift From Nuclear Power". Bloomberg.

[91] Tsuyoshi Inajima & Yuji Okada (Oct 28, 2011). "Nuclear Promotion Dropped in Japan Energy Policy After Fukushima". Bloomberg.

[92] "Areas near Japan nuclear plant may be off limits for decades". Reuters. August 27, 2011.

[93] "Kyushu restarts second reactor at Sendai plant under tighter Fukushima-inspired rules". The Japan Times Online. 2015-10-15. ISSN 0447-5763. Retrieved 2015-10-19.

[94] http://www.uic.com.au/nip58.htm

[95] "Some merchant nuclear plants could face early retirement: UBS". Platts. 9 January 2013. Retrieved 10 January 2013.

[96] "Dominion To Close, Decommission Kewaunee Power Station". Dominion. 22 October 2012. Retrieved 28 February 2013.

[97] Caroline Peachey (1 January 2013). "Why are North American plants dying?". Nuclear Engineering International. Retrieved 28 February 2013.

[98] "Crystal River Nuclear Plant to be retired; company evaluating sites for potential new gas-fueled generation". 5 February 2013.

[99] http://www.uic.com.au/nip95.htm

[100] Brazil plans to build seven nuclear reactors —MercoPress. Mercopress.com. Retrieved on 2011-06-04.

[101] "Sunday Dialogue: Nuclear Energy, Pro and Con". New York Times. February 25, 2012.

[102] MacKenzie, James J. (December 1977). "Review of The Nuclear Power Controversy by Arthur W. Murphy". The Quarterly Review of Biology. 52 (4): 467–8. doi:10.1086/410301. JSTOR 2823429.

[103] Walker, J. Samuel (10 January 2006). Three Mile Island: A Nuclear Crisis in Historical Perspective. University of California Press. pp. 10–11. ISBN 9780520246836.

[104] In February 2010 the nuclear power debate played out on the pages of the New York Times, see A Reasonable Bet on Nuclear Power and Revisiting Nuclear Power: A Debate and A Comeback for Nuclear Power?

[105] In July 2010 the nuclear power debate again played out on the pages of the New York Times, see We're Not Ready Nuclear Energy: The Safety Issues

[106] Kitschelt, Herbert P. (1986). "Political Opportunity and Political Protest: Anti-Nuclear Movements in Four Democracies" (PDF). British Journal of Political Science. 16 (1): 57. doi:10.1017/S000712340000380X.

[107] Jim Falk (1982). Global Fission: The Battle Over Nuclear Power. Oxford University Press.

[108] U.S. Energy Legislation May Be 'Renaissance' for Nuclear Power.

[109] Bernard Cohen. "The Nuclear Energy Option". Retrieved 2009-12-09.

[110] "Nuclear Energy is not a New Clear Resource.". Theworldreporter.com. 2010-09-02.

[111] Greenpeace International and European Renewable Energy Council (January 2007). Energy Revolution: A Sustainable World Energy Outlook, p. 7.

[112] Giugni, Marco (2004). *Social protest and policy change: ecology, antinuclear, and peace movements in comparative perspective*. Rowman & Littlefield. pp. 44–. ISBN 9780742518278.

[113] Stephanie Cooke (2009). *In Mortal Hands: A Cautionary History of the Nuclear Age*, Black Inc.. p. 280.

[114] Sovacool, Benjamin K. (2008). "The costs of failure: A preliminary assessment of major energy accidents, 1907–2007". *Energy Policy*. **36** (5): 1802–20. doi:10.1016/j.enpol.2008.01.040.

[115] Jim Green . Nuclear Weapons and 'Fourth Generation' Reactors *Chain Reaction*, August 2009, pp. 18-21.

[116] Kleiner, Kurt (October 2008). "Nuclear energy: assessing the emissions" (PDF). *Nature Reports*. **2**: 130–1.', **Vol** , , **pp**. .

[117] Mark Diesendorf (2007). *Greenhouse Solutions with Sustainable Energy*, University of New South Wales Press, p. 252.

[118] Mark Diesendorf. Is nuclear energy a possible solution to global warming? Archived 22 July 2012 at the Wayback Machine.

[119] Kidd, Steve (21 January 2011). "New reactors—more or less?". *Nuclear Engineering International*. Archived from the original on 12 December 2011.

[120] Ed Crooks (12 September 2010). "Nuclear: New dawn now seems limited to the east". Financial Times. Retrieved 12 September 2010.

[121] *The Future of Nuclear Power*. Massachusetts Institute of Technology. 2003. ISBN 0-615-12420-8. Retrieved 2006-11-10.

[122] Massachusetts Institute of Technology (2011). "The Future of the Nuclear Fuel Cycle" (PDF). p. xv.

[123] "Comparison of Lifecycle Greenhouse Gas Emissions of Various Electricity Generation Sources" (PDF).

[124] Economic Analysis of Various Options of Electricity Generation - Taking into Account Health and Environmental Effects, based on EU ExterneE Project data

[125] International Panel on Fissile Materials (September 2010). "The Uncertain Future of Nuclear Energy" (PDF). *Research Report 9*. p. 1.

[126] M.V. Ramana. Nuclear Power: Economic, Safety, Health, and Environmental Issues of Near-Term Technologies, *Annual Review of Environment and Resources*, 2009, 34, p. 136.

[127] Matthew Wald (February 29, 2012). "The Nuclear Ups and Downs of 2011" . *New York Times*.

[128] Benjamin K. Sovacool. A Critical Evaluation of Nuclear Power and Renewable Electricity in Asia *Journal of Contemporary Asia*, Vol. 40, No. 3, August 2010, pp. 393–400.

[129] The Worst Nuclear Disasters

[130] Arm, Stuart T. (July 2010). "Nuclear Energy: A Vital Component of Our Energy Future" (PDF). *Chemical Engineering Progress*. New York, NY: American Institute of Chemical Engineers: 27–34. ISSN 0360-7275. OCLC 1929453. Archived from the original (PDF) on 28 September 2011. Retrieved 2010-07-26.

[131] IAEA Publications

[132] Jacobson, Mark Z. & Delucchi, Mark A. (2010). "Providing all Global Energy with Wind, Water, and Solar Power. Part I: Technologies, Energy Resources, Quantities and Areas of Infrastructure, and Materials" (PDF). *Energy Policy*. p. 6.

[133] Hugh Gusterson (16 March 2011). "The lessons of Fukushima" . *Bulletin of the Atomic Scientists*.

[134] Diaz Maurin, François (26 March 2011). "Fukushima: Consequences of Systemic Problems in Nuclear Plant Design" (PDF). *Economic & Political Weekly (Mumbai)*. **46** (13): 10–12.

[135] James Paton (4 April 2011). "Fukushima Crisis Worse for Atomic Power Than Chernobyl, UBS Says" . *Bloomberg Businessweek*. Archived from the original on 15 May 2011.

[136] Benjamin K. Sovacool (January 2011). "Second Thoughts About Nuclear Power" (PDF). National University of Singapore. p. 8.

[137] Massachusetts Institute of Technology (2003). "The Future of Nuclear Power" (PDF). p. 48.

[138] http://www.inference.phy.cam.ac.uk/withouthotair/c24/page_168.shtml Dr. MacKay *Sustainable Energy without the hot air*. page 168. Data from studies by the Paul Scherrer Institute including non EU data

[139] World Nuclear Association. Safety of Nuclear Power Reactors.

[140] Ipsos 2011, p. 3

[141] Federal Ministry for the Environment (29 March 2012). *Langfristszenarien und Strategien für den Ausbau der erneuerbaren Energien in Deutschland bei Berücksichtigung der Entwicklung in Europa und global [Long-term Scenarios and Strategies for the Development of Renewable Energy in Germany Considering Development in Europe and Globally]* (PDF). Berlin, Germany: Federal Ministry for the Environment (BMU).

[142] "Advantages and Challenges of Wind Power" . *DOE*. Feb 12, 2015.

[143] John Wiseman; et al. (April 2013). "Post Carbon Pathways" (PDF). *University of Melbourne*.

[144] http://www.columbia.edu/~{}jeh1/2008/TargetCO2_20080407.pdf

[145] http://spectrum.ieee.org/energy/renewables/what-it-would-really-take-to-reverse-climate-change

[146] http://www.engerati.com/article/energy-storage-development-still-faces-obstacles

8.8 Further reading

See also: List of books about nuclear issues

- Cooke, Stephanie (2009). *In Mortal Hands: A Cautionary History of the Nuclear Age*, Black Inc.

- Cragin, Susan (2007). *Nuclear Nebraska: The Remarkable Story of the Little County That Couldn't Be Bought*, AMACOM.

- Diesendorf, Mark (2007). *Greenhouse Solutions with Sustainable Energy*, University of New South Wales Press.

- Elliott, David (2007). *Nuclear or Not? Does Nuclear Power Have a Place in a Sustainable Energy Future?*, Palgrave.

- Falk, Jim (1982). *Global Fission: The Battle Over Nuclear Power*, Oxford University Press.

- Lovins, Amory B. (1977). *Soft Energy Paths: Towards a Durable Peace*, Friends of the Earth International, ISBN 0-06-090653-7

- Lovins, Amory B. and John H. Price (1975). *Non-Nuclear Futures: The Case for an Ethical Energy Strategy*, Ballinger Publishing Company, 1975, ISBN 0-88410-602-0

- Pernick, Ron and Clint Wilder (2007). *The Clean Tech Revolution: The Next Big Growth and Investment Opportunity*, Collins, ISBN 978-0-06-089623-2

- Price, Jerome (1982). *The Antinuclear Movement*, Twayne Publishers.

- Rudig, Wolfgang (1990). *Anti-nuclear Movements: A World Survey of Opposition to Nuclear Energy*, Longman.

- Schneider, Mycle, Steve Thomas, Antony Froggatt, Doug Koplow (August 2009). *The World Nuclear Industry Status Report*, German Federal Ministry of Environment, Nature Conservation and Reactor Safety.

- Sovacool, Benjamin K. (2011). *Contesting the Future of Nuclear Power: A Critical Global Assessment of Atomic Energy*, World Scientific.

- Walker, J. Samuel (2004). *Three Mile Island: A Nuclear Crisis in Historical Perspective*, University of California Press.

- William D. Nordhaus, *The Swedish Nuclear Dilemma – Energy and the Environment*. 1997. Hardcover. ISBN 0-915707-84-5.

- Bernard Leonard Cohen, *The Nuclear Energy Option: An Alternative for the 90's*. 1990. Hardcover. ISBN 0-306-43567-5. Bernard Cohen's homepage contains the full text of the book.

8.9 External links

- German Energy Transition
- Fairewinds Energy Education

Chapter 9

Anti-nuclear movement in California

The 1970s proved to be a pivotal period for the **anti-nuclear movement in California**. Opposition to nuclear power in California coincided with the growth of the country's environmental movement. Opposition to nuclear power increased when President Richard Nixon called for the construction of 1000 nuclear plants by the year 2000.[1]

The movement succeeded in blocking plans to build a large number of facilities in the state as well as closing operating power plants. The confrontation between nuclear power advocates and environmentalists grew to include the use of non-violent civil disobedience.[2]

In 1976 the state of California placed a moratorium on new reactors until a solution to radioactive waste disposal was in place. In September 1981, over 1,900 arrests took place during a ten-day blockade at Diablo Canyon Power Plant. As part of a national anti-nuclear weapons movement Californians passed a 1982 statewide initiative calling for the end of nuclear weapons.[3] In 1984, the Davis City Council declared the city to be a nuclear free zone.

In 2013, San Onofre 2 and 3 were permanently closed.[4][5]

9.1 Early conflicts

See also: Bodega Bay Nuclear Power Plant

The birth of the anti-nuclear movement in California can be traced to controversy over Pacific Gas & Electric's attempt to build the nation's first commercially viable nuclear power plant in Bodega Bay. This conflict began in 1958 and ended in 1964, with the forced abandonment of these plans. Subsequent plans to build a nuclear power plant in Malibu were also abandoned.[6]

9.2 1970s and 1980s

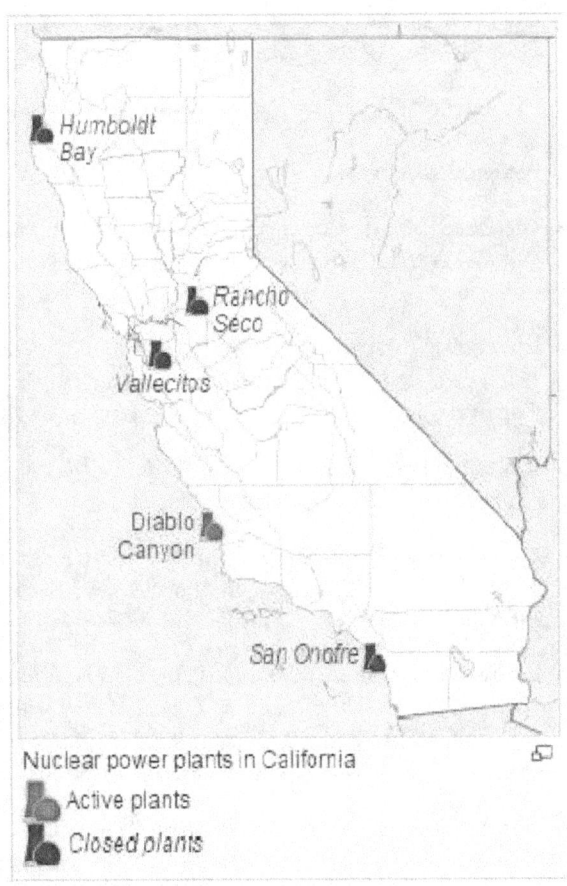

Nuclear power plants in California
Active plants
Closed plants

- Nuclear power plants in California

- Active plants

- Closed plants

The anti-nuclear movement grew in California between 1964 and 1974. It was during this period that some scientists and engineers began supporting the positions of the activists. They were influenced by the Ecology and Free Speech Movements that had inspired activists and had impacted the public consciousness.*[6] Californian's for Nuclear Safeguards would succeed at placing Proposition 15 on the June 1976 ballot which would ban new facilities and put additional safety requirements on operating reactors.*[7] The initiative failed to pass with millions of dollars spent by the nuclear industry to influence the outcome. However, as a result of the publicity which included the resignation of three General Electric nuclear engineers, the state legislature passed a moratorium on further nuclear development until a permanent solution to high level waste was in place.

Anti-nuclear groups campaigned to stop construction of several proposed plants in the seventies, especially those located on the coast and near fault lines. These proposals included the Sundesert Nuclear Power Plant, which was never built.*[2]*[8] In 1978, a year before the Three Mile Island accident, the State of California refused to allow the utility to begin construction of the Sundesert units in the "absence of federally demonstrated and approved technology for permanent disposal of radioactive wastes".*[9]*[10] The project was cancelled that year.

The discovery of an earthquake fault near General Electric's Vallecitos Nuclear Center near Pleasanton resulted in the Nuclear Regulatory Commission closing the facility down. The discovery of an earthquake fault line as well as plutonium a short distance from a school in Humboldt California resulted in the closure of Pacific Gas & Electric's.*[11]

Over a two-week period in 1981, 1,900 activists were arrested at Diablo Canyon Power Plant. It was the largest arrest in the history of the anti-nuclear movement in the United States.*[12]*[13] Specific protests included:

- August 6, 1977: The Abalone Alliance held the first blockade at Diablo Canyon Power Plant, and 47 people were arrested.*[14]

- August 1978: almost 500 people were arrested for protesting at Diablo Canyon.*[14]

- April 8, 1979: 30,000 people marched in San Francisco to support shutting down the Diablo Canyon Power Plant.*[15]

- June 30, 1979: about 40,000 people attended a protest rally at Diablo Canyon.*[16]

- September 1981: more than 1900 protesters were arrested at Diablo Canyon.*[14]*[17]

- May 1984: about 130 demonstrators showed up for start-up day at Diablo Canyon, and five were arrested.*[18]

During this period there were controversies within the Sierra Club about how to lead the anti-nuclear movement, and this led to a split over the Diablo Canyon plant which ended in success for the utilities. The split led to the formation of Friends of the Earth, led by David Brower.*[6]

In 1979, Abalone Alliance members held a 38-day sit-in in the Californian Governor Jerry Brown's office to protest continued operation of Rancho Seco Nuclear Generating Station, which was a duplicate of the Three Mile Island facility.*[19] In 1989, Sacramento voters voted to shut down the Rancho Seco power plant.*[20] The salient issues were mostly economic; the plant kept breaking down, and it had been shut from late 1985 to early 1988 for repairs, forcing the district to buy electricity from neighbors.*[21]

On June 22, 1980, about 15,000 people attended a protest near San Onofre Nuclear Generating Station.*[22] In 1977 Bechtel Corporation installed the *[23] reactor vessel backwards.

California has banned the approval of new nuclear reactors since the late 1970s because of concerns over waste disposal.*[24]*[25]

Dark Circle is a 1982 American documentary film that focuses on the connections between the nuclear weapons and the nuclear power industries, with a strong emphasis on the individual human and protracted U.S. environmental costs involved. A clear point made by the film is that while only two bombs were dropped on Japan, many hundreds were exploded in the United States. The film won the Grand Prize for documentary at the Sundance Film Festival and received a national Emmy Award for "Outstanding individual achievement in news and documentary." *[26] The film shows anti-nuclear protest activities directed at the Diablo Canyon Power Plant on the California coast in the USA. The protesters contend, and the movie supports, the assertion that the protests were responsible for delaying the licensing of the Diablo Canyon Power Plant and, as a result of the delay, the uncovering of serious construction errors was made public just before the plant went online and started producing power. For example, earthquake supports for nuclear piping had been installed backwards, and the film includes close up footage of the moment that this information became known.

9.3   1990s

On June 15, 1990 the Bureau of Land Management published the draft Environmental Impact Statement (DEIS) for the construction of a Low-level Nuclear waste repository to be located at Ward valley California. The company applying to construct and operate the repository was U.S. Ecology. An eight-year struggle between government agencies and opponents of the nuclear waste dump ended with the dump being blocked.*[27]

9.4 Nuclear-free communities

One of a set of two billboards in Davis, California advertising its nuclear-free policy

The second billboard corresponding to the one above

On November 14, 1984 the Davis, California City Council declared the city to be a nuclear free zone.*[28] Another well-known nuclear-free community is Berkeley, California, whose citizens passed the Nuclear Free Berkeley Act in 1986 which allows the city to levy fines for nuclear weapons-related activity and to boycott companies involved in the United States nuclear infrastructure.

9.5 Recent developments

PG&E announced its decision to pursue license renewal for Diablo Canyon in November 2009, and local officials "came out in support because of the economic importance of the plant and its 1,200 employees and $25 million in annual property taxes".*[29] However, local anti-nuclear activists oppose renewal and want PG&E to focus more on renewable energy. They are also concerned "about the seismic safety of the plant given the recent discovery of a new earthquake fault nearby".*[29]

In April 2011, there was demonstration of 300 people at Avila Beach calling for the closure of Diablo Canyon nuclear power plant and a halt to its relicensing application process. The event, organized by San Luis Obispo-based anti-nuclear group Mothers for Peace, was in response to the Fukushima nuclear disaster in Japan.*[30]

In 2013, San Onofre 2 and 3 were permanently closed.*[4]*[5]

9.6 See also

- Anti-nuclear groups in the United States
- Anti-nuclear protests in the United States
- California electricity crisis
- J. Samuel Walker
- List of articles associated with nuclear issues in California
- Nuclear power debate
- Nuclear-free zone
- Politics of New England
- Renewable energy in the United States
- Santa Susana Field Laboratory
- Solar power in California
- Wind power in California

9.7 References

[1] New York Times

[2] San Diego Gas & Electric, Sundesert Nuclear Power Plant Collection

[3] 1982 California Proposition 12

[4] Mark Cooper (18 June 2013). "Nuclear aging: Not so graceful". *Bulletin of the Atomic Scientists*.

[5] Matthew Wald (June 14, 2013). "Nuclear Plants, Old and Uncompetitive, Are Closing Earlier Than Expected". *New York Times*.

[6] Critical Masses: Opposition to Nuclear Power in California, 1958–1978

[7] Time Magazine

[8] August S. Carstens Collection

[9] Luther J. Carter "Political Fallout from Three Mile Island", *Science*, 204, April 13, 1979, p. 154.

[10] Critical Masses p. 176.

[11] Humboldt nuclear power plant Humboldt Bay Nuclear Power Plant

[12] Conservation Fallout: Nuclear Protest at Diablo Canyon

[13] Daniel Pope. Conservation Fallout (book review), *H-Net Reviews*, August 2007.

[14] Social Protest and Policy Change p. 44.

[15] Amplifying Public Opinion: The Policy Impact of the U.S. Environmental Movement p. 7.

[16] Gottlieb, Robert (2005). Forcing the Spring: The Transformation of the American Environmental Movement, Revised Edition, Island Press, USA, p. 240.

[17] Arrests Exceed 900 In Coast Nuclear Protest *New York Times*, September 18, 1981.

[18] Testing and Protesting *Time*, May 14, 1984.

[19] Hippy Dictionary p.559.

[20] Shutting Down Rancho Seco

[21] Matthew L. Wald. Vermont Senate Votes to Close Nuclear Plant *The New York Times*, February 24, 2010.

[22] Williams, Eesha. Wikipedia distorts nuclear history *Valley Post*, May 1, 2008.

[23] San Onofre San Onofre Nuclear Generating Station

[24] Jim Doyle. Nuclear power industry sees opening for revival *San Francisco Chronicle*, March 9, 2009.

[25] Minnesota also has a moratorium on construction of nuclear power plants, which has been in place since 1994. See Minnesota House says no to new nuclear power plants Archived May 5, 2009, at the Wayback Machine. *StarTribune.com*, April 30, 2009.

[26] *Dark Circle*, DVD release date March 27, 2007. Directors: Judy Irving, Chris Beaver, Ruth Landy. ISBN 0-7670-9304-6.

[27] Ward Valley Timeline

[28] Nuclear Free Zone

[29] Nuclear Regulatory Commission dealing with multiple issues at Diablo Canyon nuclear power plant

[30] Julia Hickey (April 17, 2001). "Anti-nuclear rally at Avila Beach". *The Tribune*. Archived from the original on 2012-03-22.

9.8 Further reading

See also: List of books about nuclear issues and List of films about nuclear issues

- Brown, Jerry and Rinaldo Brutoco (1997). *Profiles in Power: The Anti-nuclear Movement and the Dawn of the Solar Age*, Twayne Publishers.

- Lovins, Amory B. and Price, John H. (1975). *Non-Nuclear Futures: The Case for an Ethical Energy Strategy*, Ballinger Publishing Company, 1975, ISBN 0-88410-602-0

- Natti, Susanna and Acker, Bonnie (1979). *No Nukes: Everyone's Guide to Nuclear Power*, South End Press.

- Ondaatje, Elizabeth H. (c1988). *Trends in Antinuclear Protests in the United States, 1984–1987*, Rand Corporation.

- Price, Jerome (1982). *The Antinuclear Movement*, Twayne Publishers.

- Smith, Jennifer (Editor), (2002). *The Antinuclear Movement*, Cengage Gale.

- Walker, J. Samuel (2004). *Three Mile Island: A Nuclear Crisis in Historical Perspective*, University of California Press.

- Wellock, Thomas R. (1998). *Critical Masses: Opposition to Nuclear Power in California, 1958-1978*, The University of Wisconsin Press, ISBN 0-299-15850-0

- Wills, John (2006). *Conservation Fallout: Nuclear Protest at Diablo Canyon*, University of Nevada Press.

9.9 External links

- The Struggle over Nuclear Power
- Conservation Fallout: Nuclear Protest at Diablo Canyon

- Police arrest 64 at California anti-nuclear protest
- Alliance for Nuclear Responsibility
- Abalone Alliance Archives

Chapter 10

Politics of New England

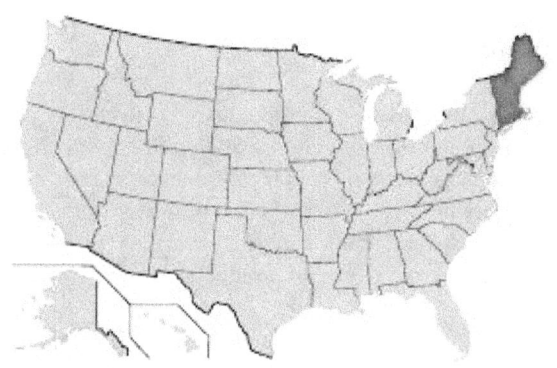

The New England region of the United States is shaded in red, above.

The **politics of New England** has long been defined by the region's political and cultural history, demographics, economy, and its loyalty to particular U.S. political parties. Within the politics of the United States, New England is sometimes viewed in terms of a single voting bloc.

Same-sex marriage is permitted in all six New England states, with Rhode Island being the final state to legalise the practice in May 2013.[1] In Maine, it was legalized by the legislature in 2009, but defeated in a referendum (53% voted to ban it versus 47% who voted to legalize it) later the same year. On January 26, 2012, supporters of a ballot initiative to repeal that decision obtained sufficient petition signatures to ensure the initiative appeared on the November 2012 ballot, where the initiative won 53–47.

The national U.S. movement against nuclear power had its roots in New England in the 1970s. In 1974, activist Sam Lovejoy toppled a weather tower at the site of the proposed Montague Nuclear Power Plant in Western Massachusetts.[2] The movement "reached critical mass" with the arrests at Seabrook Station Nuclear Power Plant on May 1, 1977, when 1,414 anti-nuclear activists from the Clamshell Alliance were arrested at the Seabrook site. Harvey Wasserman, a Clamshell spokesman at Seabrook, and Frances Crowe of Northampton, an American Friends Service Committee member, played key roles in the move-

ment.[2]

10.1 Notable laws and political movements

The New England states abolished the death penalty for robbery and burglary in the 19th century, before much of the rest of the U.S.A. As of 2012, New Hampshire is the only state in New England that has retained capital punishment.[3] Although New Hampshire currently has one death row inmate, it has not held an execution since 1939.[4]

Connecticut and Rhode Island were the only states in the union not to ratify the Eighteenth Amendment to the United States Constitution, also known as prohibition.[5] Prohibition became the law of the land on January 16, 1919.

New Hampshire has no seatbelt law for persons over 18 years of age,[6] no helmet law for motorcyclists,[7] no mandatory auto-insurance law,[8] and has neither an income tax nor a sales tax.[9]

Vermont, Maine and New Hampshire allow open carrying of firearms in public places without a permit; Vermont also allows concealed carrying without a permit.[10][11]

Massachusetts passed a ballot initiative (question 3) in the November 2012 election that legalized medical marijuana, effective January 1, 2013. Only people with debilitating diseases such as cancer, Parkinson's disease, or Alzheimer's can obtain a medical marijuana card. Massachusetts became the 18th state in the U.S.A. to legalize the medical use of marijuana.[12] Similar laws are also in place in four other New England States.[13]

10.1.1 Same-sex marriage

Main article: Same-sex marriage in New England

Same-sex marriage is legal or has been legalized in all of the New England states: Connecticut, Maine, Massachusetts, New Hampshire, Rhode Island and Vermont, as well as in the neighboring state of New York. The New England region has been noted for being the nucleus of the same-sex marriage movement in the United States,[14] with the region having among the most widespread and earliest legal support of any region. In 2004, Massachusetts became the first state in the United States to legalize same-sex marriage,[15] to be followed by three more states between October 2008 and June 2009. This followed Vermont being the first-in-the-nation with civil unions in 2000.[16] Before the 2012 election, California (2008), Iowa (2009), New York (2011) and the District of Columbia (2010) had been the only U.S. jurisdictions outside New England to have performed same-sex marriages, though same-sex marriages in California had been halted following the passage of Proposition 8.

The legalization of same-sex marriage was part of a campaign which began in November 2008, called Six by Twelve, and was organized by the Gay & Lesbian Advocates & Defenders (GLAD) to legalize same-sex marriage in all six New England states by 2012.[17][18]

The region holds a number of firsts on same-sex marriage: Vermont was the first state to enact it through legislative means and not because of a judicial ruling,[19][20] and Maine was the first state to have a governor sign a same-sex marriage bill that was not the result of a court decision.[21] Maine's first gay marriage law was repealed through a people's veto, but three years later, on November 6, 2012, the question was put to voters a second time, and Maine became one of three first US states to approve same-sex marriage at the ballot box, along with Washington and Maryland.

With Rhode Island legalizing same sex marriage, all New England states have same-sex marriage. There have been numerous[14][22][23][24][25] reasons given for why New England has found such strong legal recognition for same-sex marriages in comparison to the rest of the United States.

10.1.2 Anti-nuclear movement

See also: Paul Gunter, Harvey Wasserman, Frances Crowe, and Macy Morse

There were four targets of the anti-nuclear movement since 1974. As a result, Montague Nuclear was cancelled. Yankee Rowe closed prematurely for engineering inadequacies. Vermont Yankee was closed because it became uncompetitive. Seabrook remains operational.

Montague Nuclear Power Plant

On February 22, 1934, Sam Lovejoy took a crowbar to the weather-monitoring tower which had been erected at the Montague Nuclear Power Plant site. Lovejoy felled 349 feet (106 m) of the 550 feet (170 m) tower. He turned himself in to the local police station. He presented a statement in which he took responsibility for the action. Lovejoy's action galvanized local public opinion against the plant.[26][27] The Montague nuclear power plant proposal was canceled in 1980,[28] after $29 million was spent on the project.[26]

Seabrook

Seabrook power plant was proposed as a twin-reactor plant in 1972, at an estimated cost of $973 million. It received a commercial license in March 1990 for one reactor which cost $6.5 billion.[29] Over a period of thirteen years more than 4,000 citizens, many associated with the Clamshell Alliance anti-nuclear group, committed non-violent civil disobedience at Seabrook:[30]

- August 1, 1976: 200 residents rallied at the future Seabrook Station Nuclear Power Plant site in New Hampshire, and 18 were arrested for criminal trespass.[30]

- August 22, 1976: 188 activists from New England were arrested at the Seabrook site.[30][31]

- May 2, 1977: 1,414 protesters were arrested at Seabrook Station Nuclear Power Plant.[2][32][33] The protesters who were arrested were expected to be "released on their own recognizance", but this did not happen. Instead, they were charged with criminal trespass and asked to post bail ranging from $100 to $500. They refused and were then held in five national guard armories for 12 days. The Seabrook conflict, and role of New Hampshire Governor Meldrim Thomson, received much national media coverage.[34]

- May 13, 1977: 550 protestors were freed after being detained for thirteen days.[35]

- June 1978: some 12,000 people attended a protest at Seabrook.[32][33]

- May 25–27, 1980: Police use tear gas, riot sticks and dogs to drive 2,000 demonstrators away from the Seabrook site.[36]

- May 24, 1986: 74 anti-nuclear demonstrators were arrested in protests.[37][38]

- October 17, 1988: 84 people were arrested at the Seabrook plant.*[39]

- June 5, 1989: hundreds of demonstrators protested against the plant's first low-power testing, and the police arrested 627 people for trespassing; two state legislators, one from Massachusetts and one from New Hampshire, protested.*[30]*[40]

Yankee Rowe

The New England Coalition (NEC) is an educational nonprofit organization based in Brattleboro, Vermont. Historically, it has been part of the anti-nuclear movement in the United States.*[41] The NEC is primarily concerned with legal action more than protests. It was involved in both legal action and protests about the Yankee Rowe Nuclear Power Plant prior to its shut down in 1992, and has been involved in legal action and protests over extending the license to operate at the Vermont Yankee Nuclear Power Plant.

Vermont Yankee

In February 2010, the Vermont Senate voted 26 to 4 against allowing the PSB to consider re-certifying the Vermont Yankee Nuclear Plant after 2012, citing radioactive tritium leaks, misstatements in testimony by plant officials, a cooling tower collapse in 2007, and other problems.*[42] In January 2012, Entergy won a court case to invalidate the state's veto power on continued operation.*[43]

There were a number of anti-nuclear protests about Vermont Yankee since the 1970s. These included protests following the Japanese Fukushima nuclear disaster in March 2011 and on the date of the original operating license expiry in March 2012. On August 28, 2013 the company said economic factors, notably the low cost of electricity caused by cheap natural gas, would result in the company's decommissioning the plant in the fourth quarter of 2014.*[43]

10.2   See also

- Elections in New England

- Politics of Vermont

- Politics of New Hampshire

- Politics of Maine

- Politics of Massachusetts

- Politics of Connecticut

- Politics of Rhode Island

10.3   Notes

[1] "Same-sex marriages begin in R.I., Minnesota". Boston Globe. August 1, 2013.

[2] Michael Kenney. Tracking the protest movements that had roots in New England The Boston Globe. December 30, 2009.

[3] "Death Penalty Information Center". Deathpenaltyinfo.org. Retrieved July 19, 2006.

[4] "Supreme Court Lifts Order Blocking Connecticut Execution", Fox News, January 29, 2005. Retrieved July 19, 2006. "New Hampshire has not executed anyone since 1939 and has no one on death row. Seven inmates are waiting to die in Connecticut, which conducted New England's last execution in 1960."

[5] "CONNECTICUT BALKS AT PROHIBITION; Senate Rejects Federal Amendment—First State to Fail to Ratify". The New York Times. 1919-02-05.

[6] State Seat Belt Laws. Ghsa.org. Retrieved May 15, 2012.

[7] New Hampshire Motorcycle Helmet Law. Bikersrights.com. Retrieved May 15, 2012.

[8] State Insurance Laws. Autoinsuranceremedy.com. Retrieved May 15, 2012.

[9] Does NH have an Income Tax or Sales Tax? | Frequently Asked Questions | NH Department of Revenue Administration. Nh.gov. Retrieved May 15, 2012.

[10] State Information For New Hampshire. OpenCarry.org. Retrieved May 15, 2012.

[11] State Information For Vermont. OpenCarry.org. Retrieved May 15, 2012.

[12] "Return of Votes For Massachusetts State Election November 6, 2012" (PDF). Massachusetts Secretary of the Commonwealth. Retrieved 2013-03-24.

[13] "18 Legal Medical Marijuana States and DC Laws, Fees, and Possession Limits". procon.org. Retrieved 2013-03-24.

[14] A Push Is On for Same-Sex Marriage Rights Across New England. New York Times. April 4, 2009

[15] Burge, Kathleen (November 18, 2003). "SJC: Gay marriage legal in Mass.". The Boston Globe.

[16] Vermont legalizes gay marriage. Burlington Free Press. April 7, 2009

[17] 'Gay marriage' bill passes N.H. Senate. Baptist Press. April 24, 2009

[18] 6x12: Half-way There and Going Strong!. GLAD. April 14, 2009

[19] "Vt. legalizes same-sex marriage". The Burlington Free Press. 2009-04-07. Retrieved 2009-04-07.

[20] Goodnough, Abby (2009-04-07). "Vermont Legislature Makes Same-Sex Marriage Legal". The New York Times. Retrieved May 23, 2010.

[21] Russel, Jenna (2009-05-06). "Gay marriage law signed in Maine, advances in N.H". Boston.com. Retrieved 2009-05-06.

[22] Gay Marriage Advances in Maine, The New York Times, Abby Goodnough and Katie Zezima, May 5, 2009

[23] Lin, Joanna (2009-03-16). "New England surpasses West Coast as least religious region in America, study finds". Los Angeles Times. Retrieved 2009-05-15.

[24] New England leads on same-sex marriage. NECN. May 7, 2009

[25] N.E.'s identity bolsters gay marriage tolerance. The Boston Globe, Jenna Russell, May 11, 2009

[26] Utilities Drop Nuclear Power Plant Plans *Ocala Star-Banner*, January 4, 1981.

[27] No nukes by Anna Gyorgy pp. 393–394.

[28] Some of the Major Events in NU's History Since the 1966 Affiliation

[29] 30 years later, another nuclear struggle looms *The Daily News*, April 30, 2007.

[30] Gunter, Paul. Clamshell Alliance: Thirteen Years of Anti-Nuclear Activism at Seabrook, New Hampshire, U.S.A.*Ecologia Newsletter*, January 1990 Issue 3.

[31] Seabrook, NH Nuclear Plant Occupation Page

[32] Williams, Estha. Nuke Fight Nears Decisive Moment *Valley Advocate*, August 28, 2008.

[33] Williams, Eesha. Wikipedia distorts nuclear history *Valley Post*, May 1, 2008.

[34] William A. Gamson and Andre Modigliani. Media Coverage and Public Opinion on Nuclear Power. *American Journal of Sociology*. Vol. 95, No. 1, July 1989, p. 17.

[35] The Legacy of Seabrook

[36] Hartford Courant

[37] Anti-Nuclear Protesters Freed in New Hampshire

[38] New Hampshire / Anti-Nuclear Demonstration

[39] 84 Arrested in Protest At the Seabrook Plant

[40] Gold, Allan R. Hundreds Arrested Over Seabrook Test *New York Times*, June 5, 1989.

[41] New England Coalition.

[42] Wald, Matthew L., Vermont Senate Votes to Close Nuclear Plant *The New York Times*, February 24, 2010.

[43] "Entergy to Close, Decommission Vermont Yankee". PR Newswire.

Chapter 11

High-level radioactive waste management

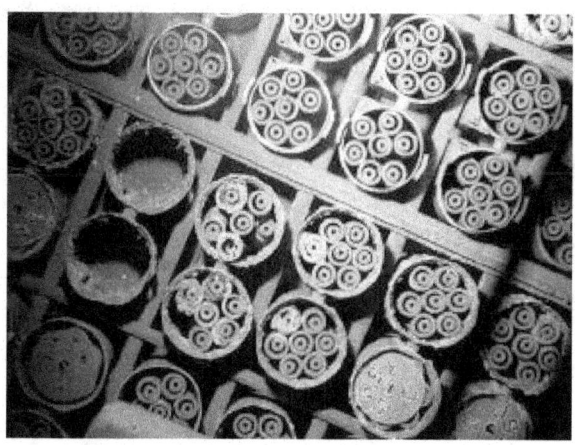

Spent nuclear fuel stored underwater and uncapped at the Hanford site in Washington, USA.

High-level radioactive waste management concerns how radioactive materials created during production of nuclear power and nuclear weapons are dealt with. Radioactive waste contains a mixture of short-lived and long-lived nuclides, as well as non-radioactive nuclides.[1] There was reported some 47,000 tonnes of high-level nuclear waste stored in the USA in 2002.

The most troublesome transuranic elements in spent fuel are neptunium-237 (half-life two million years) and plutonium-239 (half-life 24,000 years).[2] Consequently, high-level radioactive waste requires sophisticated treatment and management to successfully isolate it from the biosphere. This usually necessitates treatment, followed by a long-term management strategy involving permanent storage, disposal or transformation of the waste into a non-toxic form.[3] Radioactive decay follows the half-life rule, which means that the rate of decay is inversely proportional to the duration of decay. In other words, the radiation from a long-lived isotope like iodine-129 will be much less intense than that of short-lived isotope like iodine-131.[4]

Governments around the world are considering a range of waste management and disposal options, usually involving deep-geologic placement, although there has been limited progress toward implementing long-term waste management solutions.[5] This is partly because the timeframes in question when dealing with radioactive waste range from 10,000 to millions of years,[6][7] according to studies based on the effect of estimated radiation doses.[8]

Thus, Alfvén identified two fundamental prerequisites for effective management of high-level radioactive waste: (1) stable geological formations, and (2) stable human institutions over hundreds of thousands of years. As Alfvén suggests, no known human civilization has ever endured for so long, and no geologic formation of adequate size for a permanent radioactive waste repository has yet been discovered that has been stable for so long a period.[9] Nevertheless, avoiding confronting the risks associated with managing radioactive wastes may create countervailing risks of greater magnitude. Radioactive waste management is an example of policy analysis that requires special attention to ethical concerns, examined in the light of uncertainty and *futurity*: consideration of 'the impacts of practices and technologies on future generations'.[10]

There is a debate over what should constitute an acceptable scientific and engineering foundation for proceeding with radioactive waste disposal strategies. There are those who have argued, on the basis of complex geochemical simulation models, that relinquishing control over radioactive materials to geohydrologic processes at repository closure is an acceptable risk. They maintain that so-called "natural analogues" inhibit subterranean movement of radionuclides, making disposal of radioactive wastes in stable geologic formations unnecessary.[11] However, existing models of these processes are empirically underdetermined:[12] due to the subterranean nature of such processes in solid geologic formations, the accuracy of computer simulation models has not been verified by empirical observation, certainly not over periods of time equivalent to the lethal half-lives of high-level radioactive waste.[13][14] On the other hand, some insist deep geologic repositories in stable geologic formations are necessary. National management plans of various countries display a variety of approaches to resolving this debate.

Researchers suggest that forecasts of health detriment for such long periods *should be examined critically.*[15] Practical studies only consider up to 100 years as far as effective planning[16] and cost evaluations[17] are concerned. Long term behaviour of radioactive wastes remains a subject for ongoing research.[18] Management strategies and implementation plans of several representative national governments are described below.

11.1 Geologic disposal

The International Panel on Fissile Materials has said:

> It is widely accepted that spent nuclear fuel and high-level reprocessing and plutonium wastes require well-designed storage for periods ranging from tens of thousands to a million years, to minimize releases of the contained radioactivity into the environment. Safeguards are also required to ensure that neither plutonium nor highly enriched uranium is diverted to weapon use. There is general agreement that placing spent nuclear fuel in repositories hundreds of meters below the surface would be safer than indefinite storage of spent fuel on the surface.[19]

The process of selecting appropriate permanent repositories for high level waste and spent fuel is now under way in several countries with the first expected to be commissioned some time after 2017.[20] The basic concept is to locate a large, stable geologic formation and use mining technology to excavate a tunnel, or large-bore tunnel boring machines (similar to those used to drill the Chunnel from England to France) to drill a shaft 500–1,000 meters below the surface where rooms or vaults can be excavated for disposal of high-level radioactive waste. The goal is to permanently isolate nuclear waste from the human environment. However, many people remain uncomfortable with the immediate stewardship cessation of this disposal system, suggesting perpetual management and monitoring would be more prudent.

Because some radioactive species have half-lives longer than one million years, even very low container leakage and radionuclide migration rates must be taken into account.[21] Moreover, it may require more than one half-life until some nuclear materials lose enough radioactivity to no longer be lethal to living organisms. A 1983 review of the Swedish radioactive waste disposal program by the National Academy of Sciences found that country's estimate of several hundred thousand years—perhaps up to one million years—being necessary for waste isolation "fully justified." [22]

The proposed land-based subductive waste disposal method would dispose of nuclear waste in a subduction zone accessed from land,[23] and therefore is not prohibited by international agreement. This method has been described as a viable means of disposing of radioactive waste,[24] and as a state-of-the-art nuclear waste disposal technology.[25]

In nature, sixteen repositories were discovered at the Oklo mine in Gabon where natural nuclear fission reactions took place 1.7 billion years ago.[26] The fission products in these natural formations were found to have moved less than 10 ft (3 m) over this period,[27] though the lack of movement may be due more to retention in the uraninite structure than to insolubility and sorption from moving ground water; uraninite crystals are better preserved here than those in spent fuel rods because of a less complete nuclear reaction, so that reaction products would be less accessible to groundwater attack.[28]

11.2 Materials for geological disposal

In order to store the high level radioactive waste in long-term geological depositories, specific waste forms need to be used which will allow the radioactivity to decay away while the materials retain their integrity for thousands of years.[29] The materials being used can be broken down into a few classes: glass waste forms, ceramic waste forms, and nanostructured materials.

The glass forms include borosilicate glasses and phosphate glasses. Borosilicate nuclear waste glasses are used on an industrial scale to immobilize high level radioactive waste in many countries which are producers of nuclear energy or have nuclear weaponry. The glass waste forms have the advantage of being able to accommodate a wide variety of waste-stream compositions, they are easy to scale up to industrial processing, and they are stable against thermal, radiative, and chemical perturbations. These glasses function by binding radioactive elements to nonradioactive glass-forming elements.[30] Phosphate glasses while not being used industrially have much lower dissolution rates than borosilicate glasses, which make them a more favorable option. However, no single phosphate material has the ability to accommodate all of the radioactive products so phosphate storage requires more reprocessing to separate the waste into distinct fractions.[31] Both glasses have to be processed at elevated temperatures making them unusable for some of the more volatile radiotoxic elements.

The ceramic waste forms offer higher waste loadings than the glass options because ceramics have crystalline structure. Also, mineral analogues of the ceramic waste forms provide evidence for long term durability.[32] Due to this fact and the fact that they can be processed at lower temperatures, ceramics are often considered the next generation in high level radioactive waste forms.*[33] Ceramic waste forms offer great potential, but a lot of research remains to be done.*

11.3 National management plans

Finland, the United States and Sweden are the most advanced in developing a deep repository for high-level radioactive waste disposal. Countries vary in their plans on disposing used fuel directly or after reprocessing, with France and Japan having an extensive commitment to reprocessing. The country-specific status of high-level waste management plans are described below.

In many European countries (e.g., Britain, Finland, the Netherlands, Sweden and Switzerland) the risk or dose limit for a member of the public exposed to radiation from a future high-level nuclear waste facility is considerably more stringent than that suggested by the International Commission on Radiation Protection or proposed in the United States. European limits are often more stringent than the standard suggested in 1990 by the International Commission on Radiation Protection by a factor of 20, and more stringent by a factor of ten than the standard proposed by the U.S. Environmental Protection Agency (EPA) for Yucca Mountain nuclear waste repository for the first 10,000 years after closure. Moreover, the U.S. EPA's proposed standard for greater than 10,000 years is 250 times more permissive than the European limit.*[34]

The countries that have made the most progress towards a repository for high-level radioactive waste have typically started with public consultations and made voluntary siting a necessary condition. This consensus seeking approach is believed to have a greater chance of success than top-down modes of decision making, but the process is necessarily slow, and there is "inadequate experience around the world to know if it will succeed in all existing and aspiring nuclear nations".*[35]

Moreover, most communities do not want to host a nuclear waste repository as they are "concerned about their community becoming a de facto site for waste for thousands of years, the health and environmental consequences of an accident, and lower property values".*[36]

11.3.1   Asia

People's Republic of China

In the Peoples Republic of China, ten reactors provide about 2% of electricity and five more are under construc-

tion.*[37] China made a commitment to reprocessing in the 1980s; a pilot plant is under construction at Lanzhou, where a temporary spent fuel storage facility has been constructed. Geological disposal has been studied since 1985, and a permanent deep geological repository was required by law in 2003. Sites in Gansu Province near the Gobi desert in northwestern China are under investigation, with a final site expected to be selected by 2020, and actual disposal by about 2050.*[38]*[39]

Republic of China

In the Republic of China, nuclear waste storage facility was built at the Southern tip of Orchid Island in Taitung County, offshore of Taiwan Island. The facility was built in 1982 and it is owned and operated by Taipower. The facility receives nuclear waste from Taipower's current three nuclear power plants. However, due to the strong resistance from local community in the island, the nuclear waste has to be stored at the power plant facilities themselves.*[40]*[41]

India

Sixteen nuclear reactors produce about 3% of India's electricity, and seven more are under construction.*[37] Spent fuel is processed at facilities in Trombay near Mumbai, at Tarapur on the west coast north of Mumbai, and at Kalpakkam on the southeast coast of India. Plutonium will be used in a fast breeder reactor (under construction) to produce more fuel, and other waste vitrified at Tarapur and Trombay.*[42]*[43] Interim storage for 30 years is expected, with eventual disposal in a deep geological repository in crystalline rock near Kalpakkam.*[44]

Japan

In 2000, a Specified Radioactive Waste Final Disposal Act called for creation of a new organization to manage high level radioactive waste, and later that year the Nuclear Waste Management Organization of Japan (NUMO) was established under the jurisdiction of the Ministry of Economy, Trade and Industry. NUMO is responsible for selecting a permanent deep geological repository site, construction, operation and closure of the facility for waste emplacement by 2040.*[45]*[46] Site selection began in 2002 and application information was sent to 3,239 municipalities, but by 2006, no local government had volunteered to host the facility.*[47] Kōchi Prefecture showed interest in 2007, but its mayor resigned due to local opposition. In December 2013 the government decided to identify suitable candidate areas before approaching municipalities.*[48]

The head of the Science Council of Japan's expert panel has said Japan's seismic conditions makes it difficult to predict ground conditions over the necessary 100,000 years, so it will be impossible to convince the public of the safety of deep geological disposal.[48]

11.3.2 Europe

Belgium

Belgium has seven nuclear reactors that provide about 52% of its electricity.[37] Belgian spent nuclear fuel was initially sent for reprocessing in France. In 1993, reprocessing was suspended following a resolution of the Belgian parliament;[49] spent fuel is since being stored on the sites of the nuclear power plants. The deep disposal of high-level radioactive waste (HLW) has been studied in Belgium for more than 30 years. Boom Clay is studied as a reference host formation for HLW disposal. The Hades underground research laboratory (URL) is located at −223 m in the Boom Formation at the Mol site. The Belgian URL is operated by the Euridice Economic Interest Group, a joint organisation between SCK•CEN, the Belgian Nuclear Research Centre which initiated the research on waste disposal in Belgium in the 1970s and 1980s and ONDRAF/NIRAS, the Belgian agency for radioactive waste management. In Belgium, the regulatory body in charge of guidance and licensing approval is the Federal Agency of Nuclear Control, created in 2001.[50]

Finland

In 1983, the government decided to select a site for permanent repository by 2010. With four nuclear reactors providing 29% of its electricity,[37] Finland in 1987 enacted a Nuclear Energy Act making the producers of radioactive waste responsible for its disposal, subject to requirements of its Radiation and Nuclear Safety Authority and an absolute veto given to local governments in which a proposed repository would be located. Producers of nuclear waste organized the company Posiva, with responsibility for site selection, construction and operation of a permanent repository. A 1994 amendment to the Act required final disposal of spent fuel in Finland, prohibiting the import or export of radioactive waste.

Environmental assessment of four sites occurred in 1997–98, Posiva chose the Olkiluoto site near two existing reactors, and the local government approved it in 2000. The Finnish Parliament approved a deep geologic repository there in igneous bedrock at a depth of about 500 meters in 2001. The repository concept is similar to the Swedish model, with containers to be clad in copper and buried

below the water table beginning in 2020.[51] An underground characterization facility, Onkalo spent nuclear fuel repository, was under construction at the site in 2012.[52]

France

With 58 nuclear reactors contributing about 75% of its electricity,[37] the highest percentage of any country, France has been reprocessing its spent reactor fuel since the introduction of nuclear power there. Some reprocessed plutonium is used to make fuel, but more is being produced than is being recycled as reactor fuel.[53] France also reprocesses spent fuel for other countries, but the nuclear waste is returned to the country of origin. Radioactive waste from reprocessing French spent fuel is expected to be disposed of in a geological repository, pursuant to legislation enacted in 1991 that established a 15-year period for conducting radioactive waste management research. Under this legislation, partition and transmutation of long-lived elements, immobilization and conditioning processes, and long-term near surface storage are being investigated by the Commissariat à l' Energie Atomique (CEA). Disposal in deep geological formations is being studied by the French agency for radioactive waste management, L'Agence Nationale pour la Gestion des Déchets Radioactifs, in underground research labs.[54]

Three sites were identified for possible deep geologic disposal in clay near the border of Meuse and Haute-Marne, near Gard, and at Vienne. In 1998 the government approved the Meuse/Haute Marne Underground Research Laboratory, a site near Meuse/Haute-Marne and dropped the others from further consideration.[55] Legislation was proposed in 2006 to license a repository by 2015, with operations expected in 2025.[56]

Germany

Nuclear waste policy in Germany is in flux. German planning for a permanent geologic repository began in 1974, focused on salt dome Gorleben, a salt mine near Gorleben about 100 kilometers northeast of Braunschweig. The site was announced in 1977 with plans for a reprocessing plant, spent fuel management, and permanent disposal facilities at a single site. Plans for the reprocessing plant were dropped in 1979. In 2000, the federal government and utilities agreed to suspend underground investigations for three to ten years, and the government committed to ending its use of nuclear power, closing one reactor in 2003.[57]

Within days of the March 2011 Fukushima Daiichi nuclear disaster, Chancellor Angela Merkel "imposed a three-month moratorium on previously announced extensions for Germany's existing nuclear power plants, while shutting

seven of the 17 reactors that had been operating since 1981"
. Protests continued and, on 29 May 2011, Merkel's government announced that it would close all of its nuclear power plants by 2022.*[58]*[59]

Meanwhile, electric utilities have been transporting spent fuel to interim storage facilities at Gorleben, Lubmin and Ahaus until temporary storage facilities can be built near reactor sites. Previously, spent fuel was sent to France or the United Kingdom for reprocessing, but this practice was ended in July 2005.*[60]

Netherlands

COVRA (*Centrale Organisatie Voor Radioactief Afval*) is the Dutch interim nuclear waste processing and storage company in Vlissingen,*[61] which stores the waste produced in their only remaining nuclear power plant after it is reprocessed by Areva NC*[62] in La Hague, Manche, Basse-Normandie, France. Until the Dutch government decides what to do with the waste, it will stay at COVRA, which currently has a license to operate for one hundred years. As of early 2017, there are no plans for a permanent disposal facility.

Russia

In Russia, the Ministry of Atomic Energy (Minatom) is responsible for 31 nuclear reactors which generate about 16% of its electricity.*[37] Minatom is also responsible for reprocessing and radioactive waste disposal, including over 25,000 tons of spent nuclear fuel in temporary storage in 2001.

Russia has a long history of reprocessing spent fuel for military purposes, and previously planned to reprocess imported spent fuel, possibly including some of the 33,000 metric tons of spent fuel accumulated at sites in other countries who received fuel from the U.S., which the U.S. originally pledged to take back, such as Brazil, the Czech Republic, India, Japan, Mexico, Slovenia, South Korea, Switzerland, Taiwan, and the European Union.*[63]*[64]

An Environmental Protection Act in 1991 prohibited importing radioactive material for long-term storage or burial in Russia, but controversial legislation to allow imports for permanent storage was passed by the Russian Parliament and signed by President Putin in 2001.*[63] In the long term, the Russian plan is for deep geologic disposal.*[65] Most attention has been paid to locations where waste has accumulated in temporary storage at Mayak, near Chelyabinsk in the Ural Mountains, and in granite at Krasnoyarsk in Siberia.

Spain

Spain has five active nuclear plants with seven reactors which produced 21% of the country's electricity in 2013. Furthermore, there is legacy high-level waste from another two older, closed plants. Between 2004 and 2011, a bi-partisan initiative of the Spanish Government promoted the construction of an interim centralized storage facility (ATC, *Almacén Temporal Centralizado*), similar to the Dutch COVRA concept. In late 2011 and early 2012 the final green light was given, preliminary studies were being completed and land was purchased near Villar de Cañas (Cuenca) after a competitive tender process. The facility would be initially licensed for 60 years.

However, soon before groundbreaking was slated to begin in 2015, the project was stopped because of a mix of geological, technical, political and ecological problems. By late 2015, the Regional Government considered it "obsolete" and effectively "paralyzed." As of early 2017, the project has not been shelved but it stays frozen and no further action is expected anytime soon. Meanwhile, the spent nuclear fuel and other high-level waste is being kept in the plants' pools, as well as on-site dry cask storage (*almacenes temporales individualizados*) in Garoña and Trillo.

As of early 2017, there are no plans for a permanent high-level disposal facility either. Low- and medium-level waste is stored in the El Cabril facility (Province of Cordoba.)

Sweden

In Sweden, as of 2007 there are ten operating nuclear reactors that produce about 45% of its electricity.*[37] Two other reactors in Barsebäck were shut down in 1999 and 2005.*[66] When these reactors were built, it was expected their nuclear fuel would be reprocessed in a foreign country, and the reprocessing waste would not be returned to Sweden.*[67] Later, construction of a domestic reprocessing plant was contemplated, but has not been built.

Passage of the Stipulation Act of 1977 transferred responsibility for nuclear waste management from the government to the nuclear industry, requiring reactor operators to present an acceptable plan for waste management with "absolute safety" in order to obtain an operating license.*[68]*[69] In early 1980, after the Three Mile Island meltdown in the United States, a referendum was held on the future use of nuclear power in Sweden. In late 1980, after a three-question referendum produced mixed results, the Swedish Parliament decided to phase out existing reactors by 2010.*[70] In 2010, the Swedish government opened up for construction of new nuclear reactors. The new units can only be built at the existing nuclear power sites, Oskarshamn, Ringhals or Forsmark, and only to replace one of the existing reactors, that will have to be shut down for the new one to be able to start up.

The Swedish Nuclear Fuel and Waste Management Company, (Svensk Kärnbränslehantering AB, known as SKB) was created in 1980 and is responsible for final disposal of nuclear waste there. This includes operation of a monitored retrievable storage facility, the Central Interim Storage Facility for Spent Nuclear Fuel at Oskarshamn, about 150 miles south of Stockholm on the Baltic coast; transportation of spent fuel; and construction of a permanent repository.[71] Swedish utilities store spent fuel at the reactor site for one year before transporting it to the facility at Oskarshamn, where it will be stored in excavated caverns filled with water for about 30 years before removal to a permanent repository.

Conceptual design of a permanent repository was determined by 1983, calling for placement of copper-clad iron canisters in granite bedrock about 500 metres underground, below the water table in what is known as the KBS-3 method. Space around the canisters will be filled with bentonite clay.[71] After examining six possible locations for a permanent repository, three were nominated for further investigation, at Osthammar, Oskarshamn, and Tierp. On 3 June 2009, Swedish Nuclear Fuel and Waste Co. chose a location for a deep-level waste site at Östhammar, near Forsmark Nuclear Power plant. The application to build the repository was handed in by SKB 2011.

Switzerland

Main article: Nuclear energy in Switzerland

Switzerland has five nuclear reactors that provide about 43% of its electricity around 2007 (34% in 2015).[37] Some Swiss spent nuclear fuel has been sent for reprocessing in France and the United Kingdom; most fuel is being stored without reprocessing. An industry-owned organization, ZWILAG, built and operates a central interim storage facility for spent nuclear fuel and high-level radioactive waste, and for conditioning low-level radioactive waste and for incinerating wastes. Other interim storage facilities predating ZWILAG continue to operate in Switzerland.

The Swiss program is considering options for the siting of a deep repository for high-level radioactive waste disposal, and for low & intermediate level wastes. Construction of a repository is not foreseen until well into this century. Research on sedimentary rock (especially Opalinus Clay) is carried out at the Swiss Mont Terri rock laboratory; the Grimsel Test Site, an older facility in crystalline rock is also still active.[72]

United Kingdom

Great Britain has 19 operating reactors, producing about 20% of its electricity.[37] It processes much of its spent fuel at Sellafield on the northwest coast across from Ireland, where nuclear waste is vitrified and sealed in stainless steel canisters for dry storage above ground for at least 50 years before eventual deep geologic disposal. Sellafield has a history of environmental and safety problems, including a fire in a nuclear plant in Windscale, and a significant incident in 2005 at the main reprocessing plant (THORP).[73]

In 1982 the Nuclear Industry Radioactive Waste Management Executive (NIREX) was established with responsibility for disposing of long-lived nuclear waste[74] and in 2006 a Committee on Radioactive Waste Management (CoRWM) of the Department of Environment, Food and Rural Affairs recommended geologic disposal 200–1,000 meters underground.[75] NIREX developed a generic repository concept based on the Swedish model[76] but has not yet selected a site. A Nuclear Decommissioning Authority is responsible for packaging waste from reprocessing and will eventually relieve British Nuclear Fuels Ltd. of responsibility for power reactors and the Sellafield reprocessing plant.[77]

11.3.3 North America

Canada

The 18 operating nuclear power plants in Canada generated about 16% of its electricity in 2006.[78] A national Nuclear Fuel Waste Act was enacted by the Canadian Parliament in 2002, requiring nuclear energy corporations to create a waste management organization to propose to the Government of Canada approaches for management of nuclear waste, and implementation of an approach subsequently selected by the government. The Act defined management as "long term management by means of storage or disposal, including handling, treatment, conditioning or transport for the purpose of storage or disposal."[79]

The resulting Nuclear Waste Management Organization(NWMO) conducted an extensive three-year study and consultation with Canadians. In 2005, they recommended Adaptive Phased Management, an approach that emphasized both technical and management methods. The technical method included centralized isolation and containment of spent nuclear fuel in a deep geologic repository in a suitable rock formation, such as the granite of the Canadian Shield or Ordovician sedimentary rocks.[80] Also recommended was a phased decision making process supported by a program of continuous learning, research and development.

In 2007, the Canadian government accepted this recommendation, and NWMO was tasked with implementing the recommendation. No specific timeframe was defined for the process. In 2009, the NWMO was designing the process for site selection; siting was expected to take 10 years or more.[81]

United States

The Nuclear Waste Policy Act of 1982 established a timetable and procedure for constructing a permanent, underground repository for high-level radioactive waste by the mid-1990s, and provided for some temporary storage of waste, including spent fuel from 104 civilian nuclear reactors that produce about 19.4% of electricity there.[37] The United States in April 2008 had about 56,000 metric tons of spent fuel and 20,000 canisters of solid defense-related waste, and this is expected to increase to 119,000 metric tons by 2035.[82] The U.S. opted for Yucca Mountain nuclear waste repository, a final repository at Yucca Mountain in Nevada, but this project was widely opposed, with some of the main concerns being long distance transportation of waste from across the United States to this site, the possibility of accidents, and the uncertainty of success in isolating nuclear waste from the human environment in perpetuity. Yucca Mountain, with capacity for 70,000 metric tons of radioactive waste, was expected to open in 2017. However, the Obama Administration rejected use of the site in the 2009 United States Federal Budget proposal, which eliminated all funding except that needed to answer inquiries from the Nuclear Regulatory Commission, "while the Administration devises a new strategy toward nuclear waste disposal." [83] On March 5, 2009, Energy Secretary Steven Chu told a Senate hearing "the Yucca Mountain site no longer was viewed as an option for storing reactor waste." [82][84] Starting in 1999, military-generated nuclear waste is being entombed at the Waste Isolation Pilot Plant in New Mexico.

In a Presidential Memorandum dated January 29, 2010, President Obama established the Blue Ribbon Commission on America's Nuclear Future (the Commission).[85] The Commission, composed of fifteen members, conducted an extensive two-year study of nuclear waste disposal, what is referred to as the "back end" of the nuclear energy process.[85] The Commission established three subcommittees: Reactor and Fuel Cycle Technology, Transportation and Storage, and Disposal.[85] On January 26, 2012, the Commission submitted its final report to Energy Secretary Steven Chu.[86] In the Disposal Subcommittee's final report the Commission does not issue recommendations for a specific site but rather presents a comprehensive recommendation for disposal strategies. During their research the Commission visited Finland, France, Japan, Russia, Swe-

den, and the UK.[87] In their final report the Commission put forth seven recommendations for developing a comprehensive strategy to pursue:[87]

Recommendation #1 The United States should undertake an integrated nuclear waste management program that leads to the timely development of one or more permanent deep geological facilities for the safe disposal of spent fuel and high-level nuclear waste.[87]

Recommendation #2 A new, single-purpose organization is needed to develop and implement a focused, integrated program for the transportation, storage, and disposal 1 of nuclear waste in the United States.[87]

Recommendation #3 Assured access to the balance in the Nuclear Waste Fund (NWF) and to the revenues generated by annual nuclear waste fee payments from utility ratepayers is absolutely essential and must be provided to the new nuclear waste management organization.[87]

Recommendation #4 A new approach is needed to site and develop nuclear waste facilities in the United States in the future. We believe that these processes are most likely to succeed if they are:

- Adaptive—in the sense that process itself is flexible and produces decisions that are responsive to new information and new technical, social, or political developments.

- Staged—in the sense that key decisions are revisited and modified as necessary along the way rather than being pre-determined in advance.

- Consent-based—in the sense that affected communities have an opportunity to decide whether to accept facility siting decisions and retain significant local control.

- Transparent—in the sense that all stakeholders have an opportunity to understand key decisions and engage in the process in a meaningful way.

- Standards- and science-based—in the sense that the public can have confidence that all facilities meet rigorous, objective, and consistently-applied standards of safety and environmental protection.

- Governed by partnership arrangements or legally-enforceable agreements with host states, tribes and local communities.[87]

Recommendation #5 The current division of regulatory responsibilities for long-term repository performance between the NRC and the EPA is appropriate and should continue. The two agencies should develop new, site-independent safety standards in a formally coordinated joint process that actively engages and solicits input from all the relevant constituencies.*[87]

Recommendation #6 The roles, responsibilities, and authorities of local, state, and tribal governments (with respect to facility siting and other aspects of nuclear waste disposal) must be an element of the negotiation between the federal government and the other affected units of government in establishing a disposal facility. In addition to legally-binding agreements, as discussed in Recommendation #4, all affected levels of government (local, state, tribal, etc.) must have, at a minimum, a meaningful consultative role in all other important decisions. Additionally, states and tribes should retain—or where appropriate, be delegated—direct authority over aspects of regulation, permitting, and operations where oversight below the federal level can be exercised effectively and in a way that is helpful in protecting the interests and gaining the confidence of affected communities and citizens.*[87]

Recommendation #7 The Nuclear Waste Technical Review Board (NWTRB) should be retained as a valuable source of independent technical advice and review. *[87]

11.3.4 International repository

Although Australia does not have any nuclear power reactors, Pangea Resources considered siting an international repository in the outback of South Australia or Western Australia in 1998, but this stimulated legislative opposition in both states and the Australian national Senate during the following year.*[88] Thereafter, Pangea ceased operations in Australia but reemerged as Pangea International Association, and in 2002 evolved into the Association for Regional and International Underground Storage with support from Belgium, Bulgaria, Hungary, Japan and Switzerland.*[89] A general concept for an international repository has been advanced by one of the principals in all three ventures.*[90] Russia has expressed interest in serving as a repository for other countries, but does not envision sponsorship or control by an international body or group of other countries. South Africa, Argentina and western China have also been mentioned as possible locations.*[55]*[91]

In the EU, COVRA is negotiating a European-wide waste disposal system with single disposal sites that can be used by several EU-countries. This EU-wide storage possibility is being researched under the SAPIERR-2 program.*[92]

11.4 See also

- Radioactive waste
- Economics of new nuclear power plants
- List of nuclear waste treatment technologies
- Deep geological repository
- Nuclear reprocessing
- Decommissioning of Russian nuclear-powered vessels
- *Into Eternity*, a 2010 documentary about the construction of a Finnish waste depository
- *Journey to the Safest Place on Earth*, a 2013 documentary about the urgent need for safe depositories

11.5 Notes

[1] "Iodine-131". stoller-eser.com. Archived from the original on 2011-07-16. Retrieved 2009-01-05.

[2] Vandenbosch 2007, p. 21.

[3] Ojovan, M. I.; Lee, W.E. (2014). *An Introduction to Nuclear Waste Immobilisation*. Amsterdam: Elsevier Science Publishers. p. 362. ISBN 978-0-08-099392-8.

[4] "What about Iodine-129 - Half-Life is 15 Million Years". *Berkeley Radiological Air and Water Monitoring Forum*. University of California. 28 March 2011. Retrieved 1 December 2012.

[5] Brown, Paul (2004-04-14). "Shoot it at the sun. Send it to Earth's core. What to do with nuclear waste?". *The Guardian*.

[6] National Research Council (1995). *Technical Bases for Yucca Mountain Standards*. Washington, D.C.: National Academy Press. p. 91. ISBN 0-309-05289-0.

[7] "The Status of Nuclear Waste Disposal". The American Physical Society. January 2006. Retrieved 2008-06-06.

[8] "Public Health and Environmental Radiation Protection Standards for Yucca Mountain, Nevada; Proposed Rule" (PDF). United States Environmental Protection Agency. 2005-08-22. Retrieved 2008-06-06.

[9] Abbotts, John (October 1979). "Radioactive waste: A technical solution?". *Bulletin of the Atomic Scientists*: 12–18.

[10] Genevieve Fuji Johnson, *Deliberative Democracy for the Future: The Case of Nuclear Waste Management in Canada*. University of Toronto Press, 2008, p.9 ISBN 0-8020-9607-7

[11] Bruno, Jordi, Lara Duro, and Mireia Grivé. 2001. *The applicability and limitations of the geochemical models and tools used in simulating radionuclide behavior in natural waters: Lessons learned from the blind predictive modelling exercises performed in conjunction with natural analogue studies.* QuantiSci S. L. Parc Tecnològic del Vallès, Spain, for Swedish Nuclear Fuel and Waste Management Co.

[12] Shrader-Frechette, Kristin S. 1988. "Values and hydrogeological method: How not to site the world's largest nuclear dump" In *Planning for Changing Energy conditions*, John Byrne and Daniel Rich, eds. New Brunswick, NJ: Transaction Books, p. 101 ISBN 0-88738-713-6

[13] Shrader-Frechette, Kristin S. *Burying uncertainty: Risk and the case against geological disposal of nuclear waste* Berkeley: University of California Press (1993) p. 2 ISBN 0-520-08244-3

[14] Shrader-Frechette, Kristin S. *Expert judgment in assessing radwaste risks: What Nevadans should know about Yucca Mountain.* Carson City: Nevada Agency for Nuclear Projects, Nuclear Waste Project, 1992 ISBN 0-7881-0683-X

[15] "Issues relating to safety standards on the geological disposal of radioactive waste" (PDF). International Atomic Energy Agency. 2001-06-22. Retrieved 2008-06-06.

[16] "IAEA Waste Management Database: Report 3 – L/ILW-LL" (PDF). International Atomic Energy Agency. 2000-03-28. Retrieved 2008-06-06.

[17] "Decommissioning costs of WWER-440 nuclear power plants" (PDF). International Atomic Energy Agency. November 2002. Retrieved 2008-06-06.

[18] "Spent Fuel and High Level Waste: Chemical Durability and Performance under Simulated Repository Conditions" (PDF). International Atomic Energy Agency. October 2007. IAEA-TECDOC-1563.

[19] Harold Feiveson, Zia Mian, M.V. Ramana, and Frank von Hippel (27 June 2011). "Managing nuclear spent fuel: Policy lessons from a 10-country study". *Bulletin of the Atomic Scientists*.

[20] Vandenbosch 2007, pp. 214–248.

[21] Vandenbosch 2007, p. 10.

[22] Yates, Marshall (July 6, 1989). "DOE waste management criticized: On-site storage urged". *Public Utilities Fortnightly* (124): 33.

[23] Engelhardt, Dean; Parker, Glen. "Permanent Radwaste Solutions". San Francisco: Engelhardt, Inc. Retrieved 2008-12-24.

[24] Jack, Tricia; Robertson, Jordan. "Utah nuclear waste summary" (PDF). Salt Lake City: University of Utah Center for Public Policy and Administration. Retrieved 2008-12-24.

[25] Rao, K.R. (December 2001). "Radioactive waste: The problem and its management" (PDF). *Current Science* (81): 1534–1546. Retrieved 2008-12-24.

[26] Cowan, G. A. (1976). "Oklo, A Natural Fission Reactor". *Scientific American*. **235** (1): 36. doi:10.1038/scientificamerican0776-36. ISSN 0036-8733.

[27] "Oklo, Natural Nuclear Reactors". U.S. Department of Energy Office of Civilian Radioactive Waste Management, Yucca Mountain Project, DOE/YMP-0010. November 2004. Archived from the original on August 25, 2009. Retrieved September 15, 2009.

[28] Krauskopf, Konrad B. 1988. *Radioactive waste and geology.* New York: Chapman and Hall, 101–102. ISBN 0-412-28630-0

[29] Clark, S., Ewing, R. Panel 5 Report: Advanced Waste Forms. Basic Research Needs for Advanced Energy Systems 2006, 59–74.

[30] Grambow, B. (2006). "Nuclear Waste Glasses - How Durable?". *Elements*. **2** (6): 357. doi:10.2113/gselements.2.6.357.

[31] Oelkers, E. H.; Montel, J.-M. (2008). "Phosphates and Nuclear Waste Storage". *Elements*. **4** (2): 113. doi:10.2113/GSELEMENTS.4.2.113.

[32] Weber W. J., Navrotsky A., Stefanovsky S., Vance E.R., Vernaz E. Materials Science of High-Level Nuclear Waste Immobilization. MRS Bulletin 2009, 34, 46.

[33] Luo, S; Li, Liyu; Tang, Baolong; Wang, Dexi (1998). "Synroc immobilization of high level waste (HLW) bearing a high content of sodium". *Waste Management*. **18**: 55. doi:10.1016/S0956-053X(97)00019-6.

[34] Vandenbosch 2007, p. 248.

[35] M.V. Ramana. Nuclear Power: Economic, Safety, Health, and Environmental Issues of Near-Term Technologies, *Annual Review of Environment and Resources*, 2009, 34, p. 145.

[36] Benjamin K. Sovacool (2011). *Contesting the Future of Nuclear Power: A Critical Global Assessment of Atomic Energy.* World Scientific, p. 144.

[37] "World nuclear power reactors 2005–2007 and uranium requirements". World Nuclear Association. 2007. Retrieved 2008-12-24.

[38] Vandenbosch 2007, pp. 244–45.

[39] Tony Vince (8 March 2013). "Rock solid ambitions". Nuclear Engineering International. Retrieved 9 March 2013.

[40] http://focustaiwan.tw/news/aipl/201304030025.aspx

[41] http://www.taipeitimes.com/News/front/archives/2012/02/21/2003525985

[42] Raj, Kanwar (2005). "Commissioning and operation of high level radioactive waste vitrification and storage facilities: The Indian experience" (PDF). *International Journal of Nuclear Energy Science and Technology* (1): 148–63. Retrieved 2008-12-24.

[43] "Nuclear power in India and Pakistan". *UIC Nuclear Issues Briefing Paper #45*. World Nuclear Association. 2006. Archived from the original on 2007-12-14.

[44] Vandenbosch 2007, p. 244.

[45] Burnie, Shaun; Smith, Aileen Mioko (May–June 2001). "Japan's nuclear twilight zone". *Bulletin of the Atomic Scientists* (57): 58.

[46] "Open solicitation for candidate sites for safe disposal of high-level radioactive waste". *Nuclear Waste Management Organization of Japan*. Tokyo. 2002.

[47] Vandenbosch 2007, p. 240.

[48] "Japan's nuclear waste problem". *The Japan Times*. 21 January 2014. Retrieved 23 January 2014.

[49] "Management of irradiated fuels in Belgium". Belgian Federal Public Service Economy. Retrieved 27 January 2015.

[50] "Belgium's Radioactive Waste Management Program". U.S. Department of Energy. June 2001. Archived from the original on 2008-10-11. Retrieved 2008-12-26.

[51] *Stepwise decision making in Finland for the disposal of spent nuclear fuel*. Organization for Economic Co-operation and Development. Paris: Nuclear Energy Agency. 2002.

[52] "Posiva Oy – Nuclear Waste Management Expert".

[53] Vandenbosch 2007, p. 221.

[54] McEwen, Tim (1995). Savage, D., ed. *The scientific and regulatory basis for the geological disposal of radioactive waste. Selection of waste disposal sites*. New York: J. Wiley & Sons. ISBN 0-471-96090-X.

[55] Committee on Disposition of High-Level Radioactive Waste through Geological Isolation, Board on Radioactive Waste Management, Division on Earth and Life Studies, National Research Council. (2001). *Disposition of high-level waste and spent nuclear fuel: The continuing societal and technical challenges. U.S. National Research Council*. Washington, DC: National Academy Press. ISBN 0-309-07317-0.

[56] "Headlines: International briefs". *Radwaste Solutions* (13): 9. May–June 2006.

[57] Graham, Stephen (2003-11-15). "Germany snuffs out nuclear plant". *Seattle Times*. p. A10.

[58] Caroline Jorant (July 2011). "The implications of Fukushima: The European perspective". *Bulletin of the Atomic Scientists*. p. 15.

[59] Knight, Ben (15 March 2011). "Merkel shuts down seven nuclear reactors". Deutsche Welle. Retrieved 15 March 2011.

[60] Vandenbosch 2007, pp. 223–24.

[61] COVRA website

[62] AREVA NC - nuclear energy, nuclear fuel - La Hague

[63] Webster, Paul (May–June 2002). "Minatom: The grab for trash". *Bulletin of the Atomic Scientists* (58): 36.

[64] Vandenbosch 2007, p. 242.

[65] Bradley, Don J (1997). Payson, David R, ed. *Behind the nuclear curtain: Radioactive waste management in the former Soviet Union*. Columbus: Battelle Press. ISBN 1-57477-022-5.

[66] Vandenbosch 2007, pp. 233–34.

[67] Sundqvist, Göran (2002). *The bedrock of opinion: Science, technology and society in the siting of high-level nuclear waste*. Dordrecht: Kluwer Academic Publishers. ISBN 1-4020-0477-X.

[68] Johansson, T.B.; Steen, P. (1981). *Radioactive waste from nuclear power plants*. Berkeley: University of California Press. p. 67. ISBN 0-520-04199-2.

[69] Carter, Luther J. (1987). *Nuclear imperatives and public trust: Dealing with radioactive waste*. Washington, DC: Resources for the Future, Inc. ISBN 0-915707-29-2.

[70] Vandenbosch 2007, pp. 232–33.

[71] "Sweden's radioactive waste management program". U.S. Department of Energy. June 2001. Archived from the original on 2009-01-18. Retrieved 2008-12-24.

[72] McKie, D. "Underground Rock Laboratory Home Page". Grimsel Test Site. Retrieved 2008-12-24.

[73] Cassidy, Nick; Green, Patrick (1993). *Sellafield: The contaminated legacy*. London: Friends of the Earth. ISBN 1-85750-225-6.

[74] Openshaw, Stan; Carver, Steve; Fernie, John (1989). *Britain's nuclear waste: Siting and safety*. London: Bellhaven Press. p. 48. ISBN 1-85293-005-5.

[75] "Managing our radioactive waste safely: CoRWM's Recommendations to government" (PDF). U.K Committee on Radioactive Waste Management. 2006. Retrieved 2014-04-24.

[76] McCall, A; King, S (April 30 – May 4, 2006). "Generic repository concept development and assessment for UK high-level waste and spent nuclear fuel". *Proceedings of the 11th high-level radioactive waste management conference*. La Grange Park, IL: American Nuclear Society: 1173–79.

[77] Vandenbosch 2007, pp. 224–30.

[78] *Table 2. Generation of electric energy, 2006.* Statistics Canada (www.statcan.gc.ca). 2008.

[79] *Nuclear Fuel Waste Act.* Government of Canada. c. 23 Elizabeth II. 2002.

[80] *Choosing a way forward. Final Report.* Canada: Nuclear Waste Management Organization. 2005.

[81] *Implementing Adaptive Phased Management (2008–2012).* Canada: Nuclear Waste Management Organization. 2008 p. 8.

[82] Karen R. Olesky (2008). "Nuclear Power's Emission Reduction Potential in Utah" (PDF). Duke University. Retrieved 2009-08-01.

[83] A New Era of Responsibility, The 2010 Budget, p. 65.

[84] Hebert, H. Josef. 2009. "Nuclear waste won't be going to Nevada's Yucca Mountain, Obama official says." *Chicago Tribune.* March 6, 2009, 4. Accessed 3-6-09.

[85] "About the Commission". Retrieved 2012-18-2012. Check date values in: |access-date= (help)

[86] "Please Note". Retrieved 2012-18-2012. Check date values in: |access-date= (help)

[87] Blue Ribbon Commission on America's Nuclear Future. "Disposal Subcommittee Report to the Full Commission" (PDF). Retrieved 2012-18-2012. Check date values in: |access-date= (help)

[88] Holland, I. (2002). "Waste not want not? Australia and the politics of high-level nuclear waste". *Australian Journal of Political Science.* **37** (37): 283–301. doi:10.1080/10361140220148151.

[89] "Pangea Resources metamorphisizing into International Repository Forum". *Nuclear Waste News* (22): 41. January 31, 2002.

[90] McCombie, Charles (April 29 – May 3, 2001). "International and regional repositories: The key questions". *Proceedings of the 9th international high-level radioactive waste management conference.* La Grange Park, IL: American Nuclear Society.

[91] Vandenbosch 2007, p. 246.

[92] Nilsson, Karl Fredrik (December 10–11, 2007). *Enlargement and integration workshop: European collaboration for the management of spent nuclear fuel and radioactive waste by technology transfer and shared facilities.* Brussels: European Commission. Retrieved 2008-12-27.

11.6   References

- Vandenbosch, Robert; Vandenbosch, Susanne E. (2007). *Nuclear waste stalemate.* Salt Lake City: University of Utah Press. ISBN 0-87480-903-7.

- South Carolina Biohazard Disposal Company

11.7   Further reading

- Donald, I. W., "Waste immobilization in glass and ceramic based hosts: Radioactive, toxic and hazardous wastes", Wiley, 2010. ISBN 978-1-4443-1937-8

- Ialenti, Vincent. "Adjudicating Deep Time: Revisiting The United States' High-Level Nuclear Waste Repository Project At Yucca Mountain" (PDF). *Science & Technology Studies.* **27** (2).

- Shrader-Frechette, Kristin S. *Risk analysis and scientific method: Methodological and ethical problems with evaluating societal hazards.* Dordrecht: D. Reidel, 1985. ISBN 90-277-1836-9

11.8   External links

- International Atomic Energy Agency – Internet Directory of Nuclear Resources (links)

- Nuclear Regulatory Commission – Radioactive Waste (documents)

- Radwaste Solutions (magazine)

- "Radioactive Waste (documents and links)". UNEP Earthwatch.

- World Nuclear Association – Radioactive

Chapter 12

Lists of nuclear disasters and radioactive incidents

The Kashiwazaki-Kariwa Nuclear Power Plant, a Japanese nuclear plant with seven units, the largest single nuclear power station in the world, was completely shut down for 21 months following an earthquake in 2007.[1]

The Hanford site represents two-thirds of America's high-level radioactive waste by volume. Nuclear reactors line the riverbank at the Hanford Site along the Columbia River in January 1960.

12.1 Main lists

- List of attacks on nuclear plants
- List of Chernobyl-related articles
- List of civilian nuclear accidents
- List of civilian radiation accidents
- List of crimes involving radioactive substances
- List of criticality accidents and incidents
- List of nuclear meltdown accidents
- List of Milestone nuclear explosions
- List of military nuclear accidents
- List of nuclear and radiation accidents and incidents
- List of nuclear and radiation accidents by death toll
- List of articles about the Three Mile Island accident

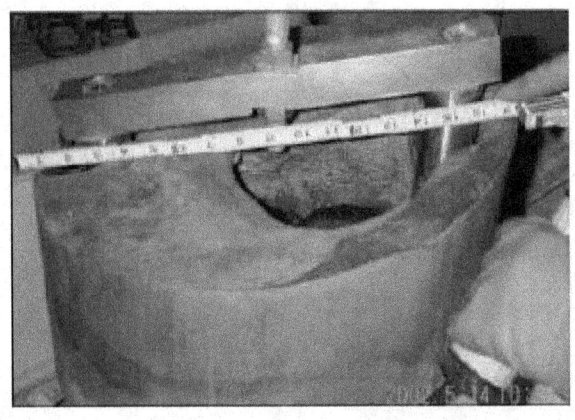

Erosion of the 150-millimetre-thick (5.9 in) carbon steel reactor head at Davis-Besse Nuclear Power Plant in 2002, caused by a persistent leak of borated water.

These are lists of nuclear disasters and radioactive incidents.

This image of the SL-1 core served as a reminder of the necessity for proper reactor practice and safeguards.

The 18,000 km^2 expanse of the Semipalatinsk Test Site (indicated in red), which covers an area the size of Wales. The Soviet Union conducted 456 nuclear tests at Semipalatinsk from 1949 until 1989 with little regard for their effect on the local people or environment. The full impact of radiation exposure was hidden for many years by Soviet authorities and has only come to light since the test site closed in 1991. [2]

12.2 Lists by country

- List of nuclear power accidents by country

- List of nuclear and radiation fatalities by country

- List of nuclear power accidents in Canada

12.3 Individual disasters, incidents and sites

- 2011 K-84 nuclear submarine incident

- 2011 Fukushima Daiichi nuclear disaster

- 2001 Instituto Oncologico Nacional radiotherapy accident

- 1997 Tokaimura nuclear accident

- 1996 San Juan de Dios radiotherapy accident

- 1990 Clinic of Zaragoza radiotherapy accident

- 1987 Goiânia accident

- 1986 Chernobyl disaster and Chernobyl disaster effects

- 1979 Church Rock uranium mill spill

- 1979 Three Mile Island accident and Three Mile Island accident health effects

- 1969 Lucens reactor

- 1962 Thor missile launch failures during nuclear weapons testing at Johnston Atoll under Operation Fishbowl

- 1961 SL-1 nuclear meltdown

- 1961 K-19 nuclear accident

- 1959 SRE partial nuclear meltdown at Santa Susana Field Laboratory

- 1957 Kyshtym disaster

- 1957 Windscale fire

- 1957 Operation Plumbbob

- 1954 Totskoye nuclear exercise

- 1950 Desert Rock exercises

- Bikini Atoll

- Hanford Site

- Rocky Flats Plant, see also radioactive contamination from the Rocky Flats Plant

- Techa River

- Pollution of Lake Karachay

- 1945 Hiroshima

- 1945 Nagasaki

12.4   See also

- List of books about nuclear issues

- List of civilian nuclear ships

- List of films about nuclear issues

- Vulnerability of nuclear plants to attack

- United States military nuclear incident terminology

- International Nuclear Event Scale

- Atomic spies

- Nuclear terrorism

- Nuclear safety

- Nuclear accident

12.5   References

[1] The north korean Parliament's Greens-EFA Group - The World Nuclear Industry Status Report 2007 p. 23.

[2] Togzhan Kassenova (28 September 2009). "The lasting toll of Semipalatinsk's nuclear testing". *Bulletin of the Atomic Scientists.*

12.6   External links

- Radiation exposures in accidents - Annex C of UN-SCEAR 2008 Report (Comprehensive list of accidents with details)

- "The world's worst nuclear power disasters". *Power Technology.* 7 October 2013.

Chapter 13

Nuclear reactor accidents in the United States

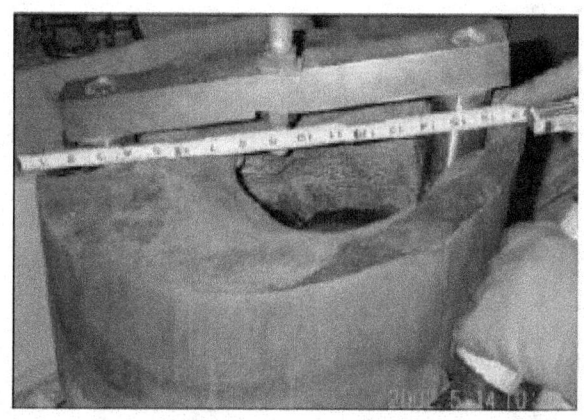

Erosion of the 6-inch-thick (150 mm) carbon steel reactor head, caused by a persistent leak of borated water, at the Davis-Besse Nuclear Power Plant.

The United States Government Accountability Office reported more than 150 incidents from 2001 to 2006 of nuclear plants not performing within acceptable safety guidelines. According to a 2010 survey of energy accidents, there have been at least 56 **accidents at nuclear reactors in the United States** (defined as incidents that either resulted in the loss of human life or more than US$50,000 of property damage). The most serious of these was the Three Mile Island accident in 1979. Davis-Besse Nuclear Power Plant has been the source of two of the top five most dangerous nuclear incidents in the United States since 1979.[1] Relatively few accidents have involved fatalities.[2]

13.1 Context

Globally, there have been at least 99 (civilian and military) recorded nuclear reactor accidents from 1952 to 2009 (defined as incidents that either resulted in the loss of human life or more than US$50,000 of property damage, the amount the US federal government uses to define major energy accidents that must be reported), totaling US$20.5 billion in property damages. The accidents involved meltdowns, explosions, fires, and loss of coolant, and occurred during both normal operation and extreme emergency conditions (such as droughts and earthquakes). Property damage costs include destruction of property, emergency response, environmental remediation, evacuation, lost product, fines, and court claims.[2] Because nuclear reactors are large and complex, accidents onsite tend to be relatively expensive.[3]

At least 56 nuclear reactor accidents have occurred in the USA. Relatively few accidents have involved fatalities.[2] The most serious of these U.S. accidents was the Three Mile Island accident in 1979. According to the Nuclear Regulatory Commission, the Davis–Besse Nuclear Power Station has been the source of two of the top five most dangerous nuclear incidents in the United States since 1979.[1]

The United States Government Accountability Office reported more than 150 incidents from 2001 to 2006 alone of nuclear plants not performing within acceptable safety guidelines. In 2006, it said: "Since 2001, the ROP has resulted in more than 4,000 inspection findings concerning nuclear power plant licensees' failure to fully comply with NRC regulations and industry standards for safe plant operation, and NRC has subjected more than 75 percent (79) of the 103 operating plants to increased oversight for varying periods".[4]

13.2 History

The Atomic Energy Act of 1954 encouraged private corporations in the United States to build nuclear reactors and a significant learning phase followed with many early partial core meltdowns and accidents at experimental reactors and research facilities.[5] This led to the introduction of the Price-Anderson Act in 1957, which was "an implicit admission that nuclear power provided risks that producers were unwilling to assume without federal backing" .*[5]

Nuclear reactor accidents continued into the 1960s with a small test reactor exploding at the Stationary Low-Power Reactor Number One in Idaho Falls in January 1961 resulting in three deaths which were the first fatalities in the history of U.S. nuclear reactor operations.*[6] There was also a partial meltdown at the Enrico Fermi Nuclear Generating Station in Michigan in 1966.*[5]

The large size of nuclear reactors ordered during the late 1960s raised new safety questions and created fears of a severe reactor accident that would send large quantities of radiation into the environment. In the early 1970s, a highly contentious debate over the performance of emergency core cooling systems in nuclear plants, designed to prevent a core meltdown that could lead to the "China syndrome", received coverage in the popular media and technical journals.*[7]*[8]

In 1976, four nuclear engineers —three from GE and one from the Nuclear Regulatory Commission—resigned, stating that nuclear power was not as safe as their superiors were claiming.*[9]*[10]*[11]*[12] They testified to the Joint Committee on Atomic Energy that:

> "the cumulative effect of all design defects and deficiencies in the design, construction and operations of nuclear power plants makes a nuclear power plant accident, in our opinion, a certain event. The only question is when, and where.*[9]

13.3   Three Mile Island accident

President Jimmy Carter leaving Three Mile Island for Middletown, Pennsylvania, April 1, 1979

On March 28, 1979, equipment failures and operator error contributed to loss of coolant and a partial core meltdown

of Unit 2's pressurized water reactor at the Three Mile Island Nuclear Power Plant in Pennsylvania.*[13] The scope and complexity of this reactor accident became clear over the course of five days, as a number of agencies at the local, state and federal levels tried to solve the problem and decide whether the ongoing accident required an emergency evacuation, and to what extent.

Cleanup started in August 1979 and officially ended in December 1993, with a total cleanup cost of about $1 billion.*[14] Benjamin K. Sovacool, in his 2007 preliminary assessment of major energy accidents, estimated that the TMI accident caused a total of $2.4 billion in property damages.*[15] The health effects of the Three Mile Island accident are widely, but not universally, agreed to be very low level.*[16]*[17]

The TMI accident forced regulatory and operational improvements on a reluctant industry, but it also increased opposition to nuclear power.*[18] The accident triggered protests around the world.*[19]

13.4   List of accidents and incidents

Main article: Nuclear power accidents by country
See also: Lists of nuclear disasters and radioactive incidents
Further information: Nuclear safety in the United States §
Emergency Classifications
This list is incomplete; you can help by expanding it.

13.5   Nuclear safety

Main article: Nuclear safety in the U.S.
 Nuclear safety in the U.S. is governed by federal regulations issued by the Nuclear Regulatory Commission (NRC). The NRC regulates all nuclear plants and materials in the U.S. except for of nuclear plants and materials controlled by the U.S. government, as well those powering naval vessels.*[24]*[25]

The 1979 Three Mile Island accident was a pivotal event that led to questions about U.S. nuclear safety.*[26] Earlier events had a similar effect, including a 1975 fire at Browns Ferry, the 1976 testimonials of three concerned GE nuclear engineers, the GE Three. In 1981, workers inadvertently reversed pipe restraints at the Diablo Canyon Power Plant reactors, compromising seismic protection systems, which further undermined confidence in nuclear safety. All of these well-publicised events, undermined public support for the U.S. nuclear industry in the 1970s and the 1980s.*[26]

A clean-up crew working to remove radioactive contamination after the Three Mile Island accident.

Recent concerns have been expressed about safety issues affecting a large part of the nuclear fleet of reactors. In 2012, the Union of Concerned Scientists, which tracks ongoing safety issues at operating nuclear plants, found that "leakage of radioactive materials is a pervasive problem at almost 90 percent of all reactors, as are issues that pose a risk of nuclear accidents".[27]

Following the Japanese Fukushima Daiichi nuclear disaster, according to Black & Veatch's annual utility survey that took place after the disaster, of the 700 executives from the US electric utility industry that were surveyed, nuclear safety was the top concern.[28] There are likely to be increased requirements for on-site spent fuel management and elevated design basis threats at nuclear power plants.[29][30] License extensions for existing reactors will face additional scrutiny, with outcomes depending on the degree to which plants can meet new requirements, and some of the extensions already granted for more than 60 of the 104 operating U.S. reactors could be revisited. On-site storage, consolidated long-term storage, and geological disposal of spent fuel is "likely to be reevaluated in a new light because of the Fukushima storage pool experience".[29]

In October 2011, the Nuclear Regulatory Commission instructed agency staff to move forward with seven of the 12 safety recommendations put forward by the federal task force in July. The recommendations include "new standards aimed at strengthening operators' ability to deal with a complete loss of power, ensuring plants can withstand floods and earthquakes and improving emergency response

capabilities". The new safety standards will take up to five years to fully implement.[31]

13.6   See also

- Nuclear power accidents by country

- Nuclear and radiation accidents by country

- Lists of nuclear disasters and radioactive incidents

- List of canceled nuclear plants in the United States

- Nuclear safety

13.7   References

[1] Nuclear Regulatory Commission (2004-09-16). "Davis-Besse preliminary accident sequence precursor analysis" (PDF). Retrieved 2006-06-14. and Nuclear Regulatory Commission (2004-09-20). "NRC issues preliminary risk analysis of the combined safety issues at Davis-Besse". Retrieved 2006-06-14.

[2] Benjamin K. Sovacool. A Critical Evaluation of Nuclear Power and Renewable Electricity in Asia, *Journal of Contemporary Asia*, Vol. 40, No. 3, August 2010, pp. 379-380.

[3] Benjamin K. Sovacool (2009). The Accidental Century - Prominent Energy Accidents in the Last 100 Years

[4] United States Government Accountability Office (2006). "Report to Congress" (PDF). p. 4.

[5] Benjamin K. Sovacool. The costs of failure: A preliminary assessment of major energy accidents, 1907–2007. *Energy Policy* 36 (2008), p. 1808.

[6] Perhaps the Worst, Not the First *TIME magazine*, May 12, 1986.

[7] Walker, J. Samuel (2004). *Three Mile Island: A Nuclear Crisis in Historical Perspective* (Berkeley: University of California Press), pp. 10-11.

[8] Wolfgang Rudig (1990). *Anti-nuclear Movements: A World Survey of Opposition to Nuclear Energy*, Longman, pp. 66-67.

[9] Mark Hertsgaard (1983). *Nuclear Inc. The Men and Money Behind Nuclear Energy*, Pantheon Books, New York, p. 72.

[10] Jim Falk (1982). *Global Fission: The Battle Over Nuclear Power*, Oxford University Press, p. 95.

[11] The San Jose Three *TIME*, Feb. 16, 1976.

[12] The Struggle over Nuclear Power *TIME*, Mar. 08, 1976.

[13] World Nuclear Association (1999). Three Mile Island: 1979 Retrieved December 24, 2008.

[14] "14-Year Cleanup at Three Mile Island Concludes". New York Times. August 15, 1993. Retrieved March 28, 2011.

[15] Benjamin K. Sovacool. The costs of failure: A preliminary assessment of major energy accidents, 1907–2007. *Energy Policy* 36 (2008). p. 1807.

[16] Mangano, Joseph (2004). Three Mile Island: Health study meltdown. *Bulletin of the atomic scientists*, 60(5), pp. 31 – 35.

[17] World Nuclear Association. Three Mile Island Accident January 2010.

[18] Wellock, Thomas R. Three Mile Island: A Nuclear Crisis in Historical Perspective (Book review) *The Historian*. 22 September 2005.

[19] Mark Hertsgaard (1983). *Nuclear Inc. The Men and Money Behind Nuclear Energy*, Pantheon Books, New York, p. 95 & 97.

[20] Benjamin K. Sovacool. A Critical Evaluation of Nuclear Power and Renewable Electricity in Asia, *Journal of Contemporary Asia*, Vol. 40, No. 3, August 2010, pp. 393–400.

[21] http://www.iaea.org/ns/tutorials/regcontrol/appendix/app96.htm

[22] Bel, Hubert T. I. "Inspector General Report - Ind" (PDF). Office of the Inspector General (OIG), U.S. Nuclear Regulatory Commission (NRC).

[23] Blade, Toledo. "Davis-Besse stirs again". Toledo Blade.

[24] About NRC, U.S. Nuclear Regulatory Commission. Retrieved 2007-6-1.

[25] Our Governing Legislation, U.S. Nuclear Regulatory Commission. Retrieved 2007-6-1.

[26] Nathan Hultman and Jonathan Koomey (1 May 2013). "Three Mile Island: The driver of US nuclear power's decline?", *Bulletin of the Atomic Scientists*.

[27] Mark Cooper (2012-68-61). "Nuclear safety and affordable reactors: Can we have both?" (PDF). *Bulletin of the Atomic Scientists*. Check date values in: |date= (help)

[28] Eric Wesoff, Greentechmedia. "Black & Veatch's 2011 Electric Utility Survey." June 16, 2011. Retrieved October 11, 2011.

[29] Massachusetts Institute of Technology (2011). "The Future of the Nuclear Fuel Cycle" (PDF). p. xv.

[30] Mark Cooper (July 2011). "The implications of Fukushima: The US perspective". *Bulletin of the Atomic Scientists*. p. 9.

[31] Andrew Restuccia (2011-10-20). "Nuke regulators toughen safety rules". *The Hill*.

13.8 Further reading

- *Conservation Fallout: Nuclear Protest at Diablo Canyon* (2006)

- *Contesting the Future of Nuclear Power* (2011)

- *Essence of Decision: Explaining the Cuban Missile Crisis* (1971)

- *Fallout: An American Nuclear Tragedy* (2004)

- *Fukushima: Japan's Tsunami and the Inside Story of the Nuclear Meltdowns* (2013)

- *Full Body Burden: Growing Up in the Nuclear Shadow of Rocky Flats* (2012)

- *Killing Our Own: The Disaster of America's Experience with Atomic Radiation* (1982)

- *In Mortal Hands: A Cautionary History of the Nuclear Age* (2009)

- *Making a Real Killing: Rocky Flats and the Nuclear West* (1999)

- *Non-Nuclear Futures: The Case for an Ethical Energy Strategy* (1975)

- *Normal Accidents: Living with High-Risk Technologies* (1984)

- *Nuclear Politics in America* (1997)

- *Nuclear Terrorism: The Ultimate Preventable Catastrophe* (2004)

- *Nuclear War Survival Skills* (1979)

- *Nuclear Weapons: The Road to Zero* (1998)

- *The Making of the Atomic Bomb* (1987)

- *Nukespeak: Nuclear Language, Visions and Mindset* (1982)

- *On Nuclear Terrorism* (2007)

- *Plutopia* (2013)

Chapter 14

Nuclear energy policy

Main article: Nuclear power

Nuclear energy policy is a national and international

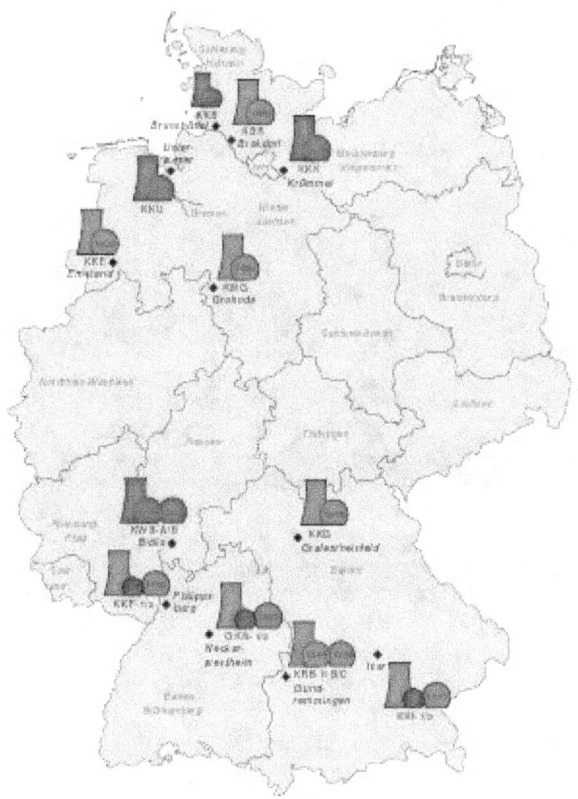

Eight German nuclear power reactors (Biblis A and B, Brunsbuettel, Isar 1, Kruemmel, Neckarwestheim 1, Philippsburg 1 and Unterweser) were permanently shutdown on 6 August 2011, following the Japanese Fukushima nuclear disaster.[1]

policy concerning some or all aspects of nuclear energy and the nuclear fuel cycle, such as uranium mining, ore concentration, conversion, enrichment for nuclear fuel, generating electricity by nuclear power, storing and reprocessing spent nuclear fuel, and disposal of radioactive waste.

Nuclear energy policies often include the regulation of energy use and standards relating to the nuclear fuel cycle.

Other measures include efficiency standards, safety regulations, emission standards, fiscal policies, and legislation on energy trading, transport of nuclear waste and contaminated materials, and their storage. Governments might subsidize nuclear energy and arrange international treaties and trade agreements about the import and export of nuclear technology, electricity, nuclear waste, and uranium.

Since about 2001 the term nuclear renaissance has been used to refer to a possible nuclear power industry revival, but nuclear electricity generation in 2012 was at its lowest level since 1999.[2][3]

Following the March 2011 Fukushima I nuclear accidents, China, Germany, Switzerland, Israel, Malaysia, Thailand, United Kingdom, and the Philippines are reviewing their nuclear power programs. Indonesia and Vietnam still plan to build nuclear power plants.[4][5][6][7] Thirty-one countries operate nuclear power stations, and there are a considerable number of new reactors being built in China, South Korea, India, and Russia.[8] As of June 2011, countries such as Australia, Austria, Denmark, Greece, Ireland, Latvia, Lichtenstein, Luxembourg, Malta, Portugal, Israel, Malaysia, and Norway have no nuclear power stations and remain opposed to nuclear power.[9][10]

Since nuclear energy and nuclear weapons technologies are closely related, military aspirations can act as a factor in energy policy decisions. The fear of nuclear proliferation influences some international nuclear energy policies.

14.1 The global picture

See also: Nuclear power by country

After 1986's Chernobyl disaster, public fear of nuclear power led to a virtual halt in reactor construction, and several countries decided to phase out nuclear power altogether.[11] However, increasing energy demand was believed to require new sources of electric power, and rising

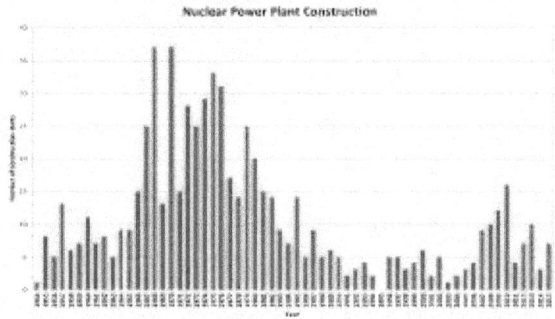

The number of nuclear power plant constructions started each year, from 1954 to 2013. Note the increase in new constructions from 2007 to 2010, before a decline following the 2011 Fukushima Dai-ichi nuclear disaster.

fossil fuel prices coupled with concerns about greenhouse gas emissions (see Climate change mitigation) have sparked heightened interest in nuclear power and predictions of a nuclear renaissance.

In 2004, the largest producer of nuclear energy was the United States with 28% of worldwide capacity, followed by France (18%) and Japan (12%).[12] In 2007, 31 countries operated nuclear power plants.[13] In September 2008 the IAEA projected nuclear power to remain at a 12.4% to 14.4% share of the world's electricity production through 2030.[14]

In 2013, almost two years after Fukushima, according to the IAEA there are 390 operating nuclear generating units throughout the world, more than 10% less than before Fukushima, and exactly the same as in Chernobyl-year 1986.[15] Asia is expected to be the primary growth market for nuclear energy in the foreseeable future, despite continued uncertainty in the energy outlooks for Japan, South Korea, and others in the region. As of 2014, 63% of all reactors under construction globally are in Asia.[16]

14.2 Policy issues

See also: Energy policy

14.2.1 Nuclear concerns

Main article: Nuclear power debate

Nuclear accidents and radioactive waste disposal are major concerns.[17] Other concerns include nuclear proliferation, the high cost of nuclear power plants, and nuclear terrorism.[17]

14.2.2 Energy security

For some countries, nuclear power affords energy independence. In the words of the French, "We have no coal, we have no oil, we have no gas, we have no choice."[18] Japan—similarly lacking in indigenous natural resources for power supply—relied on nuclear power for 1/3 of its energy mix prior to the Fukushima nuclear disaster; since March 2011, Japan has sought to offset the loss of nuclear power with increased reliance on imported liquefied natural gas, which has led to the country's first trade deficits in decades.[19] Therefore, the discussion of a future for nuclear energy is intertwined with a discussion of energy security and the use of energy mix, including renewable energy development.

Nuclear power has been relatively unaffected by embargoes, and uranium is mined in "reliable" countries, including Australia and Canada.[18][20]

14.2.3 Nuclear energy history and trends

Olkiluoto 3 under construction in 2009. It is the first EPR design, but problems with workmanship and supervision have created costly delays which led to an inquiry by the Finnish nuclear regulator STUK.[21] In December 2012, Areva estimated that the full cost of building the reactor will be about €8.5 billion, or almost three times the original delivery price of €3 billion.[22][23][24]

Proponents have long made inflated projections of the expected growth of nuclear power, but major accidents and high costs have kept growth much lower. In 1973 and 1974, the International Atomic Energy Agency predicted a worldwide installed nuclear capacity of 3,600 to 5,000 gigawatts by 2000. The IAEA's 1980 projection was for 740 to 1,075 gigawatts of installed capacity by the year 2000. Even after the 1986 Chernobyl disaster, the Nuclear Energy Agency forecasted an installed nuclear capacity of 497 to 646 gigawatts for the year 2000. The actual capacity in 2000 was 356 gigawatts. Moreover, construction costs have often been much higher, and times much longer than projected, failing to meet optimistic projections of "unlimited cheap, clean, and safe electricity."[25]

Since about 2001 the term nuclear renaissance has been

used to refer to a possible nuclear power industry revival, driven by rising fossil fuel prices and new concerns about meeting greenhouse gas emission limits.[3] However, nuclear electricity generation in 2012 was at its lowest level since 1999,[2] and new reactors under construction in Finland and France, which were meant to lead a nuclear renaissance,[26] have been delayed and are running over-budget.[26][27][28] China has 32 new reactors under construction,[29] and there are also a considerable number of new reactors being built in South Korea, India, and Russia. At the same time, at least 100 older and smaller reactors will "most probably be closed over the next 10-15 years".[8] So the expanding nuclear programs in Asia are balanced by retirements of aging plants and nuclear reactor phase-outs.[30]

In March 2011 the nuclear emergencies at Japan's Fukushima I Nuclear Power Plant and shutdowns at other nuclear facilities raised questions among some commentators over the future of the renaissance.[31][32][33][34][35] Platts has reported that "the crisis at Japan's Fukushima nuclear plants has prompted leading energy-consuming countries to review the safety of their existing reactors and cast doubt on the speed and scale of planned expansions around the world".[36] In 2011 Siemens exited the nuclear power sector following the Fukushima disaster and subsequent changes to German energy policy, and supported the German government's planned energy transition to renewable energy technologies.[37] China, Germany, Switzerland, Israel, Malaysia, Thailand, United Kingdom, Italy[38] and the Philippines have reviewed their nuclear power programs. Indonesia and Vietnam still plan to build nuclear power plants.[4][5][6][7] Countries such as Australia, Austria, Denmark, Greece, Ireland, Latvia, Liechtenstein, Luxembourg, Portugal, Israel, Malaysia, New Zealand, and Norway remain opposed to nuclear power. Following the Fukushima I nuclear accidents, the International Energy Agency halved its estimate of additional nuclear generating capacity built by 2035.[39]

The World Nuclear Association has reported that "nuclear power generation suffered its biggest ever one-year fall through 2012 as the bulk of the Japanese fleet remained offline for a full calendar year". Data from the International Atomic Energy Agency showed that nuclear power plants globally produced 2346 TWh of electricity in 2012 – seven per cent less than in 2011. The figures illustrate the effects of a full year of 48 Japanese power reactors producing no power during the year. The permanent closure of eight reactor units in Germany was also a factor. Problems at Crystal River, Fort Calhoun and the two San Onofre units in the USA meant they produced no power for the full year, while in Belgium Doel 3 and Tihange 2 were out of action for six months. Compared to 2010, the nuclear industry produced 11% less electricity in 2012.[2]

14.2.4   Reactions to Fukushima

Following the Fukushima nuclear disaster, Germany has permanently shut down eight of its reactors and pledged to close the rest by 2022.[40] The Italians have voted overwhelmingly to keep their country non-nuclear.[41] Switzerland and Spain have banned the construction of new reactors.[42] Japan's prime minister has called for a dramatic reduction in Japan's reliance on nuclear power.[43] Taiwan's president did the same. Mexico has sidelined construction of 10 reactors in favor of developing natural-gas-fired plants.[44] Belgium is considering phasing out its nuclear plants, perhaps as early as 2015.[42]

China—nuclear power's largest prospective market—suspended approvals of new reactor construction while conducting a lengthy nuclear-safety review.[35][45] Neighboring India, another potential nuclear boom market, has encountered effective local opposition, growing national wariness about foreign nuclear reactors, and a nuclear liability controversy that threatens to prevent new reactor imports. There have been mass protests against the French-backed 9900 MW Jaitapur Nuclear Power Project in Maharashtra and the 2000 MW Koodankulam Nuclear Power Plant in Tamil Nadu. The state government of West Bengal state has also refused permission to a proposed 6000 MW facility near the town of Haripur that intended to host six Russian reactors.[46]

There is little support across the world for building new nuclear reactors, a 2011 poll for the BBC indicates. The global research agency GlobeScan, commissioned by BBC News, polled 23,231 people in 23 countries from July to September 2011, several months after the Fukushima nuclear disaster. In countries with existing nuclear programmes, people are significantly more opposed than they were in 2005, with only the UK and US bucking the trend. Most believe that boosting energy efficiency and renewable energy can meet their needs.[47]

Just 22% agreed that "nuclear power is relatively safe and an important source of electricity, and we should build more nuclear power plants". In contrast, 71% thought their country "could almost entirely replace coal and nuclear energy within 20 years by becoming highly energy-efficient and focusing on generating energy from the Sun and wind". Globally, 39% want to continue using existing reactors without building new ones, while 30% would like to shut everything down now.[47]

14.3 Policies by territory

Main articles: Nuclear energy policy by country and Nuclear power by country
See also: List of nuclear reactors

Following the March 2011 Fukushima I nuclear accidents, China, Germany, Switzerland, Israel, Malaysia, Thailand, United Kingdom, and the Philippines are reviewing their nuclear power programs. Indonesia and Vietnam still plan to build nuclear power plants.*[4]*[5]*[6]*[7] Countries such as Australia, Austria, Denmark, Greece, Ireland, Luxembourg, Portugal, New Zealand, and Norway remain opposed to nuclear power.*[48]

14.4 See also

14.5 References

[1] IAEA (2011 Highlights). "Power Reactor Information System". Check date values in: ldate= (help)

[2] WNA (20 June 2013). "Nuclear power down in 2012". World Nuclear News.

[3] The Nuclear Renaissance (by the World Nuclear Association)

[4] Jo Chandler (March 19, 2011). "Is this the end of the nuclear revival?". The Sydney Morning Herald.

[5] Aubrey Belford (March 17, 2011). "Indonesia to Continue Plans for Nuclear Power". New York Times.

[6] Israel Prime Minister Netanyahu: Japan situation has "caused me to reconsider" nuclear power Piers Morgan on CNN, published 2011-03-17, accessed 2011-03-17

[7] Israeli PM cancels plan to build nuclear plant xinhuanet.com, published 2011-03-18, accessed 2011-03-17

[8] Michael Dittmar. Taking stock of nuclear renaissance that never was Sydney Morning Herald, August 18, 2010.

[9] "Nuclear power: When the steam clears". The Economist. March 24, 2011.

[10] Duroyan Fertl (June 5, 2011). "Germany: Nuclear power to be phased out by 2022". Green Left.

[11] Research and Markets: International Perspectives on Energy Policy and the Role of Nuclear Power Reuters, May 6, 2009.

[12] "Survey of energy resources" (PDF). World Energy Council. 2004. Retrieved 2007-07-13.

[13] Mycle Schneider, Steve Thomas, Antony Froggatt, Doug Koplow (August 2009). The World Nuclear Industry Status Report, German Federal Ministry of Environment, Nature Conservation and Reactor Safety, p. 6.

[14] "Energy, Electricity and Nuclear Power Estimates for the Period up to 2030" (PDF). International Atomic Energy Agency. September 2008. Retrieved 2008-09-08.

[15] Historic Move: IAEA Shifts 47 Japanese Reactors Into "Long-Term Shutdown" Category. World Nuclear Industry Status Report, 16-1-2013

[16] Multilateral Cooperation in Asia's Nuclear Sector, 2014 Pacific Energy Summit Working Paper, 8-6-14

[17] Brian Martin. Opposing nuclear power: past and present. Social Alternatives, Vol. 26, No. 2, Second Quarter 2007. pp. 43-47.

[18] "Nuclear renaissance faces realities". Platts. (subscription required). Retrieved 2007-07-13.

[19] http://www.nbr.org/research/activity.aspx?id=352 How Can Japan Compete in a Changing Global Market?, Clara Gillispie, The National Bureau of Asian Research, July 201

[20] L. Meeus; K. Purchala; R. Belmans. "Is it reliable to depend on import?" (PDF). Katholieke Universiteit Leuven. Departement of Electrical Engineering of the Faculty of Engineering. Retrieved 2007-07-13.

[21] "Olkiluoto pipe welding 'deficient', says regulator". World Nuclear News. 16 October 2009. Retrieved 8 June 2010.

[22] Kinnunen, Terhi (2010-07-01). "Finnish parliament agrees plans for two reactors". Reuters. Retrieved 2010-07-02.

[23] "Olkiluoto 3 delayed beyond 2014". World Nuclear News. 17 July 2012. Retrieved 24 July 2012.

[24] "Finland's Olkiluoto 3 nuclear plant delayed again". BBC. 16 July 2012. Retrieved 10 August 2012.

[25] Mycle Schneider and Antony Froggatt (September/October 2012 vol. 68 no. 5). "2011-2012 world nuclear industry status report". Bulletin of the Atomic Scientists. pp. 8–22. Check date values in: ldate= (help)

[26] James Kanter. Is the Nuclear Renaissance Fizzling? Green, 29 May 2009.

[27] James Kanter. In Finland, Nuclear Renaissance Runs Into Trouble New York Times, May 28, 2009.

[28] Rob Broomby. Nuclear dawn delayed in Finland BBC News, 8 July 2009.

[29] Nuclear Power in China

[30] Mark Diesendorf (2013). "Book review: Contesting the future of nuclear power" (PDF). Energy Policy.

[31] Nuclear Renaissance Threatened as Japan's Reactor Struggles Bloomberg, published March 2011, accessed 2011-03-14

[32] Analysis: Nuclear renaissance could fizzle after Japan quake Reuters, published 2011-03-14, accessed 2011-03-14

[33] Japan nuclear woes cast shadow over U.S. energy policy Reuters, published 2011-03-13, accessed 2011-03-14

[34] Nuclear winter? Quake casts new shadow on reactors MarketWatch, published 2011-03-14, accessed 2011-03-14

[35] Will China's nuclear nerves fuel a boom in green energy? Channel 4, published 2011-03-17, accessed 2011-03-17

[36] "NEWS ANALYSIS: Japan crisis puts global nuclear expansion in doubt". Platts. 21 March 2011.

[37] "Siemens to quit nuclear industry". *BBC News*. September 18, 2011.

[38] "Italy announces nuclear moratorium". World Nuclear News. 24 March 2011. Retrieved 23 May 2011.

[39] "Gauging the pressure". The Economist. 28 April 2011. Retrieved 3 May 2011.

[40] Annika Breidthardt (May 30, 2011). "German government wants nuclear exit by 2022 at latest". *Reuters*.

[41] "Italy Nuclear Referendum Results". June 13, 2011.

[42] Henry Sokolski (Nov 28, 2011). "Nuclear Power Goes Rogue". *Newsweek*.

[43] Tsuyoshi Inajima & Yuji Okada (Oct 28, 2011). "Nuclear Promotion Dropped in Japan Energy Policy After Fukushima". *Bloomberg*.

[44] Carlos Manuel Rodriguez (Nov 4, 2011). "Mexico Scraps Plans to Build 10 Nuclear Power Plants in Favor of Using Gas". *Bloomberg Businessweek*.

[45] the CNN Wire Staff. "China freezes nuclear plant approvals - CNN.com". Edition.cnn.com. Retrieved 2011-03-16.

[46] Siddharth Srivastava (27 October 2011). "India's Rising Nuclear Safety Concerns". *Asia Sentinel*.

[47] Richard Black (25 November 2011). "Nuclear power 'gets little public support worldwide'". *BBC News*.

[48] "Nuclear power: When the steam clears". *The Economist*. March 24, 2011.

14.6   Further reading

- Cooke, Stephanie (2009). *In Mortal Hands: A Cautionary History of the Nuclear Age*. Black Inc.

- Diesendorf, Mark (2007). *Greenhouse Solutions with Sustainable Energy*, University of New South Wales Press.

- Elliott, David (2007). *Nuclear or Not? Does Nuclear Power Have a Place in a Sustainable Energy Future?*, Palgrave.

- Falk, Jim (1982). *Global Fission: The Battle Over Nuclear Power*, Oxford University Press.

- Ferguson, Charles D., "Nuclear Energy: Balancing Benefits and Risks", Council on Foreign Relations, 2007

- Lovins, Amory B. (1977). *Soft Energy Paths: Towards a Durable Peace*, Friends of the Earth International, ISBN 0-06-090653-7

- Lovins, Amory B. and John H. Price (1975). *Non-Nuclear Futures: The Case for an Ethical Energy Strategy*, Ballinger Publishing Company, 1975, ISBN 0-88410-602-0

- Lowe, Ian (2007). *Reaction Time: Climate Change and the Nuclear Option*, Quarterly Essay.

- Pernick, Ron and Clint Wilder (2007). *The Clean Tech Revolution: The Next Big Growth and Investment Opportunity*, Collins, ISBN 978-0-06-089623-2

- Schneider, Mycle, Steve Thomas, Antony Froggatt, Doug Koplow (August 2009). *The World Nuclear Industry Status Report*, German Federal Ministry of Environment, Nature Conservation and Reactor Safety.

- Sovacool, Benjamin K. (2011). *Contesting the Future of Nuclear Power: A Critical Global Assessment of Atomic Energy*, World Scientific.

- Walker, J. Samuel (2004). *Three Mile Island: A Nuclear Crisis in Historical Perspective*, University of California Press.

14.7   External links

- NEI Public Policy Information

- Robert J. Duffy. *Nuclear Politics in America: A History and Theory of Government Regulation (Studies in Government and Public Policy)*. Paperback. 1997. ISBN 0-7006-0853-2.

- Carlton Stoiber, Alec Baer, Norbert Pelzer, Wolfram Tonhauser, *Handbook on Nuclear Law*, IAEA (International Atomic Energy Agency), 2003.

- Annotated bibliography for nuclear power from the Alsos Digital Library for Nuclear Issues

- Fairewinds Energy Education

- Schneider, Mycle, Steve Thomas, Antony Froggatt, Doug Koplow (2016). *The World Nuclear Industry Status Report: World Nuclear Industry Status as of 1 January 2016.*

Chapter 15

Nuclear power debate

For nuclear energy policies by nation, see Nuclear energy policy. For public protests about nuclear power, see Anti-nuclear movement. For corporate lobbying and nuclear supporters, see Pro-nuclear movement.

The **nuclear power debate** is a contro-

Stewart Brand at a 2010 debate, "Does the world need nuclear energy?"[1]

versy[2][3][4][5][6][7][8] about the deployment and use of nuclear fission reactors to generate electricity from nuclear fuel for civilian purposes. The debate about nuclear power peaked during the 1970s and 1980s, when it "reached an intensity unprecedented in the history of technology controversies", in some countries.[9][10] Observers attribute the nuclear controversy to the impossibility of generating a shared perception between social actors over the use of this technology[7] as well as systemic mismatches between expectations and experience.[8]

Proponents of nuclear energy argue that nuclear power is a sustainable energy source which reduces carbon emissions and can increase energy security if its use supplants a dependence on imported fuels.[11] Proponents advance the notion that nuclear power produces virtually no air pollution, in contrast to the chief viable alternative of fossil fuel. Proponents also believe that nuclear power is the only viable course to achieve energy independence for most Western countries. They emphasize that the risks of storing waste are small and can be further reduced by using the latest technology in newer reactors, and the operational safety record in the Western world is excellent when compared to the other major kinds of power plants.[12]

Opponents say that nuclear power poses numerous threats to people and the environment and point to studies in the literature that question if it will ever be a sustainable energy source.[13] These threats include health risks and environmental damage from uranium mining, processing and transport, the risk of nuclear weapons proliferation or sabotage, and the unsolved problem of radioactive nuclear waste.[14][15][16] They also contend that reactors themselves are enormously complex machines where many things can and do go wrong, and there have been many serious nuclear accidents.[17][18] Critics do not believe that these risks can be reduced through new technology.[19] They argue that when all the energy-intensive stages of the nuclear fuel chain are considered, from uranium mining to nuclear decommissioning, nuclear power is not a low-carbon electricity source.[20][21][22]

Three of the world's four largest economies now all generate more electricity from non-hydro renewable energy than from nuclear sources. New power generation using solar power was 33% of the global total added in 2015, wind power over 17%, and 1.3% for nuclear power, mostly due to development in China.[23]

144

15.1 Two opposing camps

Two opposing camps have evolved in society with respect to nuclear power, one supporting and promoting nuclear power and another opposing it. At the heart of this divide sit different views of risk and individual beliefs regarding public involvement in making decisions about large-scale high technology. Questions which emerge include: is nuclear power safe for humans and the environment? Could another Chernobyl disaster or Fukushima disaster happen? Can we dispose of nuclear waste in a safe manner? Can nuclear power help to reduce climate change and air pollution in a timely way?[24]

In the 2010 book *Why vs. Why: Nuclear Power*[25] Barry Brook and Ian Lowe discuss and articulate the debate about nuclear power. Brook makes the following seven arguments in favor of nuclear energy:[25]

- Renewable energy and energy efficiency may not solve the energy and climate crises

- Nuclear fuel is virtually unlimited and has extremely high specific energy

- New technology may be able to safely dispose of nuclear waste

- Nuclear power is claimed to be the safest energy option

- Advanced nuclear power may strengthen global security

- Nuclear power's true costs are claimed to be lower than either fossil fuels or renewables

- Nuclear power may lead the "clean energy" revolution

Lowe, in turn, makes the following arguments against nuclear power:[25]

- It may not be a fast enough response to climate change

- It is claimed to be too expensive

- The need for baseload electricity may be exaggerated

- The problem of waste may still remain unresolved

- It may increase the risk of nuclear war

- There are claimed to be major safety concerns

- There are claimed to be better alternatives

The Economist says that nuclear power "looks dangerous, unpopular, expensive and risky", and that "it is replaceable with relative ease and could be forgone with no huge

structural shifts in the way the world works". When asking what the world would be like without it *The Economist* notes that "(w)ithout nuclear power and with other fuels filling in its share pro rata, emissions from generation would have been about 11 billion tonnes. The difference is roughly equal to the total annual emissions of Germany and Japan combined."[26]

15.2 Electricity and energy supplied

The World Nuclear Association has reported that nuclear electricity generation in 2012 was at its lowest level since 1999. The WNA has said that "nuclear power generation suffered its biggest ever one-year fall through 2012 as the bulk of the Japanese fleet remained offline for a full calendar year".[27]

Data from the International Atomic Energy Agency showed that nuclear power plants globally produced 2346 TWh of electricity in 2012 – seven per cent less than in 2011. The figures illustrate the effects of a full year of 48 Japanese power reactors producing no power during the year. The permanent closure of eight reactor units in Germany was also a factor. Problems at Crystal River, Fort Calhoun and the two San Onofre units in the USA meant they produced no power for the full year, while in Belgium Doel 3 and Tihange 2 were out of action for six months. Compared to 2010, the nuclear industry produced 11% less electricity in 2012.[27]

Brazil, China, Germany, India, Japan, Mexico, the Netherlands, Spain and the U.K. now all generate more electricity from non-hydro renewable energy than from nuclear sources. In 2015, power generation using solar power was 33% of the global total, wind power over 17%, and 1.3% for nuclear power, exclusively due to development in China.[23]

Many studies have documented how nuclear power plants generate 16% of global electricity, but provide only 6.3% of energy production and 2.6% of final energy consumption. This mismatch stems mainly from the poor consumption efficiency of electricity compared to other energy carriers, and the transmission losses associated with nuclear plants which are usually situated far away from sources of demand.[28]

15.3 Energy security

See also: Energy security and Uranium mining

For some countries, nuclear power affords energy independence. Nuclear power has been relatively unaffected by embargoes, and uranium is mined in countries willing to export, including Australia and Canada.[29][30] However, countries now responsible for more than 30% of the world's uranium production: Kazakhstan, Namibia, Niger, and Uzbekistan, are politically unstable.[31]

One assessment from the IAEA showed that enough high-grade ore exists to supply the needs of the current reactor fleet for 40–50 years.[32] According to Sovacool (2011), reserves from existing uranium mines are being rapidly depleted, and expected shortfalls in available fuel threaten future plants and contribute to volatility of uranium prices at existing plants. Escalation of uranium fuel costs decreased the viability of nuclear projects.[32] Uranium prices rose from 2001 to 2007, before declining.[33]

The International Atomic Energy Agency and the Nuclear Energy Agency of the OCED, in their latest review of world uranium resources and demand, *Uranium 2014: Resources, Production, and Demand*, concluded that uranium resources would support "significant growth in nuclear capacity," and that: "Identified resources are sufficient for over 120 years, considering 2012 uranium requirements of 61 600 tU."[34]

According to a Stanford study, fast breeder reactors have the potential to provide power for humans on earth for billions of years, making this source sustainable.[35] But "because of the link between plutonium and nuclear weapons, the potential application of fast breeders has led to concerns that nuclear power expansion would bring in an era of uncontrolled weapons proliferation".[36]

15.4 Reliability

See also: Intermittent power sources, Energy security and renewable technology, and 100% renewable energy

In 2010, the worldwide average capacity factor was 80.1%.[37] In 2005, the global average capacity factor was 86.8%, the number of SCRAMs per 7,000 hours critical was 0.6, and the unplanned capacity loss factor was 1.6%.[38] Capacity factor is the net power produced divided by the maximum amount possible running at 100% all the time, thus this includes all scheduled maintenance/refueling outages as well as unplanned losses. The 7,000 hours is roughly representative of how long any given reactor will remain critical in a year, meaning that the scram rates translates into a sudden and unplanned shutdown about 0.6 times per year for any given reactor in the world. The unplanned capacity loss factor represents amount of power not produced due to unplanned scrams and postponed restarts.

The World Nuclear Association argues that: "Obviously sun, wind, tides and waves cannot be controlled to provide directly either continuous base-load power, or peak-load power when it is needed,...." "In practical terms non-hydro renewables are therefore able to supply up to some 15–20% of the capacity of an electricity grid, though they cannot directly be applied as economic substitutes for most coal or nuclear power, however significant they become in particular areas with favourable conditions." "If the fundamental opportunity of these renewables is their abundance and relatively widespread occurrence, the fundamental challenge, especially for electricity supply, is applying them to meet demand given their variable and diffuse nature. This means either that there must be reliable duplicate sources of electricity beyond the normal system reserve, or some means of electricity storage." "Relatively few places have scope for pumped storage dams close to where the power is needed, and overall efficiency is less than 80%. Means of storing large amounts of electricity as such in giant batteries or by other means have not been developed." [39]

According to Benjamin K. Sovacool, most studies critiquing solar and wind energy look only at individual generators and not at the system wide effects of solar and wind farms. Correlations between power swings drop substantially as more solar and wind farms are integrated (a process known as geographical smoothing) and a wider geographic area also enables a larger pool of energy efficiency efforts to abate intermittency.[40]

Sovacool says that variable renewable energy sources such as wind power and solar energy can displace nuclear resources.[40] "Nine recent studies have concluded that the variability and intermittency of wind and solar resources becomes easier to manage the more they are deployed and interconnected, not the other way around, as some utilities suggest. This is because wind and solar plants help grid operators handle major outages and contingencies elsewhere in the system, since they generate power in smaller increments that are less damaging than unexpected outages from large plants". [40]

According to a 2011 projection by the International Energy Agency, solar power generators may produce most of the world's electricity within 50 years, with wind power, hydroelectricity and biomass plants supplying much of the remaining generation. "Photovoltaic and concentrated solar power together can become the major source of electricity". [41] Renewable technologies can enhance energy security in electricity generation, heat supply, and transportation.[42]

As of 2013, the World Nuclear Association has said "There is unprecedented interest in renewable energy, particularly solar and wind energy, which provide electricity without giving rise to any carbon dioxide emission. Harnessing these for electricity depends on the cost and efficiency of the technology, which is constantly improving, thus reducing costs per peak kilowatt" .*[43]

Renewable electricity supply in the 20-50+% range has already been implemented in several European systems, albeit in the context of an integrated European grid system.*[44] In 2012 the share of electricity generated by renewable sources in Germany was 21.9%, compared to 16.0% for nuclear power after Germany shut down 7-8 of its 18 nuclear reactors in 2011.*[45] In the United Kingdom, the amount of energy produced from renewable energy is expected to exceed that from nuclear power by 2018,*[46] and Scotland plans to obtain all electricity from renewable energy by 2020.*[47] The majority of installed renewable energy across the world is in the form of hydro power, which has limited opportunity for expansion.*[48]

The IPCC has said that if governments were supportive, and the full complement of renewable energy technologies were deployed, renewable energy supply could account for almost 80% of the world's energy use within forty years.*[49] Rajendra Pachauri, chairman of the IPCC, said the necessary investment in renewables would cost only about 1% of global GDP annually. This approach could contain greenhouse gas levels to less than 450 parts per million, the safe level beyond which climate change becomes catastrophic and irreversible.*[49]

The cost of nuclear power has followed an increasing trend whereas the cost of electricity is declining in wind power.*[50] As of 2014, the wind industry in the USA is able to produce more power at lower cost by using taller wind turbines with longer blades, capturing the faster winds at higher elevations. This has opened up new opportunities and in Indiana, Michigan, and Ohio, the price of power from wind turbines built 300 feet to 400 feet above the ground can now compete with conventional fossil fuels like coal. Prices have fallen to about 4 cents per kilowatt-hour in some cases and utilities have been increasing the amount of wind energy in their portfolio, saying it is their cheapest option.*[51]

From a safety stand point, nuclear power, in terms of lives lost per unit of electricity delivered, is comparable to and in some cases, lower than many renewable energy sources.*[52]*[53] There is however no radioactive spent fuel that needs to be stored or reprocessed with conventional renewable energy sources.*[54] A nuclear plant needs to be disassembled and removed. Much of the disassembled nuclear plant needs to be stored as low level nuclear waste.*[55]

Since nuclear power plants are fundamentally heat engines, waste heat disposal becomes an issue at high ambient temperature. Droughts and extended periods of high temperature can "cripple nuclear power generation, and it is of-

ten during these times when electricity demand is highest because of air-conditioning and refrigeration loads and diminished hydroelectric capacity" .*[56] In such very hot weather a power reactor may have to operate at a reduced power level or even shut down.*[57] In 2009 in Germany, eight nuclear reactors had to be shut down simultaneously on hot summer days for reasons relating to the overheating of equipment or of rivers.*[56] Overheated discharge water has resulted in significant fish kills in the past, harming livelihood and raising public concern.*[58]

15.5 Economics

15.5.1 New nuclear plants

Main articles: Economics of new nuclear power plants and Nuclear power in the European Union
The economics of new nuclear power plants is a controver-

EDF has said its third-generation EPR Flamanville 3 project (seen here in 2010) will be delayed until 2018, due to "both structural and economic reasons," and the project's total cost has climbed to EUR 11 billion in 2012.[59] Similarly, the cost of the EPR being built at Olkiluoto, Finland has escalated dramatically, and the project is well behind schedule. The initial low cost forecasts for these megaprojects exhibited "optimism bias".*[60]*

sial subject, since there are diverging views on this topic, and multibillion-dollar investments ride on the choice of an energy source. Nuclear power plants typically have high capital costs for building the plant, but low direct fuel costs (with much of the costs of fuel extraction, processing, use and long term storage externalized). Therefore, comparison with other power generation methods is strongly dependent on assumptions about construction timescales and capital financing for nuclear plants. Cost estimates also need to take into account plant decommissioning and nuclear waste storage costs. On the other hand, measures to mitigate global warming, such as a carbon tax or carbon emissions trading, may favor the economics of nuclear power.

In recent years there has been a slowdown of electricity

demand growth and financing has become more difficult, which impairs large projects such as nuclear reactors, with very large upfront costs and long project cycles which carry a large variety of risks.[61] In Eastern Europe, a number of long-established projects are struggling to find finance, notably Belene in Bulgaria and the additional reactors at Cernavoda in Romania, and some potential backers have pulled out.[61] Where cheap gas is available and its future supply relatively secure, this also poses a major problem for nuclear projects.[61]

Analysis of the economics of nuclear power must take into account who bears the risks of future uncertainties. To date all operating nuclear power plants were developed by state-owned or regulated utility monopolies[62] where many of the risks associated with construction costs, operating performance, fuel price, and other factors were borne by consumers rather than suppliers. Many countries have now liberalized the electricity market where these risks, and the risk of cheaper competitors emerging before capital costs are recovered, are borne by plant suppliers and operators rather than consumers, which leads to a significantly different evaluation of the economics of new nuclear power plants.[63]

Following the 2011 Fukushima Daiichi nuclear disaster, costs are likely to go up for currently operating and new nuclear power plants, due to increased requirements for on-site spent fuel management and elevated design basis threats.[64]

15.5.2 Cost of decommissioning nuclear plants

Main article: nuclear decommissioning

The price of energy inputs and the environmental costs of every nuclear power plant continue long after the facility has finished generating its last useful electricity. Both nuclear reactors and uranium enrichment facilities must be decommissioned, returning the facility and its parts to a safe enough level to be entrusted for other uses. After a cooling-off period that may last as long as a century, reactors must be dismantled and cut into small pieces to be packed in containers for final disposal. The process is very expensive, time-consuming, dangerous for workers, hazardous to the natural environment, and presents new opportunities for human error, accidents or sabotage.[65]

The total energy required for decommissioning can be as much as 50% more than the energy needed for the original construction. In most cases, the decommissioning process costs between US $300 million to US$5.6 billion. Decommissioning at nuclear sites which have experienced a seri-

ous accident are the most expensive and time-consuming. In the U.S. there are 13 reactors that have permanently shut down and are in some phase of decommissioning, and none of them have completed the process.[65]

Current UK plants are expected to exceed £73bn in decommissioning costs. "Nuclear decommissioning costs exceed £73bn" .

15.5.3 Subsidies

George W. Bush signing the Energy Policy Act of 2005, which was designed to promote US nuclear reactor construction, through incentives and subsidies, including cost-overrun support up to a total of $2 billion for six new nuclear plants.[66]

U.S. 2014 Electricity Generation By Type

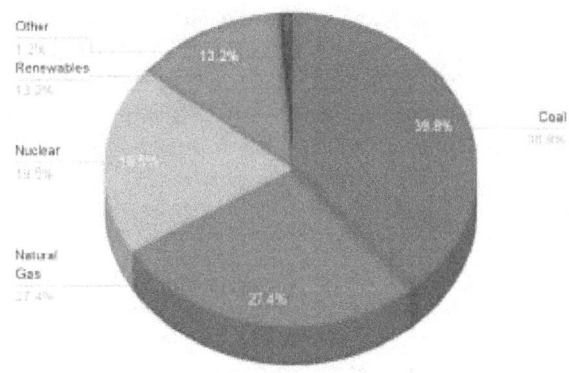

U.S. 2014 Electricity Generation By Type.[67]

Critics of nuclear power claim that it is the beneficiary of inappropriately large economic subsidies, taking the form of research and development, financing support for building new reactors and decommissioning old reactors and waste, and that these subsidies are often overlooked when comparing the economics of nuclear against other forms of power generation.[68][69] Nuclear power proponents argue that competing energy sources also receive subsidies. Fossil fuels receive large direct and indirect subsidies, such as tax benefits and not having to pay for the greenhouse gases they emit. Renewables receive proportionately large direct production subsidies and tax breaks in many nations, although in absolute terms they are often less than subsidies received by other sources.[70]

In Europe, the FP7 research program has more subsidies for nuclear than for renewable and energy efficiency together; over 70% of this is directed at the ITER fusion project.[71][72] In the US, public research money for nuclear fission declined from 2,179 to 35 million dollars between 1980 and 2000.[70]

A 2010 report by Global Subsidies Initiative compared relative subsidies of most common energy sources. It found that nuclear energy receives 1.7 US cents per kWh of energy it produces, compared to fossil fuels receiving 0.8 US cents per kWh, renewable energy receiving 5.0 US cents per kWh and biofuels receiving 5.1 US cents per kWh.[73]

Indirect nuclear insurance subsidy

Kristin Shrader-Frechette has said "if reactors were safe, nuclear industries would not demand government-guaranteed, accident-liability protection, as a condition for their generating electricity".[74] No private insurance company or even consortium of insurance companies "would shoulder the fearsome liabilities arising from severe nuclear accidents".[75]

The potential costs resulting from a nuclear accident (including one caused by a terrorist attack or a natural disaster) are great. The liability of owners of nuclear power plants in the U.S. is currently limited under the Price-Anderson Act (PAA). The Price-Anderson Act, introduced in 1957, was "an implicit admission that nuclear power provided risks that producers were unwilling to assume without federal backing".[76] The Price-Anderson Act "shields nuclear utilities, vendors and suppliers against liability claims in the event of a catastrophic accident by imposing an upper limit on private sector liability". Without such protection, private companies were unwilling to be involved. No other technology in the history of American industry has enjoyed such continuing blanket protection.[77]

The PAA was due to expire in 2002, and the former U.S. vice-president Dick Cheney said in 2001 that "nobody's going to invest in nuclear power plants" if the PAA is not renewed.[78] The U.S. Nuclear Regulatory Commission (USNRC) concluded that the liability limits placed on nuclear insurance were significant enough to constitute a subsidy, but a quantification of the amount was not attempted at that time.[79] Shortly after this in 1990, Dubin and Rothwell were the first to estimate the value to the U.S. nuclear industry of the limitation on liability for nuclear power plants under the Price Anderson Act. Their underlying method was to extrapolate the premiums operators currently pay versus the full liability they would have to pay for full insurance in the absence of the PAA limits. The size of the estimated subsidy per reactor per year was $60 million prior to the 1982 amendments, and up to $22 million following the 1988 amendments.[80] In a separate article in 2003, Anthony Heyes updates the 1988 estimate of $22 million per year to $33 million (2001 dollars).[81]

In case of a nuclear accident, should claims exceed this primary liability, the PAA requires all licensees to additionally provide a maximum of $95.8 million into the accident pool - totaling roughly $10 billion if all reactors were required to pay the maximum. This is still not sufficient in the case of a serious accident, as the cost of damages could exceed $10 billion.[82][83][84] According to the PAA, should the costs of accident damages exceed the $10 billion pool, the process for covering the remainder of the costs would be defined by Congress. In 1982, a Sandia National Laboratories study concluded that depending on the reactor size and 'unfavorable conditions' a serious nuclear accident could lead to property damages as high as $314 billion while fatalities could reach 50,000.[85] A recent study found that if only this one relatively ignored indirect subsidy for nuclear power was converted to a direct subsidy and diverted to photovoltaic manufacturing, it would result in more installed power and more energy produced by mid-century compared to the nuclear case.[86]

15.6   Environmental effects

Main article: Environmental effects of nuclear power
See also: Uranium mining debate and Lists of nuclear disasters and radioactive incidents

The primary environmental effects of nuclear power come from uranium mining, radioactive effluent emissions, and waste heat. Nuclear generation does not directly produce sulfur dioxide, nitrogen oxides, mercury or other pollutants associated with the combustion of fossil fuels.

Nuclear plants require slightly more cooling water than fossil-fuel power plants due to their slightly lower generation efficiencies. Uranium mining can use large amounts of water —for example, the Roxby Downs mine in South Australia uses 35 million litres of water each day and plans to increase this to 150 million litres per day.[87]

15.6.1   Effect on greenhouse gas emissions

Main article: Life-cycle greenhouse-gas emissions of energy sources

Carbon emissions from nuclear power
Sovacool life cycle study survey, 2008

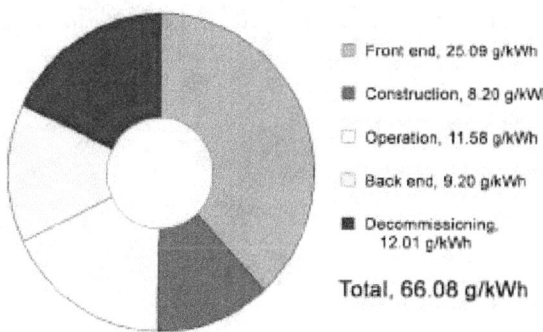

- Front end, 25.09 g/kWh
- Construction, 8.20 g/kWh
- Operation, 11.58 g/kWh
- Back end, 9.20 g/kWh
- Decommissioning, 12.01 g/kWh

Total, 66.08 g/kWh

Mean value of carbon dioxide emissions from qualified life cycle studies among 103 surveyed. Includes results of 1997 Vattenfall study.

According to Sovacool, nuclear power plants produce electricity with about 66 g equivalent lifecycle carbon dioxide emissions per kWh, while renewable power generators produce electricity with 9.5-38 g carbon dioxide per kWh.[88] A 2012 study by Yale University disputed this estimate, and found that the mean value from nuclear power ranged from 11- 25 g/kWh of total life cycle CO2 emissions[89]

While nuclear power does not directly emit greenhouse gases, emissions occur, as with every source of energy, over a facility's life cycle: mining and fabrication of construction materials, plant construction, operation, uranium mining and milling, and plant decommissioning. A literature survey by the Intergovernmental Panel on Climate Change of 32 greenhouse gas emissions studies, found a median value of 16 g equivalent lifecycle carbon dioxide emissions per kWh for nuclear power.[90]

Climate and energy scientists James Hansen, Ken Caldeira, Kerry Emanuel and Tom Wigley have released an open letter[91] stating, in part, that

> Renewables like wind and solar and biomass will certainly play roles in a future energy economy, but those energy sources cannot scale up fast enough to deliver cheap and reliable power at the scale the global economy requires. While it may be theoretically possible to stabilize the climate without nuclear power, in the real world there is no credible path to climate stabilization that does not include a substantial role for nuclear power.

In a published rebuttal to Hansen's analyses, eight energy and climate scholars say that "nuclear power reactors are less effective at displacing greenhouse gas emissions than

energy efficiency initiatives and renewable energy technologies". They go on to argue "that (a) its near-term potential is significantly limited compared to energy efficiency and renewable energy; (b) it displaces emissions and saves lives only at high cost and at the enhanced risk of nuclear weapons proliferation; (c) it is unsuitable for expanding access to modern energy services in developing countries; and (d) Hansen's estimates of cancer risks from exposure to radiation are flawed".[92] James Hansen's rebuttal can be found here.

Mark Diesendorf and B.K. Sovacool review the "little-known research which shows that the life-cycle CO2 emissions of nuclear power may become comparable with those of fossil power as the 5.4 million tonnes of high-grade uranium ore is used up over the next several decades and low-grade uranium is mined and milled using fossil fuels".[93][94] Critics calculate that if nuclear energy were used to rapidly replace existing energy sources, there would be an energy cannibalism effect, which would affect the carbon neutral growth rate of the technology.[95]

As the nuclear power debate continues, greenhouse gas emissions are not decreasing, they are increasing. Predictions estimate that even with draconian emission reductions within the ten years, the world will still pass 650ppm of carbon dioxide and a catastrophic 4C average rise in temperature.[96] Public perception is that renewable energies such as wind, solar, biomass and geothermal are significantly affecting global warming.[97] All of these sources combined only supplied 1.3% of global energy in 2013 as 8 billion tonnes of coal was burned annually.[98] This too little, too late may be a mass form of climate change denial, or an idealistic pursuit of green energy.

15.6.2 High-level radioactive waste

Main article: High-level radioactive waste management
The world's nuclear fleet creates about 10,000 metric tons of high-level spent nuclear fuel each year.[99] High-level radioactive waste management concerns management and disposal of highly radioactive materials created during production of nuclear power. The technical issues in accomplishing this are daunting, due to the extremely long periods radioactive wastes remain deadly to living organisms. Of particular concern are two long-lived fission products, technetium-99 (half-life 220,000 years) and iodine-129 (half-life 15.7 million years),[100] which dominate spent nuclear fuel radioactivity after a few thousand years. The most troublesome transuranic elements in spent fuel are neptunium-237 (half-life two million years) and plutonium-239 (half-life 24,000 years).[101] Consequently, high-level radioactive waste requires sophisticated treatment and management to successfully isolate it from the biosphere.

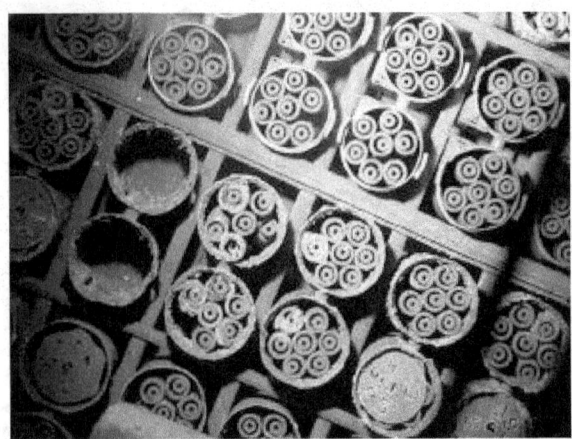

Spent nuclear fuel stored underwater and uncapped at the Hanford site in Washington, USA.

This usually necessitates treatment, followed by a long-term management strategy involving permanent storage, disposal or transformation of the waste into a non-toxic form.[102]

Governments around the world are considering a range of waste management and disposal options, usually involving deep-geologic placement, although there has been limited progress toward implementing long-term waste management solutions.[103] This is partly because the timeframes in question when dealing with radioactive waste range from 10,000 to millions of years,[104][105] according to studies based on the effect of estimated radiation doses.[106]

Since the fraction of a radioisotope's atoms decaying per unit of time is inversely proportional to its half-life, the relative radioactivity of a quantity of buried human radioactive waste would diminish over time compared to natural radioisotopes (such as the decay chain of 120 trillion tons of thorium and 40 trillion tons of uranium which are at relatively trace concentrations of parts per million each over the crust's 3×10^{19} ton mass).[107][108][109] For instance, over a timeframe of thousands of years, after the most active short half-life radioisotopes decayed, burying U.S. nuclear waste would increase the radioactivity in the top 2000 feet of rock and soil in the United States (10 million km^2) by ≈ 1 part in 10 million over the cumulative amount of natural radioisotopes in such a volume, although the vicinity of the site would have a far higher concentration of artificial radioisotopes underground than such an average.[110]

Nuclear waste disposal is one of the most controversial facets of the nuclear power debate. Presently, waste is mainly stored at individual reactor sites and there are over 430 locations around the world where radioactive material continues to accumulate. Experts agree that centralized underground repositories which are well-managed, guarded, and monitored, would be a vast improvement.[111] There

is an international consensus on the advisability of storing nuclear waste in deep underground repositories.[112] but no country in the world has yet opened such a site.[112][113][114][115] There are dedicated waste storage sites at the Waste Isolation Pilot Plant in New Mexico and two in German salt mines, the Morsleben Repository and the Schacht Asse II.

15.6.3 Prevented mortality

In March 2013, climate scientists Pushker Kharecha and James Hansen published a paper in Environmental Science & Technology, entitled *Prevented mortality and greenhouse gas emissions from historical and projected nuclear power.*[116] It estimated an average of 1.8 million lives saved worldwide by the use of nuclear power instead of fossil fuels between 1971 and 2009. The paper examined mortality levels per unit of electrical energy produced from fossil fuels (coal and natural gas) as well as nuclear power. Kharecha and Hansen assert that their results are probably conservative, as they analyze only deaths and do not include a range of serious but non-fatal respiratory illnesses, cancers, hereditary effects and heart problems, nor do they include the fact that fossil fuel combustion in developing countries tends to have a higher carbon and air pollution footprint than in developed countries.[117] The authors also conclude that the emission of some 64 billion tonnes of carbon dioxide equivalent have been avoided by nuclear power between 1971 and 2009, and that between 2010 and 2050, nuclear could additionally avoid up to 80 to 240 billion tonnes.

15.7 Accidents and safety

See also: Nuclear safety, Nuclear and radiation accidents, and Lists of nuclear disasters and radioactive incidents

Benjamin K. Sovacool has reported that worldwide there have been 99 accidents at nuclear power plants.[118] Fifty-seven accidents have occurred since the Chernobyl disaster, and 57% (56 out of 99) of all nuclear-related accidents have occurred in the USA.[118] Serious nuclear power plant accidents include the Fukushima Daiichi nuclear disaster (2011), Chernobyl disaster (1986), Three Mile Island accident (1979), and the SL-1 accident (1961).[119] Nuclear-powered submarine mishaps include the K-19 reactor accident (1961),[120] the K-27 reactor accident (1968),[121] and the K-431 reactor accident (1985).[119]

The effect of nuclear accidents has been a topic of debate practically since the first nuclear reactors were constructed.

It has also been a key factor in public concern about nuclear facilities.[122] Some technical measures to reduce the risk of accidents or to minimize the amount of radioactivity released to the environment have been adopted. Despite the use of such measures, "there have been many accidents with varying effects as well near misses and incidents" .[122]

Nuclear power plants are a complex energy system[123][124] and opponents of nuclear power have criticized the sophistication and complexity of the technology. Helen Caldicott has said: "... in essence, a nuclear reactor is just a very sophisticated and dangerous way to boil water -- analogous to cutting a pound of butter with a chain saw." [125] The 1979 Three Mile Island accident inspired Charles Perrow's book *Normal Accidents*, where a nuclear accident occurs, resulting from an unanticipated interaction of multiple failures in a complex system. TMI was an example of a normal accident because it was "unexpected, incomprehensible, uncontrollable and unavoidable" .[126]

> Perrow concluded that the failure at Three Mile Island was a consequence of the system's immense complexity. Such modern high-risk systems, he realized, were prone to failures however well they were managed. It was inevitable that they would eventually suffer what he termed a 'normal accident'. Therefore, he suggested, we might do better to contemplate a radical redesign, or if that was not possible, to abandon such technology entirely.[127]

15.7.1   Chernobyl explosion

Main article: Chernobyl explosion

The Chernobyl explosion was a nuclear accident that occurred on 26 April 1986 at the Chernobyl Nuclear Power Plant in Ukraine. An explosion and fire released large quantities of radioactive contamination into the atmosphere, which spread over much of Western USSR and Europe. It is considered the worst nuclear power plant accident in history, and is one of only two classified as a level 7 event on the International Nuclear Event Scale (the other being the Fukushima Daiichi nuclear disaster).[128] The battle to contain the contamination and avert a greater catastrophe ultimately involved over 500,000 workers and cost an estimated 18 billion rubles, crippling the Soviet economy.[129] The accident raised concerns about the safety of the nuclear power industry, slowing its expansion for a number of years.[130]

UNSCEAR has conducted 20 years of detailed scientific and epidemiological research on the effects of the Chernobyl accident. Apart from the 57 direct deaths in the

Map showing caesium-137 contamination in Belarus, Russia, and Ukraine as of 1996.

accident itself, UNSCEAR predicted in 2005 that up to 4,000 additional cancer deaths related to the accident would appear "among the 600 000 persons receiving more significant exposures (liquidators working in 1986–87, evacuees, and residents of the most contaminated areas)".[131] Russia, Ukraine, and Belarus have been burdened with the continuing and substantial decontamination and health care costs of the Chernobyl disaster.[132]

15.7.2   Fukushima disaster

The 2011 Fukushima Daiichi nuclear disaster, the worst nuclear incident in 25 years, displaced 50,000 households after radioactive material leaked into the air, soil and sea.[133] Whereas the radiation level never was an immediate life hazard outside the plant, the displacement was the direct cause of over 1500 deaths[134] [135] Radiation checks led to bans on some shipments of vegetables and fish.[136]

Main article: Fukushima nuclear disaster

Following an earthquake, tsunami, and failure of cooling systems at Fukushima I Nuclear Power Plant and issues concerning other nuclear facilities in Japan on March 11, 2011, a nuclear emergency was declared. This was the first time a nuclear emergency had been declared in Japan, and 140,000 residents within 20 km (12 mi) of the plant were evacuated.[137] Explosions and a fire resulted in dangerous levels of radiation, sparking a stock market collapse and panic-buying in supermarkets.[138] The UK, France and some other countries advised their nationals to consider leaving Tokyo, in response to fears of spreading nuclear contamination. The accidents drew attention to ongoing concerns over Japanese nuclear seismic design standards and caused other governments to re-evaluate their nuclear programs. John Price, a former member of the Safety Policy Unit at the UK's National Nuclear Corporation, said that it "might be 100 years before melting fuel rods can be safely removed from Japan's Fukushima nuclear plant".[139]

15.7.3 Three Mile Island accident

Main article: Three Mile Island accident
The Three Mile Island accident was a core meltdown

President Jimmy Carter leaving Three Mile Island for Middletown, Pennsylvania, April 1, 1979.

in Unit 2 (a pressurized water reactor manufactured by Babcock & Wilcox) of the Three Mile Island Nuclear Generating Station in Dauphin County, Pennsylvania near Harrisburg, United States in 1979. It was the most significant accident in the history of the USA commercial nuclear power generating industry, resulting in the release of approximately 2.5 million curies of radioactive noble gases, and approximately 15 curies of iodine-131.[140] Cleanup started in August 1979 and officially ended in December 1993, with a total cleanup cost of about $1 billion.[141] The incident was rated a five on the seven-point International Nuclear Event Scale: Accident With Wider

Consequences.[142][143]

The health effects of the Three Mile Island nuclear accident are widely, but not universally, agreed to be very low level. However, there was an evacuation of 140,000 pregnant women and pre-school age children from the area.[144][145][146] The accident crystallized anti-nuclear safety concerns among activists and the general public, resulted in new regulations for the nuclear industry, and has been cited as a contributor to the decline of new reactor construction that was already underway in the 1970s.[147]

15.7.4 New reactor designs

The nuclear power industry has improved the safety and performance of reactors, and has proposed new safer (but generally untested) reactor designs but there is no guarantee that the reactors will be designed, built and operated correctly.[148] Mistakes do occur and the designers of reactors at Fukushima in Japan did not anticipate that a tsunami generated by an earthquake would disable the backup systems that were supposed to stabilize the reactor after the earthquake.[149] According to UBS AG, the Fukushima I nuclear accidents have cast doubt on whether even an advanced economy like Japan can master nuclear safety.[150] Catastrophic scenarios involving terrorist attacks are also conceivable.[148] An interdisciplinary team from MIT have estimated that given a three-fold increase in nuclear power from 2005 to 2055, and an unchanged accident frequency, four core damage accidents would be expected in that period [151]

Proponents of nuclear power argue that in comparison to any other form of power, nuclear power is the safest form of energy, accounting for all the risks from mining to production to storage, including the risks of spectacular nuclear accidents. Accidents in the nuclear industry have been less damaging than accidents in the hydro industry, and less damaging than the constant, incessant damage from air pollutants from fossil fuels. Coal plants release more radioactivity into the environment than nuclear plants, through the release of thorium and uranium in coal ash.[152] The World Nuclear Association provides a comparison of deaths from accidents in course of different forms of energy production. In their comparison, deaths per TW-yr of electricity produced from 1970 to 1992 are quoted as 885 for hydropower, 342 for coal, 85 for natural gas, and 8 for nuclear.[153] Nuclear power plant accidents rank first in terms of their economic cost, accounting for 41 percent of all property damage attributed to energy accidents.[18]

The Union of Concerned Scientists supports a national renewable energy standard which would require utilities to produce a certain percentage of their energy from sources such as wind power, solar energy and geothermal energy. The group also supports a national energy efficiency standard for home appliances.[154] The UCS also acknowledges that nuclear power can reduce greenhouse gas emissions, but maintains that it must become much safer and cheaper before it can be considered a workable solution to global warming. They support increased safety enforcement from the Nuclear Regulatory Commission among other steps to improve nuclear power.[155] UCS has been critical of proposed Generation III reactor designs. Edwin Lyman, a senior staff scientist at UCS, has challenged specific cost-saving design choices made for both the AP1000 and ESBWR. Lyman is concerned about the strength of the steel containment vessel and the concrete shield building around the AP1000. The AP1000 containment vessel does not have sufficient safety margins, says Lyman.[156]

15.8 Whistleblowers

See also: Nuclear whistleblowers and List of nuclear whistleblowers

This is a list of nuclear whistleblowers. They are mainly former employees of nuclear power facilities who have spoken out about safety concerns.

15.8.1 Health effects on population near nuclear power plants and workers

See also: Environmental impact of nuclear power § Risk of cancer

A major concern in the nuclear debate is what the long-term effects of living near or working in a nuclear power station are. These concerns typically center around the potential for increased risks of cancer. However, studies conducted by non-profit, neutral agencies have found no compelling evidence of correlation between nuclear power and risk of cancer.[169]

There has been considerable research done on the effect of low-level radiation on humans. Debate on the applicability of Linear no-threshold model versus Radiation hormesis and other competing models continues, however, the predicted low rate of cancer with low dose means that large sample sizes are required in order to make meaningful conclusions. A study conducted by the National Academy of Science found that carcinogenic effects of radiation does increase with dose.[170] The largest study on nuclear industry workers in history involved nearly a half-million individuals and concluded that a 1–2% of cancer deaths were

Fishermen near the now-dismantled Trojan Nuclear Power Plant in Oregon. The reactor dome is visible on the left, and the cooling tower on the right.

likely due to occupational dose. This was on the high range of what theory predicted by LNT, but was "statistically compatible".[171]

The Nuclear Regulatory Commission (NRC) has a factsheet that outlines 6 different studies. In 1990 the United States Congress requested the National Cancer Institute to conduct a study of cancer mortality rates around nuclear plants and other facilities covering 1950 to 1984 focusing on the change after operation started of the respective facilities. They concluded in no link. In 2000 the University of Pittsburgh found no link to heightened cancer deaths in people living within 5 miles of plant at the time of the Three Mile Island accident. The same year, the Illinois Public Health Department found no statistical abnormality of childhood cancers in counties with nuclear plants. In 2001 the Connecticut Academy of Science and Engineering confirmed that radiation emissions were negligibly low at the Connecticut Yankee Nuclear Power Plant. Also that year, the American Cancer Society investigated cancer clusters around nuclear plants and concluded no link to radiation noting that cancer clusters occur regularly due to unrelated reasons. Again in 2001, the Florida Bureau of Environmental Epidemiology reviewed claims of increased cancer rates in counties with nuclear plants, however, using the same data as the claimants, they observed no abnormalities.[172]

Scientists learned about exposure to high level radiation from studies of the effects of bombing populations at Hiroshima and Nagasaki. However, it is difficult to trace the relationship of low level radiation exposure to resulting cancers and mutations. This is because the latency period between exposure and effect can be 25 years or more for cancer and a generation or more for genetic damage. Since nuclear generating plants have a brief history, it is early to judge the effects.[173]

Most human exposure to radiation comes from natural background radiation. Natural sources of radiation amount to an average annual radiation dose of 295 mrem. The average person receives about 53 mrem from medical procedures and 10 mrem from consumer products.[174] According to the National Safety Council, people living within 50 miles of a nuclear power plant receive an additional 0.01 mrem per year. Living within 50 miles of a coal plant adds 0.03 mrem per year.[175]

In its 2000 report, "Sources and effects of ionizing radiation",[176] the UNSCEAR also gives some values for areas where the radiation background is very high.[177] You can for example have some value like 370 nGy/h on average in Yangjiang (meaning 3.24 mSv per year or 324 mrem), or 1,800 gGy/h in Kerala (meaning 15.8 mSv per year or 1580 mrem). They are also some other "hot spots", with some maximum values of 17,000 nGy/h in the hot springs of Ramsar (that would be equivalent to 149 mSv per year pr 14,900 mrem per year). The highest background seem to be in Guarapari with a reported 175 mSv per year (or 17,500 mrem per year), and 90,000 nGy/h maximum value given in the UNSCEAR report (on the beaches).[177] A study made on the Kerala radiation background, using a cohort of 385,103 residents, concludes that "showed no excess cancer risk from exposure to terrestrial gamma radiation" and that "Although the statistical power of the study might not be adequate due to the low dose, our cancer incidence study [...]suggests it is unlikely that estimates of risk at low doses are substantially greater than currently believed." [178]

Current guidelines established by the NRC, require extensive emergency planning, between nuclear power plants, Federal Emergency Management Agency (FEMA), and the local governments. Plans call for different zones, defined by distance from the plant and prevailing weather conditions and protective actions. In the reference cited, the plans detail different categories of emergencies and the protective actions including possible evacuation.[179]

A German study on childhood cancer in the vicinity of nuclear power plants, the KiKK study[180] was published in December 2007. According to Ian Fairlie, it "resulted in a public outcry and media debate in Germany which has received little attention elsewhere". It has been established "partly as a result of an earlier study by Körblein and Hoffmann[181] which had found statistically significant increases in solid cancers (54%), and in leukemia (76%) in children aged less than 5 within 5 km of 15 German nuclear power plant sites. It red a 2.2-fold increase in leukemias and a 1.6-fold increase in solid (mainly embryonal) cancers among children living within 5 km of all German nuclear power stations." [182] In 2011 a new study of the KiKK data was incorporated into an assessment by the Committee on Medical Aspects of Radiation in the Environment (COMARE) of the incidence of childhood leukemia around British nuclear power plants. It found that the control sample of population used for comparison in the German study may have been incorrectly selected and other possible contributory factors, such as socio-economic ranking, were not taken into consideration. The committee concluded that there is no significant evidence of an association between risk of childhood leukemia (in under 5 year olds) and living in proximity to a nuclear power plant.[183]

15.8.2   Safety culture in host nations

Some developing countries which plan to go nuclear have very poor industrial safety records and problems with political corruption.[184] Inside China, and outside the country, the speed of the nuclear construction program has raised safety concerns. Prof He Zuoxiu, who was involved with China's atomic bomb program, has said that plans to expand production of nuclear energy twentyfold by 2030 could be disastrous, as China was seriously underprepared on the safety front. China's fast-expanding nuclear sector is opting for cheap technology that "will be 100 years old by the time dozens of its reactors reach the end of their lifespans", according to diplomatic cables from the US embassy in Beijing.[185] The rush to build new nuclear power plants may "create problems for effective management, operation and regulatory oversight" with the biggest potential bottleneck being human resources – "coming up with enough trained personnel to build and operate all of these new plants, as well as regulate the industry".[185] The challenge for the government and nuclear companies is to "keep an eye on a growing army of contractors and subcontractors who may be tempted to cut corners".[186] China is advised to maintain nuclear safeguards in a business culture where quality and safety are sometimes sacrificed in favor of cost-cutting, profits, and corruption. China has asked for international assistance in training more nuclear power plant inspectors.[186]

15.9 Nuclear proliferation and terrorism concerns

See also: Nuclear proliferation and List of crimes involving radioactive substances

According to Mark Z. Jacobson, the growth of nuclear power has "historically increased the ability of nations to obtain or enrich uranium for nuclear weapons, and a large-scale worldwide increase in nuclear energy facilities would exacerbate this problem, putting the world at greater risk of a nuclear war or terrorism catastrophe". [148] The historic link between energy facilities and weapons is evidenced by the secret development or attempted development of weapons capabilities in nuclear power facilities in Pakistan, India, Iraq (prior to 1981), Iran, and to some extent in North Korea. [148]

Four AP1000 reactors, which were designed by the American Westinghouse Electric Company are currently, as of 2011, being built in China [187] and a further two AP1000 reactors are to be built in the USA. [188] Hyperion Power Generation, which is designing modular reactor assemblies that are proliferation resistant, is a privately owned US corporation, as is Terrapower which has the financial backing of Bill Gates. [189]

15.9.1 Vulnerability of plants to attack

See also: Vulnerability of nuclear plants to attack

Nuclear reactors become preferred targets during military conflict and, over the past three decades, have been repeatedly attacked during military air strikes, occupations, invasions and campaigns: [190]

- In September 1980, Iran bombed the Al Tuwaitha nuclear complex in Iraq.

- In June 1981, an Israeli air strike completely destroyed Iraq' s Osirak nuclear research facility.

- Between 1984 and 1987, Iraq bombed Iran' s Bushehr nuclear plant six times.

- In Iraq in 1991, the U.S. bombed three nuclear reactors and an enrichment pilot facility.

- In 1991, Iraq launched Scud missiles at Israel' s Dimona nuclear power plant.

- In September 2003, Israel bombed a Syrian reactor under construction. [190]

According to a 2004 report by the U.S. Congressional Budget Office, "The human, environmental, and economic costs from a successful attack on a nuclear power plant that results in the release of substantial quantities of radioactive material to the environment could be great." [191] The United States 9/11 Commission has said that nuclear power plants were potential targets originally considered for the September 11, 2001 attacks. If terrorist groups could sufficiently damage safety systems to cause a core meltdown at a nuclear power plant, and/or sufficiently damage spent fuel pools, such an attack could lead to a widespread radioactive contamination. [192]

If nuclear power use is to expand significantly, nuclear facilities will have to be made extremely safe from attacks that could release massive quantities of radioactivity into the environment and community. New reactor designs have features of passive safety, such as the flooding of the reactor core without active intervention by reactor operators. But these safety measures have generally been developed and studied with respect to accidents, not to the deliberate reactor attack by a terrorist group. However, the US Nuclear Regulatory Commission does now also require new reactor license applications to consider security during the design stage. [192]

15.9.2 Use of waste byproduct as a weapon

An additional concern with nuclear power plants is that if the by-products of nuclear fission (the nuclear waste generated by the plant) were to be left unprotected it could be stolen and used as a radiological weapon, colloquially known as a "dirty bomb". There were incidents in post-Soviet Russia of nuclear plant workers attempting to sell nuclear materials for this purpose (for example, there was such an incident in Russia in 1999 where plant workers attempted to sell 5 grams of radioactive material on the open market, [193] and an incident in 1993 where Russian workers were caught attempting to sell 4.5 kilograms of enriched uranium. [194] [195] [196]), and there are additional concerns that the transportation of nuclear waste along roadways or railways opens it up for potential theft. The United Nations has since called upon world leaders to improve security in order to prevent radioactive material falling into the hands of terrorists, [197] and such fears have been used as justifications for centralized, permanent, and secure waste repositories and increased security along transportation routes. [198]

Proponents state that the spent fissile fuel is not radioactive enough to create any sort of effective nuclear weapon, in a traditional sense where the radioactive material is the means of explosion.

15.10 Public opinion

Main article: Public opinion on nuclear issues
There is little support across the world for building new nu-

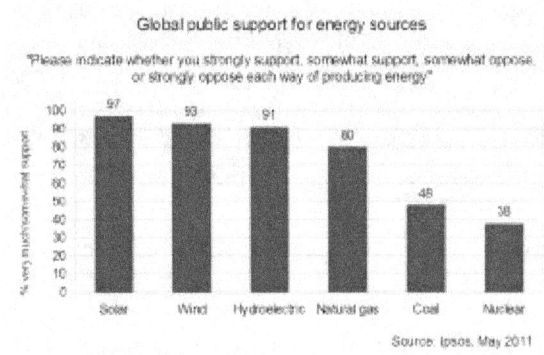

Global public support for energy sources, based on a survey by Ipsos (2011).[199]

clear reactors, a 2011 poll for the BBC indicates. The global research agency GlobeScan, commissioned by BBC News, polled 23,231 people in 23 countries from July to September 2011, several months after the Fukushima nuclear disaster. In countries with existing nuclear programmes, people are significantly more opposed than they were in 2005, with only the UK and US bucking the trend. Most believe that boosting energy efficiency and renewable energy can meet their needs.[200]

> Just 22% agreed that "nuclear power is relatively safe and an important source of electricity, and we should build more nuclear power plants". In contrast, 71% thought their country "could almost entirely replace coal and nuclear energy within 20 years by becoming highly energy-efficient and focusing on generating energy from the Sun and wind". Globally, 39% want to continue using existing reactors without building new ones, while 30% would like to shut everything down now.[200]

In 2011, Deutsche Bank analysts concluded that "the global impact of the Fukushima accident is a fundamental shift in public perception with regard to how a nation prioritizes and values its populations health, safety, security, and natural environment when determining its current and future energy pathways". As a consequence, "renewable energy will be a clear long-term winner in most energy systems, a conclusion supported by many voter surveys conducted over the past few weeks. At the same time, we consider natural gas to be, at the very least, an important transition fuel, especially in those regions where it is considered secure".[201]

15.10.1 European Union

A poll in the European Union for February–March 2005 showed 37% were in favour of nuclear energy and 55% opposed, leaving 8% undecided.[202] The same agency ran another poll in Oct-Nov 2006 that showed 14% favoured building new nuclear plants, 34% favoured maintaining the same number, and 39% favoured reducing the number of operating plants, leaving 13% undecided. This poll showed that respondents with a lower level of education and that women were less likely to approve.[203]

15.10.2 Japan

In June 2011, both Ipsos Mori and the Japanese Asahi Shimbun newspaper found drops in support for nuclear power technology in most countries, with support continuing in a number including the US. The Ipsos Mori poll found that nuclear had the lowest support of any established technology for generating electricity, with 38%. Coal was at 48% support while solar energy, wind power and hydro all found favour with more than 90% of those surveyed.[200]

15.10.3 Sweden

A 2011 poll found that skepticism over nuclear power had grown in Sweden following Japan's nuclear crisis. 36 percent of respondents wanted to phase-out nuclear power, up from 15 percent two years previous. An equal percentage of 36 percent were in favor of keeping nuclear power at its present level, and another 21 percent favored increasing nuclear power, with 7% undecided.[204]

15.10.4 United States

What had been growing acceptance of nuclear power in the United States was eroded sharply following the 2011 Japanese nuclear accidents, with support for building nuclear power plants in the U.S. dropping slightly lower than it was immediately after the Three Mile Island accident in 1979, according to a CBS News poll. Only 43 percent of those polled after the Fukushima nuclear emergency said they would approve building new power plants in the United States.[205]

A Gallup poll in the US in March 2015 found support for nuclear power at 51%, with 43% opposed. This was the lowest level of support for nuclear since 2001, and significantly down from the 2010 peak of 62% in favor, versus 33% opposed.[206] Similarly, a Roper poll in 2013 found support for new nuclear power plants at 55%, with 41% opposed, down from the peak level of support in 2010 of 70%

in favor versus 27% opposed.*[207]

The two energy sources that attracted the highest levels of support in the *2007 MIT Energy Survey* were solar power and wind power. Outright majorities would choose to "increase a lot" use of these two fuels, and better than three out of four Americans would like to increase these fuels in the U. S. energy portfolio. Fourteen per cent of respondents would like to see nuclear power "increase a lot". .*[208]

15.11   Trends and future prospects

See also: Nuclear renaissance, Pro-nuclear movement, and Anti-nuclear movement

As of May 15, 2011, a total of 438 nuclear reactors were operating in 30 countries, six fewer than the historical maximum of 444 in 2002. Since 2002, utilities have started up 26 units and disconnected 32 including six units at the Fukushima Daiichi nuclear power plant in Japan. The current world reactor fleet has a total nominal capacity of about 372 gigawatts (or thousand megawatts). Despite six fewer units operating in 2011 than in 2002, the capacity is still about 9 gigawatts higher.*[209] The numbers of new operative reactors, final shutdowns and new initiated constructions according to International Atomic Energy Agency (IAEA) in recent years are as follows: *[210]

Stephanie Cooke has argued that the cost of building new reactors is extremely high, as are the risks involved. Most utilities have said that they won't build new plants without government loan guarantees. There are also bottlenecks at factories that produce reactor pressure vessels and other equipment, and there is a shortage of qualified personnel to build and operate the reactors,*[211] although the recent acceleration in nuclear power plant construction is drawing a substantial expansion of the heavy engineering capability.*[212]

Following the Fukushima Daiichi nuclear disaster, the International Energy Agency halved its estimate of additional nuclear generating capacity to be built by 2035.*[213] Platts has reported that "the crisis at Japan's Fukushima nuclear plants has prompted leading energy-consuming countries to review the safety of their existing reactors and cast doubt on the speed and scale of planned expansions around the world".*[214] In 2011, *The Economist* reported that nuclear power "looks dangerous, unpopular, expensive and risky", and that "it is replaceable with relative ease and could be forgone with no huge structural shifts in the way the world works".*[26]

In September 2011, German engineering giant Siemens announced it will withdraw entirely from the nuclear industry, as a response to the Fukushima nuclear disaster in Japan.*[215] The company is to boost its work in the renewable energy sector.*[216] Commenting on the German government's policy to close nuclear plants, Werner Sinn, president of the Ifo Institute for Economic Research at the University of Munich, stated: "It is wrong to shut down the atomic power plants, because this is a cheap source of energy, and wind and solar power are by no means able to provide a replacement. They are much more expensive, and the energy that comes out is of inferior quality. Energy-intensive industries will move out, and the competitiveness of the German manufacturing sector will be reduced or wages will be depressed." *[217]

In 2011, Mycle Schneider spoke of a global downward trend in the nuclear power industry:

> The international nuclear lobby has pursued a 10-year-long, massive propaganda strategy aimed at convincing decision-makers that atomic technology has a bright future as a low-carbon energy option... however, most of the high-flying nuclear plans never materialized. The historic maximum of reactors operating worldwide was achieved in 2002 with 444 units. In the European Union the historic peak was reached as early as 1988 with 177 reactors, of which only 134 are left. The only new projects underway in Europe are heavily over budget and much delayed.

> As *Time* magazine rightly stated in March, "Nuclear power is expanding only in places where taxpayers and ratepayers can be compelled to foot the bill." China is building 27 -- or more than 40 percent -- of the 65 units officially under construction around the world. Even there, though, nuclear is fading as an energy option. While China has invested the equivalent of about $10 billion per year into nuclear power in recent years, in 2010 it spent twice as much on wind energy alone and some $54.5 billion on all renewables combined.*[218]

In contrast, proponents of nuclear power argue that nuclear power has killed by far the fewest number of people per terawatt hour of any type of power generation, and it has a very small effect on the environment with effectively zero emissions of any kind. And this even taking into account the Chernobyl and Fukushima accidents, in which few people were killed directly and few excess cancers will be caused by releases of radioactivity to the environment. Some proponents acknowledge that most people will not accept this sort of statistical argument nor will they believe reassuring statements from industry or government. Indeed, the industry itself has created fear of nuclear power by pointing out that radioactivity can be dangerous. Improved communication by industry might help to overcome current fears regarding nuclear power, but it will be a difficult task to change current perceptions in the general population.[219]

But with regard to the proposition that "Improved communication by industry might help to overcome current fears regarding nuclear power", M.V. Ramana says that the basic problem is that there is "distrust of the social institutions that manage nuclear energy", and a 2001 survey by the European Commission found that "only 10.1 percent of Europeans trusted the nuclear industry". This public distrust is periodically reinforced by safety violations by nuclear companies, or through ineffectiveness or corruption on the part of nuclear regulatory authorities. Once lost, says Ramana, trust is extremely difficult to regain.[220] Faced with public antipathy, the nuclear industry has "tried a variety of strategies to persuade the public to accept nuclear power", including the publication of numerous "fact sheets" that discuss issues of public concern. Ramana says that none of these strategies have been very successful.[220]

In March 2012, E.ON UK and RWE npower announced they would be pulling out of developing new nuclear power plants in the UK, placing the future of nuclear power in the UK in doubt.[221] More recently, Centrica (who own British Gas) pulled out of the race on 4 February 2013 by letting go its 20% option on four new nuclear plants.[222] Cumbria county council (a local authority) turned down an application for a final waste repository on 30 January 2013 —there is currently no alternative site on offer.[223]

In terms of current nuclear status and future prospects:[224]

- Ten new reactors were connected to the grid, In 2015, the highest number since 1990, but expanding Asian nuclear programs are balanced by retirements of aging plants and nuclear reactor phase-outs.[225] Seven reactors were permanently shut down.

- 441 operational reactors had a worldwide net capacity of 382,855 megawatts of electricity in 2015. However, some reactors are classified as operational, but are not producing any power.[226]

- 67 new nuclear reactors were under construction in 2015, including four EPR units.[227] The first two EPR projects, in Finland and France, were meant to lead a nuclear renaissance[228] but both are facing costly construction delays. Construction commenced on two Chinese EPR units in 2009 and 2010.[229] The Chinese units were to start operation in 2014 and 2015,[230] but the Chinese government halted construction because of safety concerns.[231] China's

National Nuclear Safety Administration carried out on-site inspections and issued a permit to proceed with function tests in 2016. Taishan 1 is expected to start up in the first half of 2017 and Taishan 2 is scheduled to begin operating by the end of 2017.[232]

Brazil, China, India, Japan and the Netherland now all generate more electricity from wind energy than from nuclear sources. New power generation using solar power grew by 33% in 2015, wind power over 17%, and 1.3% for nuclear power, exclusively due to development in China.[23]

15.12   See also

15.13   Footnotes

[1] "Stewart Brand + Mark Z. Jacobson: Debate: Does the world need nuclear energy?". *TED* (published June 2010). February 2010. Retrieved 21 October 2013.

[2] "Sunday Dialogue: Nuclear Energy, Pro and Con". *New York Times*. February 25, 2012.

[3] MacKenzie, James J. (December 1977). "*The Nuclear Power Controversy* by Arthur W. Murphy". *The Quarterly Review of Biology*. **52** (4): 467–8. doi:10.1086/410301. JSTOR 2823429.

[4] Walker, J. Samuel (10 January 2006). *Three Mile Island: A Nuclear Crisis in Historical Perspective*. University of California Press. pp. 10–11. ISBN 9780520246836.

[5] In February 2010 the nuclear power debate played out on the pages of the *New York Times*, see A Reasonable Bet on Nuclear Power and Revisiting Nuclear Power: A Debate and A Comeback for Nuclear Power?

[6] In July 2010 the nuclear power debate again played out on the pages of the *New York Times*, see We're Not Ready Nuclear Energy: The Safety Issues

[7] Diaz-Maurin, François (2014). "Going beyond the Nuclear Controversy". *Environmental Science & Technology*. **48** (1): 25–26. doi:10.1021/es405282z.

[8] Diaz-Maurin, François; Kovacic, Zora (2015). "The unresolved controversy over nuclear power: A new approach from complexity theory". *Global Environmental Change*. **31** (C): 207–216. doi:10.1016/j.gloenvcha.2015.01.014.

[9] Kitschelt, Herbert P. (2009). "Political Opportunity Structures and Political Protest: Anti-Nuclear Movements in Four Democracies". *British Journal of Political Science*. **16**: 57. doi:10.1017/S000712340000380X.

[10] Jim Falk (1982). *Global Fission: The Battle Over Nuclear Power*. Oxford University Press, pages 323-340.

[11] U.S. Energy Legislation May Be 'Renaissance' for Nuclear Power.

[12] Bernard Cohen. "The Nuclear Energy Option". Retrieved 2009-12-09.

[13] J.M. Pearce."Limitations of Nuclear Power as a Sustainable Energy Source, *Sustainability* 4(6), pp.1173-1187 (2012).

[14] "Nuclear Energy is not a New Clear Resource". Theworldreporter.com. 2010-09-02.

[15] Greenpeace International and European Renewable Energy Council (January 2007). *Energy Revolution: A Sustainable World Energy Outlook*, p. 7.

[16] Giugni, Marco (2004). *Social protest and policy change: ecology, antinuclear, and peace movements in comparative perspective*. Rowman & Littlefield. pp. 44–. ISBN 9780742518278.

[17] Stephanie Cooke (2009). *In Mortal Hands: A Cautionary History of the Nuclear Age*. Black Inc., p. 280.

[18] Sovacool, Benjamin K. (2008). "The costs of failure: A preliminary assessment of major energy accidents, 1907–2007". *Energy Policy*. **36** (5): 1802. doi:10.1016/j.enpol.2008.01.040.

[19] Jim Green . Nuclear Weapons and 'Fourth Generation' Reactors *Chain Reaction*, August 2009, pp. 18-21.

[20] Kleiner, Kurt (2008). "Nuclear energy: Assessing the emissions". *Nature Reports Climate Change* (810): 130. doi:10.1038/climate.2008.99.

[21] Mark Diesendorf (2007). *Greenhouse Solutions with Sustainable Energy*, University of New South Wales Press, p. 252.

[22] Mark Diesendorf (July 2007). "Is nuclear energy a possible solution to global warming?" (PDF).

[23] Mycle Schneider, The World Nuclear Industry Status Report 2016: Summary and Conclusions, 13 July 2016, p.12.

[24] John R. Parkins & Randolph Haluza-DeLay (2011). "Social and Ethical Considerations of Nuclear Power Development" (PDF). *University of Alberta*.

[25] Brook, B.W. & Lowe, I. (2010). *Why vs Why: Nuclear Power*. Pantera Press. ISBN 978-0-9807418-5-8

[26] "Nuclear power: When the steam clears". *The Economist*. March 24, 2011.

[27] WNA (20 June 2013). "Nuclear power down in 2012" . *World Nuclear News*.

[28] Benjamin K. Sovacool (2011). *Contesting the Future of Nuclear Power: A Critical Global Assessment of Atomic Energy*. World Scientific, p. 90.

[29] "Nuclear renaissance faces realities" . Platts. Retrieved 2007-07-13.

[30] L. Meeus; K. Purchala; R. Belmans. "Is it reliable to depend on import?" (PDF). Katholieke Universiteit Leuven, Department of Electrical Engineering of the Faculty of Engineering. Retrieved 2007-07-13.

[31] Benjamin K. Sovacool (January 2011). "Second Thoughts About Nuclear Power" (PDF). National University of Singapore. pp. 5–6.

[32] Benjamin K. Sovacool (2011). *Contesting the Future of Nuclear Power*, World Scientific, p. 88 and 122-123.

[33] Commodities Price History, International Monetary Fund, accessed 6 July 2016.

[34] *Uranium 2014: Resources, Production, and Demand*, International Atomic Energy Agency/OCED Nuclear Energy Agency, 2014, p.130.

[35] John McCarthy (2006). "Facts From Cohen and Others" . *Progress and its Sustainability*. Stanford. Retrieved 2008-01-18.

[36] Benjamin K. Sovacool (2011). *Contesting the Future of Nuclear Power: A Critical Global Assessment of Atomic Energy*, World Scientific, p. 113-114.

[37] http://web.archive.org/web/20110705134219/http://www.iaea.org/cgi-bin/db.page.pl/pris.factors3y.htm?faccve=EAF&facname=Energy%20Availability%20Factor&group=Country

[38] "15 years of progress" (PDF). World Nuclear Association.

[39] "Renewable Energy and Electricity" . World Nuclear Association. June 2010. Retrieved 2010-07-04.

[40] Benjamin K. Sovacool (2011). *Contesting the Future of Nuclear Power: A Critical Global Assessment of Atomic Energy*, World Scientific, p. 220.

[41] Ben Sills (August 29, 2011). "Solar May Produce Most of World's Power by 2060, IEA Says" . *Bloomberg*.

[42] Contribution of Renewables to Energy Security

[43] World Nuclear Association (September 2013). "Renewable Energy and Electricity" .

[44] Amory Lovins (2011). *Reinventing Fire*, Chelsea Green Publishing, p. 199.

[45] Entwicklungen in der deutschen Strom- und Gaswirtschaft 2012 BDEW (german)

[46] Harvey, Fiona (2012-10-30). "Renewable energy will overtake nuclear power by 2018, research says" . *The Guardian*. London.

[47] Scotland aims for 100% renewable energy by 2020

[48] Proliferation of Hydroelectric Dams in the Andean Amazon and Implications for Andes-Amazon Connectivity Matt Finer, Clinton N. Jenkins

[49] Fiona Harvey (9 May 2011). "Renewable energy can power the world, says landmark IPCC study". *The Guardian*. London.

[50] http://eetd.lbl.gov/ea/ems/reports/wind-energy-costs-2-2012.pdf

[51] Diane Cardwell (March 20, 2014). "Wind Industry's New Technologies Are Helping It Compete on Price". *New York Times*.

[52] "Dr. MacKay *Sustainable Energy without the hot air*". Data from studies by the Paul Scherrer Institute including non-EU data. p. 168. Retrieved 15 September 2012.

[53] Nils Starfelt; Carl-Erik Wikdahl. "Economic Analysis of Various Options of Electricity Generation - Taking into Account Health and Environmental Effects" (PDF). Retrieved 2012-09-08

[54] Spent Nuclear Fuel: A Trash Heap Deadly for 250,000 Years or a Renewable Energy Source?

[55] "Closing and Decommissioning Nuclear Power Plants" (PDF). March 7, 2012.

[56] Benjamin K. Sovacool (2011). *Contesting the Future of Nuclear Power: A Critical Global Assessment of Atomic Energy*. World Scientific. p. 146.

[57] "TVA reactor shut down; cooling water from river too hot"

[58] http://www.startribune.com/sudden-shutdown-of-monticello-nuclear-power-plant-causes-fish-kill/354007091/

[59] EDF raises French EPR reactor cost to over $11 billion. *Reuters*, Dec 3, 2012.

[60] Mancini, Mauro and Locatelli, Giorgio and Sainati, Tristano (2015). The divergence between actual and estimated costs in large industrial and infrastructure projects: is nuclear special? In: *Nuclear new build: insights into financing and project management*. Nuclear Energy Agency. pp. 177-188.

[61] Kidd, Steve (January 21, 2011). "New reactors—more or less?". *Nuclear Engineering International*.

[62] Ed Crooks (12 September 2010). "Nuclear: New dawn now seems limited to the east". Financial Times. Retrieved 12 September 2010.

[63] *The Future of Nuclear Power*. Massachusetts Institute of Technology. 2003. ISBN 0-615-12420-8. Retrieved 2006-11-10.

[64] Massachusetts Institute of Technology (2011). "The Future of the Nuclear Fuel Cycle" (PDF). p. xv.

[65] Benjamin K. Sovacool (2011). *Contesting the Future of Nuclear Power: A Critical Global Assessment of Atomic Energy*. World Scientific. p. 118-119.

[66] John Quiggin (8 November 2013). "Reviving nuclear power debates is a distraction. We need to use less energy". *The Guardian*.

[67] http://www.eia.gov/electricity/monthly/epm_table_grapher.cfm?t=epmt_1_01

[68] "Nuclear Power: Still Not Viable without Subsidies". Union of Concerned Scientists. Retrieved 2012-02-04.

[69] "Billions of Dollars in Subsidies for the Nuclear Power Industry Will Shift Financial Risks to Taxpayers" (PDF). Union of Concerned Scientists. Retrieved 4 February 2012.

[70] "Energy Subsidies and External Costs". *Information and Issue Briefs*. World Nuclear Association. 2005. Retrieved 2006-11-10.

[71] FP7 budget breakdown

[72] FP7 Euratom spending

[73] http://www.iisd.org/gsi/sites/default/files/relative_energy_subsidies.pdf

[74] Kristin Shrader-Frechette (19 August 2011). "Cheaper, safer alternatives than nuclear fission". *Bulletin of the Atomic Scientists*.

[75] Arjun Makhijani (21 July 2011). "The Fukushima tragedy demonstrates that nuclear energy doesn't make sense". *Bulletin of the Atomic Scientists*.

[76] Benjamin K. Sovacool. The costs of failure: A preliminary assessment of major energy accidents, 1907–2007. *Energy Policy* 36 (2008), p. 1808.

[77] John Byrne and Steven M. Hoffman (1996). *Governing the Atom: The Politics of Risk*, Transaction Publishers, p. 136.

[78] Reuters, 2001. "Cheney says push needed to boost nuclear power", Reuters News Service, May 15, 2001.

[79] United States Nuclear Regulatory Commission, 1983. The Price-Anderson Act: the Third Decade, NUREG-0957

[80] Dubin, Jeffrey A.; Rothwell, Geoffrey S. (1990). "Subsidy to Nuclear Power Through Price-Anderson Liability Limit". *Contemporary Economic Policy*. **8** (3): 73. doi:10.1111/j.1465-7287.1990.tb00645.x.

[81] Heyes, Anthony (2003). "Determining the Price of Price-Anderson". *Regulation*. **25** (4): 105–10.

[82] U.S. Department of Energy. 1999. Department of Energy Report to Congress on the Price-Anderson Act, Prepared by the U.S. Department of Energy, Office of General Council. Accessed 20 August 2010. Available: http://www.gc.energy.gov/documents/paa-rep.pdf

[83] Reuters, 2001. "Cheney says push needed to boost nuclear power", *Reuters News Service*, May 15, 2001.

[84] Bradford, Peter A. (January 23, 2002). "Testimony before the United States Senate Committee on Environment and Public Works Subcommittee on Transportation, Infrastructure and Nuclear Safety" (PDF). *Renewal of the Price Anderson Act.*

[85] Wood, W.C. 1983. Nuclear Safety: Risks and Regulation. American Enterprise Institute for Public Policy Research. Washington, D.C. pp. 40-48.

[86] Zelenika-Zovko, I.; Pearce, J.M. (2011). "Diverting indirect subsidies from the nuclear industry to the photovoltaic industry: Energy and financial returns". *Energy Policy.* **39** (5): 2626. doi:10.1016/j.enpol.2011.02.031.

[87] Nuclear power and water scarcity, ScienceAlert, 28 October 2007. Retrieved 2008-08-08

[88] Benjamin K. Sovacool. A Critical Evaluation of Nuclear Power and Renewable Electricity in Asia, *Journal of Contemporary Asia,* Vol. 40, No. 3, August 2010, p. 386.

[89] Warner, Ethan S.; Heath, Garvin A. (2012). "Life Cycle Greenhouse Gas Emissions of Nuclear Electricity Generation". *Journal of Industrial Ecology.* **16**: S73. doi:10.1111/j.1530-9290.2012.00472.x.

[90] Moomaw, W., P. Burgherr, G. Heath, M. Lenzen, J. Nyboer, A. Verbruggen, 2011: Annex II: Methodology. In IPCC: Special Report on Renewable Energy Sources and Climate Change Mitigation (ref. page 10)

[91] Patterson, Thom (November 3, 2013). "Climate change warriors: It's time to go nuclear". *CNN.*

[92] Sovacool, Benjamin K.; Parenteau, Patrick; Ramana, M.V.; Valentine, Scott V.; Jacobson, Mark Z.; Delucchi, Mark A.; Diesendorf, Mark (2013). "Comment on 'Prevented Mortality and Greenhouse Gas Emissions from Historical and Projected Nuclear Power'". *Environmental Science & Technology.* **47** (12): 6715–7. doi:10.1021/es401667h.

[93] Mark Diesendorf (2013). "Book review: Contesting the future of nuclear power" (PDF). *Energy Policy.*

[94] Benjamin K. Sovacool (2011). "The 'Self-Limiting' Future of Nuclear Power" (PDF). *Contesting the Future of Nuclear Power.* World Scientific.

[95] Kenny, R.; Law, C.; Pearce, J.M. (2010). "Towards real energy economics: Energy policy driven by life-cycle carbon emission". *Energy Policy.* **38** (4): 1969. doi:10.1016/j.enpol.2009.11.078.

[96] Adam, David (9 December 2008). "Too late? Why scientists say we should expect the worst". *The Guardian.* ISSN 0261-3077. Retrieved 2016-10-07.

[97] "Nine out of 10 people want more renewable energy". *The Guardian.* 23 April 2012. ISSN 0261-3077. Retrieved 2016-10-07.

[98] *Renewables 2015: Global Status Report* (PDF). Renewable Energy Policy Network for the 21st Century. p. 27. Archived from the original (PDF) on 19 June 2015.

[99] Benjamin K. Sovacool (2011). *Contesting the Future of Nuclear Power: A Critical Global Assessment of Atomic Energy.* World Scientific, p. 141.

[100] "Environmental Surveillance, Education and Research Program". Idaho National Laboratory. Retrieved 2009-01-05.

[101] Vandenbosch, Robert; Vandenbosch, Susanne E. (2007). *Nuclear Waste Stalemate: Political and Scientific Controversies.* University of Utah Press. p. 21. ISBN 978-0-87480-903-9.

[102] Ojovan, M. I.; Lee, W.E. (2005). *An Introduction to Nuclear Waste Immobilisation.* Amsterdam: Elsevier Science Publishers. p. 315. ISBN 0-08-044462-8.

[103] Brown, Paul (2004-04-14). "Shoot it at the sun. Send it to Earth's core. What to do with nuclear waste?". *The Guardian.* London.

[104] National Research Council (1995). *Technical Bases for Yucca Mountain Standards.* Washington, D.C.: National Academy Press. p. 91. ISBN 0-309-05289-0.

[105] "The Status of Nuclear Waste Disposal". The American Physical Society. January 2006. Retrieved 2008-06-06.

[106] "Public Health and Environmental Radiation Protection Standards for Yucca Mountain, Nevada; Proposed Rule" (PDF). United States Environmental Protection Agency. 2005-08-22. Retrieved 2008-06-06.

[107] Sevior, Martin (2006). "Considerations for nuclear power in Australia". *International Journal of Environmental Studies.* **63** (6): 859. doi:10.1080/00207230601047255.

[108] Ragheb. M. (October 7, 2013). "Thorium Resources In Rare Earth Elements" (PDF).

[109] Peterson, B. T.; Depaolo, D. J. (2007). "Mass and Composition of the Continental Crust Estimated Using the CRUST2.0 Model". *American Geophysical Union.* **33**: 1161. Bibcode:2007AGUFM.V33A1161P.

[110] Cohen, Bernard L. (1998). "Perspectives on the High Level Waste Disposal Problem". *Interdisciplinary Science Reviews.* **23** (3): 193–203. doi:10.1179/030801898789764480.

[111] Montgomery, Scott L. (2010). *The Powers That Be,* University of Chicago Press, p. 137.

[112] Al Gore (2009). *Our Choice.* Bloomsbury, pp. 165-166.

[113] "A Nuclear Power Renaissance?". *Scientific American.* April 28, 2008. Retrieved 2008-05-15.

[114] von Hippel, Frank N. (April 2008). "Nuclear Fuel Recycling: More Trouble Than It's Worth". *Scientific American.* Retrieved 2008-05-15.

[115] Is the Nuclear Renaissance Fizzling?

[116] Kharecha, Pushker A.; Hansen, James E. (2013). "Prevented Mortality and Greenhouse Gas Emissions from Historical and Projected Nuclear Power". *Environmental Science & Technology*. **47** (9): 4889. doi:10.1021/es3051197.

[117] "Nuclear Power Prevents More Deaths Than It Causes | Chemical & Engineering News". Cen.acs.org. Retrieved 2013-06-18.

[118] Sovacool, Benjamin K. (2010). "A Critical Evaluation of Nuclear Power and Renewable Electricity in Asia". *Journal of Contemporary Asia*. **40** (3): 369. doi:10.1080/00472331003798350.

[119] The Worst Nuclear Disasters

[120] Strengthening the Safety of Radiation Sources p. 14.

[121] Johnston, Robert (September 23, 2007). "Deadliest radiation accidents and other events causing radiation casualties". Database of Radiological Incidents and Related Events.

[122] Ramana, M.V. (2009). "Nuclear Power: Economic, Safety, Health, and Environmental Issues of Near-Term Technologies". *Annual Review of Environment and Resources*. **34**: 127. doi:10.1146/annurev.environ.033108.092057.

[123] Storm van Leeuwen, Jan (2008). Nuclear power – the energy balance

[124] Wolfgang Rudig (1990). *Anti-nuclear Movements: A World Survey of Opposition to Nuclear Energy*. Longman. p. 53 & p. 61.

[125] Helen Caldicott (2006). *Nuclear power is not the answer to global warming or anything else*. Melbourne University Press. ISBN 0-522-85251-3. p.xvii

[126] Perrow, C. (1982). 'The President's Commission and the Normal Accident', in Sils, D., Wolf, C. and Shelanski, V. (Eds), *Accident at Three Mile Island: The Human Dimensions*, Westview, Boulder, pp.173–184.

[127] Pidgeon, Nick (2011). "In retrospect: Normal Accidents". *Nature*. **477** (7365): 404. doi:10.1038/477404a.

[128] Black, Richard (2011-04-12). "'Fukushima: As Bad as Chernobyl?'". Bbc.co.uk. Retrieved 2011-08-20.

[129] From interviews with Mikhail Gorbachev, Hans Blix and Vassili Nesterenko. *The Battle of Chernobyl*. Discovery Channel. Relevant video locations: 31:00, 1:10:00.

[130] Kagarlitsky, Boris (1989). "Perestroika: The Dialectic of Change". In Mary Kaldor; Gerald Holden; Richard A. Falk. *The New Detente: Rethinking East-West Relations*. United Nations University Press. ISBN 0-86091-962-5.

[131] "IAEA Report". *In Focus: Chernobyl*. International Atomic Energy Agency. Archived from the original on 17 December 2007. Retrieved 29 March 2006.

[132] Hallenbeck, William H (1994). *Radiation Protection*. CRC Press. p. 15. ISBN 0-87371-996-4. Reported thus far are 237 cases of acute radiation sickness and 31 deaths.

[133] Tomoko Yamazaki & Shunichi Ozasa (27 June 2011). "Fukushima Retiree Leads Anti-Nuclear Shareholders at Tepco Annual Meeting'. *Bloomberg*.

[134] Evacuation-related deaths now more than quake/tsunami toll in Fukushima Prefecture, Japan Daily Press, 18 December 2013.

[135] Fukushima evacuation has killed more than earthquake and tsunami, survey says, NBC News, 10 septembre 2013.

[136] Mari Saito (7 May 2011). "Japan anti-nuclear protesters rally after PM call to close plant". Reuters.

[137] Weisenthal, Joe (11 March 2011). "Japan Declares Nuclear Emergency, As Cooling System Fails At Power Plant". *Business Insider*. Retrieved 11 March 2011.

[138] "Blasts escalate Japan's nuclear crisis". *World News Australia*. March 16, 2011.

[139] David Mark; Mark Willacy (April 1, 2011). "Crews 'facing 100-year battle' at Fukushima". *ABC News*.

[140] Rogovin. pp. 153.

[141] "14-Year Cleanup at Three Mile Island Concludes". New York Times. August 15, 1993. Retrieved March 28, 2011.

[142] Spiegelberg-Planer, Rejane. "A Matter of Degree: A revised International Nuclear and Radiological Event Scale (INES) extends its reach". IAEA.org. Retrieved March 19, 2011.

[143] King, Laura; Kenji Hall; Mark Magnier (March 18, 2011). "In Japan, workers struggling to hook up power to Fukushima reactor". Los Angeles Times. Retrieved March 19, 2011.

[144] Susan Cutter and Barnes. Evacuation behavior and Three Mile Island, Disasters, vol.6, 1982, p 116-124.

[145] A Decade Later, TMI's Legacy Is Mistrust *The Washington Post*, March 28, 1989, p. A01.

[146] People & Events: Dick Thornburgh

[147] Michael Levi on Nuclear Policy, in video "Tea with the Economist", 1:55-2:10, on http://audiovideo.economist.com/, retrieved April 6th 2011, 3.24pm.

[148] Jacobson, Mark Z.; Delucchi, Mark A. (2011). "Providing all global energy with wind, water, and solar power. Part I: Technologies, energy resources, quantities and areas of infrastructure, and materials". *Energy Policy*. **39** (3): 1154. doi:10.1016/j.enpol.2010.11.040.

[149] Hugh Gusterson (16 March 2011). "The lessons of Fukushima". *Bulletin of the Atomic Scientists*.

[150] James Paton (April 4, 2011). "Fukushima Crisis Worse for Atomic Power Than Chernobyl, UBS Says". *Bloomberg Businessweek*.

[151] Massachusetts Institute of Technology (2003). "The Future of Nuclear Power" (PDF). p. 48.

[152] Hvistendahl, Mara. "Coal Ash Is More Radioactive than Nuclear Waste". Scientific American. Retrieved 2013-06-18.

[153] "Safety of Nuclear Power Reactors".

[154] Brower, Michael. Ph. D. and Leon, Warren, Ph. D. *The Consumer's Guide to Effective Environmental Choices: Practical Advice from the Union of Concerned Scientists* 1999. Three Rivers Press.

[155] Nuclear Power and Global Warming

[156] Adam Piore (June 2011). "Nuclear energy: Planning for the Black Swan". *Scientific American*.

[157] Whistleblower on Nuclear Plant Safety

[158] The San Jose Three

[159] The Struggle over Nuclear Power

[160] A book chapter which discusses the whistleblowing, written by Vivian Weil, was published in 1983 as "The Browns Ferry Case" in *Engineering Professionalism and Ethics*, edited by James H. Schaub and Karl Pavlovic, and published by John Wiley & Sons.

[161] Julie Miller (February 12, 1995). "Paying The Price For Blowing The Whistle". *The New York Times*.

[162] Boughton, Katherine (10 December 1999). "The Whistleblower: Arnold Gundersen of Goshen". *Litchfield County Times*. Retrieved 10 September 2013.

[163] Eric Pooley. Nuclear Warriors *Time Magazine*. March 4, 1996.

[164] Adam Bowles. A Cry in the Nuclear Wilderness *Christianity Today*, October 2, 2000.

[165] George Galatis, Nuclear Whistleblower *Time Magazine*. March 4, 1996.

[166] NRC Failure to Adequately Regulate - Millsone Unit 1, 1995

[167] National Security Whistleblowers in the Post-September 11th Era pp.177-178.

[168] Nuclear power and antiterrorism: obscuring the policy contradictions

[169] "No Excess Mortality Risk Found in Counties with Nuclear Facilities". National Cancer Institute. Retrieved 2009-02-06.

[170] Clapp, Richard (November 2005). "Nuclear Power and Public Health". *Environmental Health Perspectives*. Retrieved 2009-01-28.

[171] Cardis, E; Vrijheid, M; Blettner, M; Gilbert, E; Hakama, M; Hill, C; Howe, G; Kaldor, J; Muirhead, CR; Schubauer-Berigan, M; Yoshimura, T; Bermann, F; Cowper, G; Fix, J; Hacker, C; Heinmiller, B; Marshall, M; Thierry-Chef, I; Utterback, D; Ahn, YO; Amoros, E; Ashmore, P; Auvinen, A; Bae, JM; Solano, JB; Biau, A; Combalot, E; Deboodt, P; Diez Sacristan, A; Eklof, M (2005). "Risk of cancer after low doses of ionising radiation: Retrospective cohort study in 15 countries". *BMJ*. **331** (7508): 77. doi:10.1136/bmj.38499.599861.E0. PMC 558612⮌. PMID 15987704.

[172] Nuclear Regulatory Commission. Backgrounder on Radiation Protection and the "Tooth Fairy" Issue. December 2004

[173] "Low-Level Radiation: How the Linear No-Threshold Model Keeps Canadians Safe". Canadian Nuclear Safety Commission. Retrieved 2010-06-27.

[174] "Average Annual Radiation Exposure". Lbl.gov. 2011-05-04. Retrieved 2013-06-18.

[175] "National Safety Council". Nsc.org. Retrieved 2013-06-18.

[176] "Sources and effects of ionizing radiation". UNSCEAR. Retrieved 2013-11-08.

[177] "Appendix B, page 121, Table 11 Areas of high natural radiation background" (PDF). UNSCEAR. Retrieved 2013-11-08.

[178] Nair, Raghu Ram K.; Rajan, Balakrishnan; Akiba, Suminori; Jayalekshmi, P; Nair, M Krishnan; Gangadharan, P; Koga, Taeko; Morishima, Hiroshige; Nakamura, Seiichi; Sugahara, Tsutomu (2009). "Background Radiation and Cancer Incidence in Kerala, India—Karanagappally Cohort Study". *Health Physics*. **96** (1): 55–66. doi:10.1097/01.HP.0000327646.54923.11. PMID 19066487.

[179] "NRC: Backgrounder on Emergency Preparedness at Nuclear Power Plants". Nrc.gov. Retrieved 2013-06-18.

[180] Kinderkrebs in der Umgebung von KernKraftwerken

[181] Körblein A, Hoffmann W; . Childhood Cancer in the Vicinity of German Nuclear Power Plants, Medicine & Global Survival 1999, 6(1):18-23.

[182] Fairlie, Ian (2009). "Commentary: Childhood cancer near nuclear power stations". *Environmental Health*. **8**: 43. doi:10.1186/1476-069X-8-43. PMC 2757021⮌. PMID 19775438.

[183] "Further consideration of the incidence of childhood leukemia around nuclear power plants in Great Britain" (Press release). COMARE. 6 May 2011. Retrieved 7 May 2011.

[184] Safety issues cloud nuclear renaissance: Developing nations' track record gives cause for concern

[185] Jonathan Watts (25 August 2011). "WikiLeaks cables reveal fears over China's nuclear safety". *The Guardian*. London.

[186] Keith Bradsher (December 15, 2009). "Nuclear Power Expansion in China Stirs Concerns". New York Times. Retrieved 2010-01-21.

[187] "China Nuclear Power | Chinese Nuclear Energy". World-nuclear.org. Retrieved 2013-06-18.

[188] "Obama Administration Announces Loan Guarantees to Construct New Nuclear Power Reactors in Georgia | The White House". Whitehouse.gov. 2010-02-16. Retrieved 2013-06-18.

[189] TED2010. "Bill Gates on energy: Innovating to zero! | Video on". Ted.com. Retrieved 2013-06-18.

[190] Benjamin K. Sovacool (2011). *Contesting the Future of Nuclear Power: A Critical Global Assessment of Atomic Energy*. World Scientific, p. 192.

[191] "Congressional Budget Office Vulnerabilities from Attacks on Power Reactors and Spent Material".

[192] Charles D. Ferguson & Frank A. Settle (2012). "The Future of Nuclear Power in the United States" (PDF). *Federation of American Scientists*.

[193] Vadim Nesvizhskiy (1999). "Neutron Weapon from Underground". *Research Library*. Nuclear Threat Initiative. Retrieved 2006-11-10.

[194] "Information on Nuclear Smuggling Incidents". *Nuclear Almanac*. Nuclear Threat Initiative. Retrieved 2006-11-10.

[195] Amelia Gentleman & Ewen MacAskill (2001-07-25). "Weapons-grade Uranium Seized". London: Guardian Unlimited. Retrieved 2006-11-10.

[196] Pavel Simonov (2005). "The Russian Uranium That is on Sale for the Terrorists". *Global Challenges Research*. Axis. Retrieved 2006-11-10.

[197] "Action Call Over Dirty Bomb Threat". BBC News. 2003-03-11. Retrieved 2006-11-10.

[198] For an example of the former, see the quotes in Erin Neff, Cy Ryan, and Benjamin Grove. "Bush OKs Yucca Mountain waste site", *Las Vegas Sun* (2002 February 15). For an example of the latter, see ""Dirty Bomb" Plot spurs Schumer to call for US Marshals to guard Nuclear waste that would go through New York", press release of Senator Charles E. Shumer (13 June 2002).

[199] Ipsos (23 June 2011). *Global Citizen Reaction to the Fukushima Nuclear Plant Disaster (theme: environment / climate) Ipsos Global @dvisor* (PDF). Survey website: Ipsos MORI: Poll: Strong global opposition towards nuclear power.

[200] Richard Black (25 November 2011). "Nuclear power 'gets little public support worldwide'". *BBC News*.

[201] Deutsche Bank Group (2011). The 2011 inflection point for energymarkets: Health, safety, security and the environment. *DB Climate Change Advisors*, May 2.

[202] EurActiv.com - Majority of Europeans oppose nuclear power | EU - European Information on EU Priorities & Opinion

[203] "Europeans and Nuclear Safety Report" (PDF). Special Eurobarometer 271. European Commission. February 2007.

[204] "Poll shows anti-nuclear sentiment up in Sweden". *Businessweek*. 22 March 2011.

[205] Michael Cooper (March 22, 2011). "Nuclear Power Loses Support in New Poll". *The New York Times*.

[206] Rebecca Rifkin. U.S. Support for Nuclear Energy at 51%. Gallup, 30 March 2015.

[207] Roper Center, , 2013.

[208] Stephen Ansolabehere. Public Attitudes Toward America's Energy Options Report of the 2007 MIT Energy Survey. Center for Energy and Environmental Policy research, March 2007, p. 3.

[209] Schneider, M.; Froggatt, A.; Thomas, S. (2011). "2010-2011 world nuclear industry status report". *Bulletin of the Atomic Scientists*. **67** (4): 60. doi:10.1177/0096340211413539.

[210] IAEA Pris. Power reactor information system

[211] Stephanie Cooke (2009). *In Mortal Hands: A Cautionary History of the Nuclear Age*. Black Inc. p. 387.

[212] Heavy Manufacturing of Power Plants

[213] "Gauging the pressure". The Economist. 28 April 2011.

[214] "NEWS ANALYSIS: Japan crisis puts global nuclear expansion in doubt". Platts. 21 March 2011.

[215] "Siemens to quit nuclear industry". *BBC News*. 18 September 2011.

[216] "Siemens to Exit Nuclear Energy Business". *Spiegel Online*. 19 September 2011.

[217] David Talbot (July–August 2012). "The Great German Energy Experiment". *Technology Review*. Massachusetts Institute of Technology. Retrieved 25 July 2012.

[218] Mycle Schneider (9 September 2011). "Fukushima crisis: Can Japan be at the forefront of an authentic paradigm shift?". *Bulletin of the Atomic Scientists*.

[219] Steve Kidd (19 January 2012). "Nuclear as the last resort – why and how?". Nuclear Engineering International. Retrieved 22 January 2012.

[220] Ramana, M. V. (2011). "Nuclear power and the public". *Bulletin of the Atomic Scientists*. **67** (4): 43. doi:10.1177/0096340211413358.

[221] David Maddox (30 March 2012). "Nuclear disaster casts shadow over future of UK's energy plans". *The Scotsman*.

[222] Carrington, Damian (4 February 2013). "Centrica withdraws from new UK nuclear projects". *The Guardian*. Retrieved 13 February 2013.

[223] Wainwright, Martin (30 January 2013). "Cumbria rejects underground nuclear storage dump". *The Guardian*. Retrieved 13 February 2013.

[224] "Ten New Nuclear Power Reactors Connected to Grid in 2015, Highest Number Since 1990". Retrieved May 22, 2016.

[225] Mark Diesendorf (2013). "Book review: Contesting the future of nuclear power" (PDF). *Energy Policy*.

[226] "Japan approves two reactor restarts". Taipei Times. 2013-06-07. Retrieved 2013-06-14.

[227] *Pub.iaea.org* (PDF). May 9, 2015 http://www-pub.iaea.org/MTCD/Publications/PDF/RDS_2-36_web.pdf. Retrieved May 22, 2016. Missing or empty |title= (help)

[228] James Kanter. In Finland, Nuclear Renaissance Runs Into Trouble *New York Times*, May 28, 2009.

[229] Geert De Clercq (31 July 2014). "EDF hopes French EPR will launch before Chinese reactors". Reuters. Retrieved 9 December 2014.

[230] Symbolic milestone for Finnish EPR, *World Nuclear News*, 24 October 2013.

[231] Mycle Schneider, Antony Froggatt. "China dialogue: World nuclear industry in decline". 3 February 2016.

[232] http://www.world-nuclear-news.org/NN-First-Taishan-EPR-completes-cold-tests-0102164.html

15.14   Further reading

- Ferguson, Charles D. (June 2007). *Nuclear energy: balancing benefits and risks*. Council on Foreign Relations. ISBN 978-0-87609-400-6.

- Ferguson, Charles D.; Marburger, Lindsey E.; Farmer, J. Doyne; Makhijani, Arjun (2010). "A US nuclear future?". *Nature*. **467** (7314): 391–3. doi:10.1038/467391a. PMID 20864972.

- Diaz-Maurin, François (2014). "Going beyond the Nuclear Controversy". *Environmental Science & Technology*. **48** (1): 25–26. doi:10.1021/es405282z.

- Schneider, Mycle, Steve Thomas, Antony Froggatt, Doug Koplow (2016). *The World Nuclear Industry Status Report: World Nuclear Industry Status as of 1 January 2016*.

15.15   External links

- The World Nuclear Industry Status Reports website

- Beyond Nuclear at Nuclear Policy Research Institute advocacy organization

- Greenpeace Nuclear Campaign

- World Information Service on Energy (WISE)

- 1 million europeans against nuclear power

- Nuclear Files

- "Critical Hour: Three Mile Island, The Nuclear Legacy, And National Security" (PDF). (929 KB) Online book

- "Natural Resources Defense Council" (PDF). (158 KB)

- The New York Times Finally Reports the Economic Disaster of New Nukes

- American Nuclear Society (ANS)

- Representing the People and Organisations of the Global Nuclear Profession

- Environmentalists for Nuclear Power

- SCK.CEN Belgian Nuclear Research Centre

- Nuclear Energy Institute (NEI)

- Atomic Insights

- Freedom for Fission

- The Nuclear Energy Option, online book by Bernard L. Cohen. Emphasis on risk estimates of nuclear.

- Fairewinds Energy Education

- Should we use nuclear energy? - Wikidebate on Wikiversity

Chapter 16

World Association of Nuclear Operators

The **World Association of Nuclear Operators** (WANO) is an international group of nuclear power plant operators, dedicated to nuclear safety.

It was formed in 1989, following the world's worst nuclear disaster at Chernobyl (Ukraine), which was blamed on faulty design, wrong use of procedures and poor management control. As public confidence in the nuclear industry had been shaken, nuclear operators worldwide began to work together to prevent recurrences.

WANO promotes effective communication and open information sharing, via its five main programmes: Peer Review, Performance Analysis, Technical Support, Professional Development, and Corporate Communications. It is not a regulatory or advisory body, or a lobbyist for the industry.

16.1 Mission

"To maximise the safety and reliability of nuclear power plants worldwide by working together to assess, benchmark and improve performance through mutual support, exchange of information and emulation of best practice".[1]

16.2 WANO Programmes

- The Peer Review programme provides a critical assessment of station performance by an experiences team of global industry peers against nuclear industry standards of excellence as defined by WANO Performance Objectives and Criteria.[2]

- The Performance Analysis programme collects, screens and analyses operating experience and performance data, providing members with lessons learnt and industry performance insight reports. Fundamental to its success is the willingness of WANO members to openly share operating experience and performance data for the benefit of nuclear operators worldwide.[3][4]

- The Technical Support programme works with members to improve safety and reliability. Activities include technical support missions; new unit assistance; principles, guidelines and good practices; and plant of focus. Together, they help members learn from the experiences of their peers.[5]

- The Training & Development programme provides assistance to WANO members through workshops, seminars and training. This includes new entrants as well as operating stations. Specific activities include workshops, seminars, training courses and leadership courses.[6]

- The Corporate Communications programme ensures WANO's mission, vision and activities are shared with all internal and external audiences, including WANO members, industry vendors, new entrants, nations considering adding nuclear to their energy mix, other interested parties and the media. A variety of channels are used to promote access to WANO products and services.

16.3 See also

- World Institute for Nuclear Security

- Institute of Nuclear Power Operations

16.4 References

[1] "World Association of Nuclear Operators (WANO):: Our mission".

[2] "World Association of Nuclear Operators (WANO):: Peer reviews".

[3] "World Association of Nuclear Operators (WANO):: Operating Experience".

[4] "World Association of Nuclear Operators (WANO):: About us".

[5] "World Association of Nuclear Operators (WANO):: Technical Support and Exchange".

[6] "World Association of Nuclear Operators (WANO):: Professional and Technical Development".

16.5   External links

- Official website

- Inside WANO Magazine

- International Atomic Energy Agency

- WANO Media Announcements

Chapter 17

Anti-nuclear organizations

Main article: Anti-nuclear movement

Anti-nuclear organizations may oppose uranium mining, nuclear power, and/or nuclear weapons. Anti-nuclear groups have undertaken public protests and acts of civil disobedience which have included occupations of nuclear plant sites. Some of the most influential groups in the anti-nuclear movement have had members who were elite scientists, including several Nobel Laureates and many nuclear physicists.

17.1 Types of organizations

Various types of organizations have identified themselves with the anti-nuclear movement:[1]

- direct action groups, such as the Clamshell Alliance and Shad Alliance;

- environmental groups, such as Friends of the Earth and Greenpeace;

- consumer protection groups, such as Ralph Nader's Critical Mass;

- professional organizations,[2] such as Union of Concerned Scientists and International Physicians for the Prevention of Nuclear War; and

- political parties such as European Free Alliance.

Some of the most influential groups in the anti-nuclear movement have had members who were elite scientists, including several Nobel Laureates and many nuclear physicists. In the United States, these scientists have belonged primarily to three groups: the Union of Concerned Scientists, the Federation of American Scientists, and the Committee for Nuclear Responsibility.[3]

17.2 Activities

Anti-nuclear groups have undertaken public protests and acts of civil disobedience which have included occupations of nuclear plant sites. Other salient strategies have included lobbying, petitioning government authorities, influencing public policy through referendum campaigns and involvement in elections. Anti-nuclear groups have also tried to influence policy implementation through litigation and by participating in licensing proceedings.[4]

17.3 International organizations

- The ATOM Project, an International nonprofit organization seeking entry into force of the Nuclear Nonproliferation Treaty and the limitation of all nuclear arsenals.[5]

- European Nuclear Disarmament, which held annual conventions in the 1980s involving thousands of anti-nuclear weapons activists mostly from Western Europe but also from Eastern Europe, the United States, and Australia.[6]

- Friends of the Earth International, a network of environmental organizations in 77 countries.[7]

- Global Zero, an international non-partisan group of 300 world leaders dedicated to achieving the elimination of nuclear weapons.[8]

- Global Initiative to Combat Nuclear Terrorism, an international partnership of 83 nations.

- Greenpeace International, a non-governmental environmental organization[9] with offices in over 41 countries and headquarters in Amsterdam, Netherlands.[10]

- International Campaign to Abolish Nuclear Weapons

- International Network of Engineers and Scientists for Global Responsibility
- International Physicians for the Prevention of Nuclear War, which had affiliates in 41 nations in 1985, representing 135,000 physicians;*[6] IPPNW was awarded the UNESCO Peace Education Prize in 1984 and the Nobel Peace Prize in 1985.*[11]
- Nuclear Free World Policy
- Nuclear Information and Resource Service
- OPANAL
- Parliamentarians for Nuclear Non-Proliferation and Disarmament, a global network of over 700 parliamentarians from more than 75 countries working to prevent nuclear proliferation.*[12]
- Pax Christi International, a Catholic group which took a "sharply anti-nuclear stand" .*[6]
- Ploughshares Fund
- Pugwash Conferences on Science and World Affairs
- Socialist International, the world body of social democratic parties.*[13]
- Sōka Gakkai, a peace-orientated Buddhist organisation, which held anti-nuclear exhibitions in Japanese cities during the late 1970s, and gathered 10 million signatures on petitions calling for the abolition of nuclear weapons.*[13]*[14]
- The Ribbon International, a United Nations Non-Governmental Organization promoting nuclear disarmament.
- United Nations Office for Disarmament Affairs
- World Disarmament Campaign*[13]
- World Information Service on Energy, based in Amsterdam, The Netherlands
- World Union for Protection of Life

17.4   List of other organizations

Many of these groups are listed at "Protest movements against nuclear energy" in Wolfgang Rudig (1990). *Antinuclear Movements: A World Survey of Opposition to Nuclear Energy*, Longman, pp. 381–403.

- Alliance for Nuclear Accountability
- Alliance for Nuclear Responsibility
- Arms Control Association
- Australian Conservation Foundation
- Bellona Foundation
- Beyond Nuclear
- Campaign Against Nuclear Energy
- Campaign for Nuclear Disarmament
- Campaign for Nuclear Disarmament (NZ)
- Canadian Coalition for Nuclear Responsibility
- Canadian Voice of Women for Peace
- Christian CND
- Citizens' Nuclear Information Center
- Clamshell Alliance
- Coalition for Nuclear Power Postponement
- Committee for Non-Violent Action
- Committee for a Nuclear Free Island
- Committee for Nuclear Responsibility
- Council for a Livable World
- Critical Mass
- Cumbrians Opposed to a Radioactive Environment
- Don't Make a Wave Committee
- Earthlife Africa
- East Coast Solidarity for Anti-Nuke Group
- Economists for Peace and Security
- Energy Fair
- Energy Probe
- European Nuclear Disarmament
- Friends of the Earth (EWNI)
- Friends of the Earth Scotland
- Global Security Institute
- Greenpeace Aotearoa New Zealand
- Greenpeace Australia Pacific
- INFORSE-Europe

- Institute for Energy and Environmental Research
- International Campaign to Abolish Nuclear Weapons
- International Physicians for the Prevention of Nuclear War
- Koeberg Alert
- Klimaat -en vredesactiegroep Pimpampoentje - President Andy Vermaut
- Labour CND
- Legambiente
- Low Level Radiation Campaign
- MEDACT
- Musicians United for Safe Energy
- Natural Resources Defense Council
- Nevada Desert Experience
- Nevada Semipalatinsk

| width=1200 align=left |

- New England Coalition
- No Nukes group
- No New Nukes Y'all
- No to Nuclear Weapons
- One Less Nuclear Power Plant
- Nuclear Age Peace Foundation
- Nuclear Control Institute
- Nuclear Disarmament Party
- Nuclear Free World Policy
- Nuclear Information and Research Service
- Nuclear Threat Initiative
- NukeWatch
- Oak Ridge Peace and Environmental Alliance
- Operation Gandhi
- Peace Action
- Peace Boat
- Peace Organisation of Australia

- Pembina Institute
- People's Movement Against Nuclear Energy
- Performers and Artists for Nuclear Disarmament
- Physicians for Social Responsibility
- Plowshares Movement
- Public Citizen Energy Program
- Rocky Flats Truth Force
- Sayonara Nuclear Power Plants
- Scientists against Nuclear Arms
- Scottish Campaign for Nuclear Disarmament
- Seeds of hope
- Shad Alliance
- Sierra Club
- Sortir du nucléaire (Canada)
- Sortir du nucléaire (France)
- Stop Rokkasho
- The Seneca Women's Encampment for a Future of Peace and Justice
- The Wilderness Society (Australia)
- Top Level Group
- Trident Ploughshares
- Two Futures Project
- White House Peace Vigil
- Women from Fukushima Against Nukes
- Women Strike for Peace
- Women's Action for New Directions (WAND) previously called Women's Action for Nuclear Disarmament, forerunner organization: Women's Party for Survival
- Women's International League for Peace and Freedom

17.5   See also

- Anti-nuclear groups in the United States

- List of anti-nuclear power groups

- Anti-nuclear protests in the United States

- List of books about nuclear issues

- List of companies in the nuclear sector

- List of nuclear power groups

- List of Nuclear-Free Future Award recipients

- List of renewable energy organizations

- List of anti-war organizations

- List of peace activists

- Non-nuclear future

- Nuclear organizations (Wikipedia category)

17.6   References

[1] William A. Gamson and Andre Modigliani. Media Coverage and Public Opinion on Nuclear Power Archived March 24, 2012, at the Wayback Machine.. *American Journal of Sociology*, Vol. 95, No. 1, July 1989, p. 7.

[2] Fox Butterfield. Professional Groups Flocking to Antinuclear Drive, *The New York Times*, March 27, 1982.

[3] Jerome Price (1982). *The Anti-nuclear Movement*, Twayne Publishers, p. 65.

[4] Herbert P. Kitschelt. Political Opportunity and Political Protest: Anti-Nuclear Movements in Four Democracies *British Journal of Political Science*, Vol. 16, No. 1, 1986, p. 67.

[5] "The ATOM Project". Friends of the Earth International. Retrieved 2015-06-09.

[6] Lawrence S. Wittner (2009). *Confronting the Bomb: A Short History of the World Nuclear Disarmament Movement*, Stanford University Press, pp. 164-165.

[7] "About Friends of the Earth International". Friends of the Earth International. Retrieved 2009-06-25.

[8] http://www.globalzero.org?name=2.htm&id=2

[9] United Nations, Department of Public Information, Non-Governmental Organizations

[10] Greenpeace International: Greenpeace worldwide

[11] Profile from Helix Magazine

[12] Henry Mhara (Oct 17, 2011). "Coltart elected anti-nuclear organisation president". *News Day*.

[13] Lawrence S. Wittner (2009). *Confronting the Bomb: A Short History of the World Nuclear Disarmament Movement*, Stanford University Press, p. 128.

[14] Lawrence S. Wittner (2009). *Confronting the Bomb: A Short History of the World Nuclear Disarmament Movement*, Stanford University Press, p. 125.

Chapter 18

List of anti–nuclear power groups

Anti-nuclear power groups have emerged in every country that has had a nuclear power programme. Protest movements against nuclear power first emerged in the USA, at the local level, and spread quickly to Europe and the rest of the world. National nuclear campaigns emerged in the late 1970s. Fuelled by the Three Mile Island accident and the Chernobyl disaster, the anti-nuclear power movement mobilised political and economic forces which for some years "made nuclear energy untenable in many countries" .*[1]

Some of these anti-nuclear power organisations are reported to have developed considerable expertise on nuclear power and energy issues.*[2] In 1992, the chairman of the Nuclear Regulatory Commission said that "his agency had been pushed in the right direction on safety issues because of the pleas and protests of nuclear watchdog groups" .*[3]

18.1   International

- Friends of the Earth International, a network of environmental organizations in 77 countries.*[4]

- Greenpeace International, a non-governmental environmental organization*[5] with offices in over 41 countries and headquarters in Amsterdam, Netherlands.*[6]

- International Network of Engineers and Scientists for Global Responsibility

- Nuclear Information and Resource Service

- Pax Christi International, a Catholic group which took a "sharply anti-nuclear stand" .*[7]

- Pugwash Conferences on Science and World Affairs

- Socialist International, the world body of social democratic parties.*[8]

- Sōka Gakkai, a peace-orientated Buddhist organisation, which held anti-nuclear exhibitions in Japanese

cities during the late 1970s, and gathered 10 million signatures on petitions calling for the abolition of nuclear weapons.*[8]*[9]

- World Information Service on Energy, based in Amsterdam, The Netherlands

- World Union for Protection of Life

18.2   Australia

- Campaign Against Nuclear Energy

- Greenpeace Australia Pacific

18.3   Canada

- Pembina Institute

- Sortir du nucléaire (Canada)

18.4   France

- Sortir du nucléaire (France)

- TchernoBlaye

18.5   Japan

- Citizens' Nuclear Information Center

- Green Action Japan

18.6   New Zealand

- Greenpeace Aotearoa New Zealand

18.7   South Africa

- Koeberg Alert

18.8   United Kingdom

- Cumbrians Opposed to a Radioactive Environment

- Friends of the Earth (EWNI)

- Friends of the Earth Scotland

- Kick Nuclear

- Sustainable Development Commission

18.9   United States

- Arms Control Association

- Abalone Alliance

- Clamshell Alliance

- Institute for Energy and Environmental Research

- Musicians United for Safe Energy

- Natural Resources Defense Council

- New England Coalition

- Nuclear Watch South

- Shad Alliance

18.10   See also

- List of nuclear power groups

- Non-nuclear future

- List of anti-nuclear protests in the United States

- List of anti-nuclear groups in the United States

18.11   References

[1] Wolfgang Rudig (1990). *Anti-nuclear Movements: A World Survey of Opposition to Nuclear Energy*, Longman, p. 1.

[2] Lutz Mez, Mycle Schneider and Steve Thomas (Eds.) (2009). *International Perspectives of Energy Policy and the Role of Nuclear Power*, Multi-Science Publishing Co. Ltd. p. 279.

[3] Matthew L. Wald. Nuclear Agency's Chief Praises Watchdog Groups, *The New York Times*, June 23, 1992.

[4] "About Friends of the Earth International". Friends of the Earth International. Retrieved 2009-06-25.

[5] United Nations, Department of Public Information, Non-Governmental Organizations

[6] Greenpeace International: Greenpeace worldwide

[7] Lawrence S. Wittner (2009). *Confronting the Bomb: A Short History of the World Nuclear Disarmament Movement*, Stanford University Press, pp. 164-165.

[8] Lawrence S. Wittner (2009). *Confronting the Bomb: A Short History of the World Nuclear Disarmament Movement*, Stanford University Press, p. 128.

[9] Lawrence S. Wittner (2009). *Confronting the Bomb: A Short History of the World Nuclear Disarmament Movement*, Stanford University Press, p. 125.

Chapter 19

Campaign for Nuclear Disarmament

"CND" redirects here. For other uses, see CND (disambiguation).

The **Campaign for Nuclear Disarmament (CND)** is

The CND symbol, designed by Gerald Holtom in 1958. It has become a nearly universal peace symbol used in many different versions worldwide. [1]

an organisation that advocates unilateral nuclear disarmament by the United Kingdom, international nuclear disarmament and tighter international arms regulation through agreements such as the Nuclear Non-Proliferation Treaty. It opposes military action that may result in the use of nuclear, chemical or biological weapons and the building of nuclear power stations in the UK.

CND was formed in 1957 and since that time has periodically been at the forefront of the peace movement in the UK. It claims to be Europe's largest single-issue peace campaign. Between 1959 and 1965 it organised the Aldermaston March, which was held over the Easter weekend from the Atomic Weapons Establishment near Aldermaston to Trafalgar Square, London.

19.1 Campaigns

CND's current strategic objectives are:

- The elimination of British nuclear weapons and global abolition of nuclear weapons. It campaigns for the cancellation of Trident by the British government and against the deployment of nuclear weapons in Britain.

- The abolition of weapons of mass destruction, in particular chemical and biological weapons. CND wants a ban on the manufacture, testing and use of depleted uranium weapons

- A nuclear-free, less militarised and more secure Europe. It supports the Organisation for Security and Co-operation in Europe (OSCE). It opposes US military bases and nuclear weapons in Europe and British membership of NATO.

- The closure of the nuclear power industry. [2]

In recent years CND has extended its campaigns to include opposition to U.S. and British policy in the Middle East, rather as it broadened its anti-nuclear campaigns in the 1960s to include opposition to the Vietnam War. In collaboration with the Stop the War Coalition and the Muslim Association of Britain, CND has organised anti-war marches under the slogan "Don't Attack Iraq", including protests on 28 September 2002 and 15 February 2003. It also organised a vigil for the victims of the 2005 London bombings.

CND campaigns against the Trident missile. In March 2007 it organised a rally in Parliament Square to coincide with the Commons motion to renew the weapons system. The rally was attended by over 1,000 people. It was addressed by Labour MPs Jon Trickett, Emily Thornberry, John McDonnell, Michael Meacher, Diane Abbott and Jeremy Corbyn, and Elfyn Llwyd of Plaid Cymru and Angus MacNeil of the Scottish National Party. In the House of Commons, 161 MPs (88 of them Labour) voted against the renewal of

Trident and the Government motion was carried only with the support of Conservatives.[3]

In 2006 CND launched a campaign against nuclear power. Its membership, which had fallen to 32,000 from a peak of 110,000 in 1983, increased threefold after Prime Minister Tony Blair made a commitment to nuclear energy.[4]

19.2   Structure

CND is based in London and has national groups in Wales, Ireland and Scotland, regional groups in Cambridgeshire, Cumbria, the East Midlands, Kent, London, Manchester, Merseyside, Mid Somerset, Norwich, South Cheshire and North Staffordshire, Southern England, South West England, Suffolk, Surrey, Sussex, Tyne and Wear, the West Midlands and Yorkshire, and local branches.

There are five "specialist sections": Trade Union CND, Christian CND, Labour CND, Green CND and Ex-Services CND,[5] which have rights of representation on the governing council. There are also parliamentary, youth and student groups.

19.3   History

19.3.1   The First Wave: 1957–63

Bertrand Russell (centre), alongside his wife Edith and Ralph Schoenman with Michael Randle (second left), leading an anti-nuclear march in London, 18 February 1961

The Campaign for Nuclear Disarmament was founded in 1957 in the wake of widespread fear of nuclear conflict and the effects of nuclear tests. In the early 1950s Britain had become the third atomic power, after the USA and the USSR, and had recently tested an H-bomb.[6]

In November 1957 J. B. Priestley wrote an article for the

CND rally, in Aberystwyth, Wales May 25, 1961

New Statesman magazine, "Britain and the Nuclear Bombs",[7] advocating unilateral nuclear disarmament by Britain. In it he said:

> "In plain words: now that Britain has told the world she has the H-bomb she should announce as early as possible that she has done with it, that she proposes to reject, in all circumstances, nuclear warfare."

The article prompted many letters of support and at the end of the month the editor of the *New Statesman*, Kingsley Martin, chaired a meeting in the rooms of Canon John Collins in Amen Court to launch the Campaign for Nuclear Disarmament. Collins was chosen as its Chairman, Bertrand Russell as its President and Peggy Duff as its organising secretary. The other members of its executive committee were Martin, Priestley, Ritchie Calder, journalist James Cameron, Howard Davies, Michael Foot, Arthur Goss, and Joseph Rotblat. The Campaign was launched at a public meeting at Central Hall, Westminster, on 17 February 1958 attended by 5,000 people. After the meeting a few hundred left to demonstrate at Downing Street.[8][9]

The new organisation attracted considerable public interest and drew support from a range of interests, including scientists, religious leaders, academics, journalists, writers, actors and musicians. Its sponsors included John Arlott, Peggy Ashcroft, the Bishop of Birmingham Dr J. L. Wilson, Benjamin Britten, Viscount Chaplin, Michael de la Bédoyère, Bob Edwards, MP, Dame Edith Evans, A.S.Frere, Gerald Gardiner, QC, Victor Gollancz, Dr I.Grunfeld, E.M.Forster, Barbara Hepworth, Patrick

Heron, Rev. Trevor Huddleston, Sir Julian Huxley, Edward Hyams, the Bishop of Llandaff Dr Glyn Simon, Doris Lessing, Sir Compton Mackenzie, the Very Rev George McLeod, Miles Malleson, Denis Matthews, Sir Francis Meynell, Henry Moore, John Napper, Ben Nicholson, Sir Herbert Read, Flora Robson, Michael Tippett, the cartoonist 'Vicky', Professor C. H. Waddington and Barbara Wootton.[10] Other prominent founding members of CND were Fenner Brockway, E. P. Thompson, A. J. P. Taylor, Anthony Greenwood, Lord Simon, D. H. Pennington, Eric Baker and Dora Russell. Organisations that had previously opposed British nuclear weapons supported CND, including the British Peace Committee, the Direct Action Committee,[11] the National Committee for the Abolition of Nuclear Weapons Tests[10] and the Quakers.[12]

In the same year, a branch of CND was also set in the Republic of Ireland by John de Courcy Ireland, and his wife Beatrice, aiming to campaign for the Irish government to support international efforts to achieve nuclear disarmament and to keep Ireland free of nuclear power.[13] Notable supporters of the Irish CND included Peadar O'Donnell, Owen Sheehy-Skeffington and Hubert Butler.[14]

The formation of CND marked a significant change in the international peace movement, which from the late 1940s had been dominated by the World Peace Council (WPC), an anti-western organisation directed by the Soviet Communist Party. Because the WPC had a large budget and organised high-profile international conferences, the peace movement became identified with the Communist cause.[15] CND represented the growth of the unaligned peace movement and its detachment from the WPC.

With a general election due in 1959, which Labour was widely expected to win,[16] CND's founders envisaged a campaign by eminent individuals to secure a government that would adopt its policies: the unconditional renunciation of the use, production of or dependence upon nuclear weapons by Britain and the bringing about of a general disarmament convention; halting the flight of planes armed with nuclear weapons; ending nuclear testing; not proceeding with missile bases; and not providing nuclear weapons to any other country.[10]

In Easter 1958, CND, after some initial reluctance, supported a march from London to the Atomic Weapons Research Establishment at Aldermaston (a distance of 52 miles), that had been organised by a small pacifist group, the Direct Action Committee. Thereafter, CND organised annual Easter marches from Aldermaston to London that became the main focus for supporters' activity. 60,000 people participated in the 1959 march and 150,000 in the 1961 and 1962 marches.[17][18] The 1958 march was the subject of a documentary by Lindsay Anderson, *March to Al-*

dermaston.

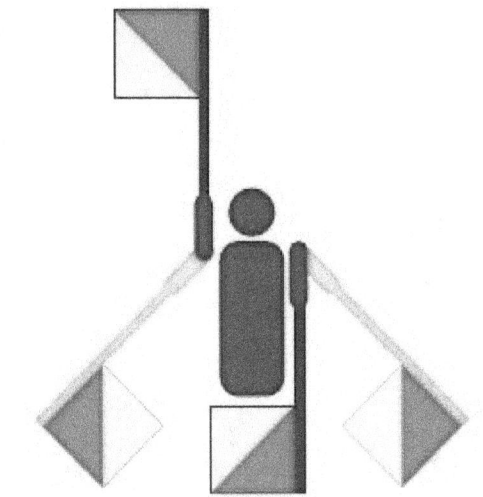

The flag semaphore symbols for letters "N" (green) and "D" (blue)

The symbol adopted by CND, designed for them in 1958 by Gerald Holtom,[10] became the international peace symbol. It is based on the semaphore symbols for "N" (two flags held 45 degrees down on both sides, forming the triangle at the bottom) and "D" (two flags, one above the head and one at the feet, forming the vertical line) (for Nuclear Disarmament) within a circle.[19] Holtom later said that it also represented "an individual in despair, with hands palm outstretched outwards and downwards in the manner of Goya's peasant before the firing squad," (although in that painting, *The Third of May 1808*, the peasant is actually holding his hands *upwards*).[20] The CND symbol, the Aldermaston march, and the slogan "Ban the Bomb" became icons and part of the youth culture of the 1960s.

CND's supporters were generally left of centre in politics. About three-quarters were Labour voters[12] and many of the early executive committee were Labour Party members.[10] The ethos of CND at that time was described as "essentially that of middle-class radicalism".[21]

In the event, Labour lost the 1959 election, but it voted at its 1960 Conference for unilateral nuclear disarmament, which represented CND's greatest influence and coincided with the highest level of public support for its programme.[22] The resolution was passed against the wishes of the party's leaders, who refused to be bound by it and proceeded to organise to have it overturned at the next conference.[23] Hugh Gaitskell, the Labour Party leader, promised to "fight, fight, and fight again" against the decision, which was duly overturned at the 1961 Conference. Labour's failure to win the election and its rejection of unilateralism upset CND's plans, and from about 1961 its prospects of success began to fade. It was said that from that time onward it lacked any clear idea of how nuclear disarmament was to be implemented and that its demonstrations had become an end in themselves.ᵃ[24] The sociologist Frank Parkin said that, for many supporters, the question of implementation was of secondary importance anyway because, for them, involvement in the campaign was "an expressive activity in which the defence of principles was felt to have higher priority than 'getting things done'." ᵃ[12] He suggested CND's survival in the face of its failure was explained by the fact that it provided "a rallying point and symbol for radicals", which was more important for them than "its manifest function of attempting to change the government's nuclear weapons policy." ᵃ[12] Despite setbacks, it retained the support of a significant minority of the population and became a mass movement, with a network of autonomous branches and specialist groups and an increased participation in demonstrations until about 1963.

In 1960 Bertrand Russell resigned from the Campaign in order to form the Committee of 100, which became, in effect, the direct action wing of CND. Russell argued that direct action was necessary because the press was losing interest in CND and because the danger of nuclear war was so great that it was necessary to obstruct government preparations for it.ᵃ[25] In 1958 CND had cautiously accepted direct action as a possible method of campaigning,ᵃ[10] but, largely under the influence of its chairman, Canon Collins, the CND leadership opposed any sort of unlawful protest. The Committee of 100 was created as a separate organisation partly for that reason and partly because of personal animosity between Collins and Russell. Although the Committee was supported by many in CND, it has been suggestedᵃ[26] that the campaign against nuclear weapons was weakened by the friction between the two organisations. The Committee organised large sit-down demonstrations in London and at military bases. It later diversified into other political campaigns, including Biafra, the Vietnam war and housing in the UK. It was dissolved in 1968. When direct action came to the fore again in the 1980s, it was generally accepted by the peace movement as a normal part of protest.ᵃ[27]

CND's executive committee did not give its supporters a voice in the Campaign until 1961, when a national council was formed and until 1966 it had no formal membership. The relationship between supporters and leaders was unclear, as was the relationship between the executive and the local branches. The executive committee's lack of authority made possible the inclusion within CND of a wide range of views, but it resulted in lengthy internal discussions and the adoption of contradictory resolutions at conferences.ᵃ[24] There was friction between the founders, who conceived of CND as a campaign by eminent individuals focused on the Labour Party, and CND's supporters (including the more radical members of the executive committee), who saw it as an extra-parliamentary mass move-

ment. Collins was unpopular with many supporters because of his strictly constitutional approach and found himself increasingly out of sympathy with the direction the movement was taking.ᵃ[28] He resigned in 1964 and put his energies into the International Confederation for Disarmament and Peace.ᵃ[29]

The Cuban Missile Crisis in the Autumn of 1962, in which the United States blockaded a Soviet attempt to put nuclear missiles on Cuba, created widespread public anxiety about imminent nuclear war and CND organised demonstrations on the issue. But six months after the crisis, a Gallup Poll found that public concern about nuclear weapons had fallen to its lowest point since 1957,ᵃ[10] and there was a view (disputed by some CND supporters)ᵃ[30] that U.S. President John F. Kennedy's success in facing down Soviet premier Nikita Khrushchev turned the British public away from the idea of unilateral nuclear disarmament.

On the 1963 Aldermaston march, a clandestine group calling itself Spies for Peace distributed leaflets about a secret government establishment, RSG 6, that the march was passing. The people behind Spies for Peace remain unknown, except for Nicholas Walter, a leading member of the Committee of 100.ᵃ[31] The leaflet said that RSG 6 was to be the local HQ for a military dictatorship after nuclear war. A large group left the march, against the wishes of the CND leadership, to demonstrate at RSG 6. Later, when the march reached London, there were disorderly demonstrations in which anarchists were prominent, quickly deprecated in the press and in parliament.ᵃ[10] In 1964 there was only a one-day march, partly because of the events of 1963 and partly because the logistics of the march, which had grown beyond all expectation, had exhausted the organisers.ᵃ[8] The Aldermaston March was resumed in 1965.

Support for CND dwindled after the 1963 Test Ban Treaty, one of the things it had been campaigning for. From the mid-1960s, the anti-war movement's preoccupation with the Vietnam War tended to eclipse concern about nuclear weapons but CND continued to campaign against both.

Although CND has never formally allied itself to any political party and has never been an election campaigning body, CND members and supporters have stood for election at various times on a nuclear disarmament ticket. The nearest CND has come to having an electoral arm was the Independent Nuclear Disarmament Election Campaign (INDEC) which stood candidates in a few local elections during the 1960s. INDEC was never endorsed by CND nationally and candidates were generally put up by local branches as a means of raising the profile of the nuclear threat.

19.3.2   The Second Wave: 1980–83

In the 1980s, CND underwent a major revival in response to the resurgence of the Cold War.[21] There was increasing tension between the superpowers following the deployment of SS20s in the Soviet Bloc countries, American Pershing missiles in Western Europe, and Britain's replacement of the Polaris armed submarine fleet with Trident missiles.[21] The NATO exercise Able Archer 83 also added to international tension.

CND's membership increased rapidly, and in the early 1980s it claimed 90,000 national members and a further 250,000 in local branches.[21]   "This made it one of the largest political organisations in Britain and probably the largest peace movement in the world (outside the state-sponsored movements of the Communist bloc)."[21] Public support for unilateralism reached its highest level since the 1960s.[32] In October 1981, 250,000 people joined an anti-nuclear demonstration in London. CND's demonstration on the eve of Cruise missile deployment in October 1983 was one of the largest in British history,[21] with 300,000 taking part in London as three million protested across Europe.[33]

1983 Easter CND march around the Atomic Weapons Research Establishment (AWRE) at Aldermaston

New sections were formed, including Ex-services CND, Green CND, Student CND, Tories Against Cruise and Trident (TACT), Trade Union CND, and Youth CND. More women than men supported CND.[8] The campaign attracted supporters who opposed the Government's civil defence plans as outlined in an official booklet, *Protect and Survive*. This publication was ridiculed in a popular pamphlet, *Protest and Survive*, by E. P. Thompson, a leading anti-nuclear campaigner of the period.

The British anti-nuclear movement at this time differed from that of the 1960s. Many groups sprang up independently of CND, some affiliating later. CND's previous objection to civil disobedience was dropped and it became a normal part of anti-nuclear protest. The women's move-

ment had a strong influence, much of it emanating from the Greenham Common Women's Peace Camp,[8] followed by Molesworth People's Peace Camp.

A network of protesters, calling itself Cruise Watch, tracked and harassed Cruise missiles whenever they were carried on public roads. After a while, the missiles traveled only at night under police escort.

At its 1982 conference, the Labour Party adopted a policy of unilateral nuclear disarmament. It lost the 1983 general election "in which, following the Falklands war, foreign policy was high on the agenda. Election defeats under, first, Michael Foot, then Neil Kinnock, led Labour to abandon the policy in the late 1980s."[34] The re-election of a Conservative government in 1983 and the defeat of left-wing parties in continental Europe "made the deployment of Cruise missiles inevitable and the movement again began to lose steam."[21]

19.4   Extent of support for CND policies

In 2006, CND had 32,000 members.[35] As it did not have a national membership until 1966, the strength of public support in its early days can be estimated only from the numbers of those attending demonstrations or expressing approval in opinion polls. Polls on a number of related issues have been taken over the past fifty years.

- Between 1955 and 1962, between 19% and 33% of people in Britain expressed disapproval of the manufacture of nuclear weapons.[36]

- Public support for unilateralism in September 1982 was 31%, falling to 21% in January 1983, but it is hard to say whether this decline was a result of the contemporary propaganda campaign against CND or not.[32]

- Support for CND fell after the end of the Cold war. It had not succeeded in converting the British public to unilateralism and even after the collapse of the Soviet Union British nuclear weapons still have majority support.[32] "Unilateral disarmament has always been opposed by a majority of the British public, with the level of support for unilateralism remaining steady at around one in four of the population."[22][37]

- In 2005, MORI conducted an opinion poll which asked about attitudes to Trident and the use of nuclear weapons. When asked whether the UK should replace Trident, without being told of the cost, 44% of respondents said "Yes" and 46% said "No". When asked the same question and told of the cost, the proportion saying "Yes" fell to 33% and the proportion saying "No" increased to 54%.*[38]

- In the same poll, MORI asked "Would you approve or disapprove of the UK using nuclear weapons against a country we are at war with?". 9% approved if that country did not have nuclear weapons, and 84% disapproved. 16% approved if that country had nuclear weapons but never used them, and 72% disapproved. 53% approved if that country used nuclear weapons against the UK, and 37% disapproved.*[38]

- CND's policy of opposing American nuclear bases is said to be in tune with public opinion.*[21]

On three occasions the Labour Party, when in opposition, has been significantly influenced by CND in the direction of unilateral nuclear disarmament. Between 1960-1961 it was official Party policy although the Labour leader Hugh Gaitskell opposed the decision and succeeded in quickly reversing it. In 1980 long time CND supporter Michael Foot became Labour Party leader and in 1982 succeeded in changing official Labour policy in line with his views. After losing the 1983 and 1987 general elections Labour leader Neil Kinnock persuaded the party to abandon unilateralism in 1989.*[39] In 2015 another long time CND supporter Jeremy Corbyn was elected leader of the Labour Party, although the official Labour policy has not changed as yet in line with his views.*[40]

19.5 Organised opposition to CND

CND's growing support in the 1980s provoked opposition from several sources, including Peace Through Nato, the British Atlantic Committee (which received government funding).*[41] Women and Families for Defence (set up by Conservative journalist and later MP Lady Olga Maitland to oppose the Greenham Common Peace Camp), the Conservative Party's Campaign for Defence and Multilateral Disarmament, the Coalition for Peace through Security, the Foreign Affairs Research Institute, and The 61, a private sector intelligence agency. The British government also took direct steps to counter the influence of CND. Secretary of State for Defence Michael Heseltine setting up Defence Secretariat 19 "to explain to the public the facts about the Government's policy on deterrence and multilateral disarmament" .*[42] The activities of anti-CND organisations are said to have included research, publication, mobilising public opinion, counter-demonstrations, working within the Churches, smears against CND leaders and spying.

In an article on anti-CND groups, Stephen Dorril reported that in 1982 Eugene V. Rostow, Director of the US Arms Control and Disarmament Agency, became concerned about the growing unilateralist movement. According to Dorril, Rostow helped to initiate a propaganda exercise in Britain, "aimed at neutralising the efforts of CND. It would take three forms: mobilising public opinion, working within the Churches, and a 'dirty tricks' operation against the peace groups." *[43]

One of the groups set up to carry out this work was the Coalition for Peace through Security (CPS), modelled on the US Coalition for Peace through Strength. The CPS was founded in 1981. Its main activists were Julian Lewis, Edward Leigh and Francis Holihan.*[43] Amongst the activities of the CPS were commissioning Gallup polls*[44] which showed the levels of support for British possession of nuclear weapons, providing speakers at public meetings, highlighting the left-wing affiliations of leading CND figures and mounting counter-demonstrations against CND. These including haranguing CND marchers from the roof of the CPS's Whitehall office and flying a plane over a CND festival with a banner reading, "Help the Soviets, Support CND!"*[45] The CPS attracted criticism for refusing to say where its funding came from while alleging that the anti-nuclear movement was funded by the Soviet Union.*[46] Although the CPS called itself a grass-roots movement, it had no members and was financed by The 61,*[45] "a private sector operational intelligence agency" *[47] said by its founder, Brian Crozier, to be funded by "rich individuals and a few private companies" .*[48] It is said to have also received funding from the Heritage Foundation.*[49]

The CPS claimed that Bruce Kent, the general secretary of CND and a Catholic priest, was a supporter of IRA terrorism.*[45] Kent alleged in his autobiography that Francis Holihan spied on CND. Dorril claimed*[43]

> that Holihan had organised aerial propaganda, had entered CND offices under false pretences, and that CPS workers had joined CND in order to gain access to the Campaign's 1982 Annual Conference. When Bruce Kent went on a speaking tour of America, Holihan followed him around. Offensive material on Kent was sent to newspapers and radio stations, and demonstrations were organised against him with support from the College Republican Committee.

Gerald Vaughan, a government minister, tried to halve government funding for the Citizens Advice Bureau, apparently because Joan Ruddock, CND's chair, was employed part-time at his local bureau. Bruce Kent was warned by Cardinal Basil Hume not to become too involved in politics.

19.6 Allegations of Communist influence and intelligence surveillance

CND's opponents claimed that CND was a Communist or Soviet-dominated organisation, a charge its supporters denied. In the 1960s MI5 designated it "communist controlled", but there was no evidence of direct Soviet influence or funding, and an MI5 officer who went public said there was no evidence that it was controlled by any other extreme left-wing elements either.[50]

In 1981, the Foreign Affairs Research Institute, which shared an office with the CPS, was said by *Sanity*, the CND newspaper, to have published a booklet claiming that Russian money was being used by CND.[43] In response to Lord Chalfont's claim in that the Soviet Union was giving the European peace movement £100 million a year, Bruce Kent said: "If they were, it was certainly not getting to our grotty little office in Finsbury Park." [51] In the 1980s, the Federation of Conservative Students (FCS) claimed that one of CND's elected officers, Dan Smith, was a communist. CND sued for defamation and the FCS settled on the second day of the trial, apologised and paid damages and costs.[52]

The security service (MI5) carried out surveillance of CND members it considered to be subversive. From the late 1960s until the mid-1970s, MI5 designated CND as subversive by virtue of its being "communist controlled". Communists have played an active role in the organisation, and John Cox, its chairman from 1971 to 1977, was a member of the Communist Party of Great Britain.[53] From the late 1970s, MI5 downgraded CND to "communist-penetrated". MI5 says it has no current investigations in this area.[54]

The British journalist Charles Moore reported a conversation he had with the Soviet double agent Oleg Gordievsky after the death of leading Labour politician Michael Foot. As editor of the newspaper *Tribune*, says Moore, Foot was regularly visited by KGB agents who identified themselves as diplomats and gave him money. "A leading supporter of the Campaign for Nuclear Disarmament, Foot ... passed on what he knew about debates over nuclear weapons. In return, the KGB gave him drafts of articles encouraging British disarmament which he could then edit and publish, unattributed to their real source, in *Tribune*." [55] Foot had received libel damages from the *Sunday Times* for a similar claim made earlier.[56]

In 1985, Cathy Massiter, an MI5 officer who had been responsible for the surveillance of CND from 1981 to 1983, resigned and made disclosures to a Channel 4 *20/20 Vision* programme, "MI5's Official Secrets".[57][58] She said that her work was determined more by the political importance of CND than by any security threat posed by subversive elements within it. In 1983, she analysed telephone intercepts on John Cox that gave her access to conversations with Joan Ruddock and Bruce Kent. MI5 also placed a spy, Harry Newton, in the CND office. According to Massiter, Newton believed that CND was controlled by extreme left-wing activists and that Bruce Kent might be a crypto-communist, but Massiter found no evidence to support either opinion.[50] On the basis of Ruddock's contacts, MI5 suspected her of being a communist sympathiser. Speaking in the House of Commons, Dale Campbell-Savours, MP, said:

> "it was felt within the service that officers were likely to be questioned about the true political affiliation of Mrs. Joan Ruddock, who became chair of CND in 1983. It was fully recognised by the service that she had no subversive affiliations and therefore should not be recorded under any of the usual subversive categories. In fact, she was recorded as a contact of a hostile intelligence service after giving an interview to a Soviet journalist based in London who was suspected of being a KGB intelligence officer. In Joan Ruddock's file, MI5 recorded special branch references to her movements—usually public meetings—and kept press cuttings and the products of mail and telephone intercepts obtained through active investigation of other targets, such as the Communist party and John Cox. There were police reports recording her appearances at demonstrations or public meetings. There were references to her also in reports from agents working, for example, in the Communist party. These would also appear in her file." [58]

According to Stephen Dorril, at about the same time, Special Branch officers recruited an informant within CND, Stanley Bonnett, on the instructions of MI5.[49] MI5 is also said to have suspected CND's treasurer, Cathy Ashton, of being a communist sympathiser[53] because she shared a house with a communist.[49] When Michael Heseltine became Secretary of State for Defence in 1983, Massiter was asked to provide information for Defence Secretariat 19 (DS19) about leading CND personnel but was instructed to include only information from published sources. Ruddock claims that DS19 released distorted information regarding her political party affiliations to the media and Conservative Party candidates.[59]

Brian Crozier claimed in his book *Free Agent: The Unseen War 1941-1991* (Harper Collins, 1993) that The 61 infiltrated a mole into CND in 1979.[49]

In 1990, it was discovered in the archive of the Stasi (the state security service of the former German Democratic Republic) that a member of CND's governing council, Vic Allen, had passed information to them about CND. This discovery was made public in a BBC TV programme in 1999, reviving debate about Soviet links to CND. Allen stood against Joan Ruddock for the leadership of CND in 1985, but was defeated. Ruddock responded to the Stasi revelations by saying that Allen "certainly had no influence on national CND, and as a pro-Soviet could never have succeeded to the chair," and that "CND was as opposed to Soviet nuclear weapons as Western ones." *[60]*[61]

19.7   Chairs of CND since 1958

- Canon John Collins 1958–64
- Olive Gibbs 1964–67
- Sheila Oakes 1967–68
- Malcolm Caldwell 1968–70
- April Carter 1970–71
- John Cox 1971–77
- Bruce Kent 1977–79
- Hugh Jenkins 1979–81
- Joan Ruddock 1981–85
- Paul Johns 1985–87
- Bruce Kent 1987–90
- Marjorie Thompson 1990–93
- Janet Bloomfield 1993–96
- David Knight 1996–2001
- Carol Naughton 2001–03
- Kate Hudson 2003–10
- Dave Webb 2010– present

19.8   General Secretaries of CND since 1958

- Peggy Duff 1958–67
- Dick Nettleton 1967–73
- Dan Smith 1974–75

- Duncan Rees 1976–79
- Bruce Kent 1979–85
- Meg Beresford 1985–90
- Gary Lefley, 1990–94

The post was abolished in 1994, and reinstated in 2010.

- Kate Hudson, 2010–

19.9   Archives

Much of National CNDs historical archive is at the London School of Economics and the *Modern Records Centre* at the University of Warwick. Records of local and regional groups are spread throughout the country in public and private collections.

- Catalogue of CNDs papers held at LSE Archives.

19.10   See also

- Anti-nuclear movement in the United Kingdom
- Anti-war
- Campaign for Nuclear Disarmament (NZ)
- Counterculture of the 1960s
- European Nuclear Disarmament
- European Peace Marches
- Greenham Common Women's Peace Camp
- Independent Nuclear Disarmament Election Committee
- Koeberg Alert
- List of anti-war organizations
- List of peace activists
- Nuclear disarmament
- Nuclear-Free Future Award
- Nuclear-free zone
- Nuclear Information Service
- Nuclear proliferation

- Nuclear weapons and the United Kingdom
- Peace movement
- Peace symbols
- World Peace Council
- Youth for Multilateral Disarmament (YMD)

19.11 References

[1] 'BBC NEWS : Magazine : World's best-known protest symbol turns 50'. *BBC News*. London. 20 March 2008. Retrieved 2008-05-25.

[2] "CND aims and policies". Cnduk.org. Retrieved 2011-01-09.

[3]

[4] Herbert, Ian (2006-07-17). "CND membership booms after nuclear U-turn". London: Independent.co.uk. Retrieved 2011-01-09.

[5] "CND Constitution" (PDF). Retrieved 2011-01-09.

[6] CND. *The history of CND*

[7] J. B. Priestley. "Britain and the Nuclear Bombs". *New Statesman*, 2 November 1957.

[8] John Minnion and Philip Bolsover (eds). *The CND Story*, Allison and Busby, 1983. ISBN 0-85031-487-9

[9] "Campaign for Nuclear Disarmament (CND)". Spartacus.schoolnet.co.uk. Retrieved 2011-01-09.

[10] Christopher Driver, *The Disarmers: A Study in Protest*, Hodder and Stoughton, 1964

[11] "The history of CND". Cnduk.org. 1945-08-06. Retrieved 2011-01-09.

[12] Frank Parkin, *Middle Class Radicalism: The Social Bases of the Campaign for Nuclear Disarmament*, Manchester University Press, 1968, p. 39.

[13] Fagan, Kieran (9 April 2006). "John de Courcy Ireland". Obituary. *Irish Independent*. Retrieved 4 March 2011.

[14] Richard S. Harrison, *Irish Anti-War Movements*. Dublin : Irish Peace 1986 (pp. 59-61).

[15] Rainer Santi, *100 years of peace making*. International Peace Bureau, January 1991

[16] David E. Butler and Richard Rose, *The British General Election of 1959* (1960)

[17] Peter Barberis, John McHugh, Mike Tyldesley, *Encyclopedia of British and Irish Political Organizations*, Continuum International Publishing, 2005. ISBN 0-8264-5814-9

[18] April Carter, "Campaign for Nuclear Disarmament", in Linus Pauling, Ervin László, and Jong Youl Yoo (eds), *The World Encyclopedia of Peace*, Oxford: Pergamon, 1986. ISBN 0-08-032685-4, (vol. 1, pp. 109-113).

[19] "Early Defections in March", *Manchester Guardian*, 5 April 1958

[20] Information, Campaign for Nuclear Disarmament

[21] James Hinton "Campaign for Nuclear Disarmament", in Roger S.Powers, *Protest, Power and Change*, Taylor and Francis, 1997, p. 63, ISBN 0-8153-0913-9

[22] April Carter, *Direct Action and Liberal Democracy*, London: Routledge and Kegan Paul, 1973, p. 64.

[23] Robert McKenzie, "Power in the Labour Party: The Issue of 'Intra-Party Democracy'", in Dennis Kavanagh, *The Politics of the Labour Party*, Routledge, 2013.

[24] Peers, Dave, "The impasse of CND", *International Socialism*, No. 12, Spring 1963, pp. 6-11.

[25] Russell, B., "Civil Disobedience", *New Statesman*, 17 February 1961

[26] Taylor, R., *Against the Bomb*, Oxford University Press, 1988.

[27] "A brief history of CND". Cnduk.org. 1945-08-06. Retrieved 2011-01-09.

[28] "Collins, (Lewis) John", *Oxford Dictionary of National Biography*

[29] Oxford Conference of Non-aligned Peace Organizations

[30] Nigel Young, "Cuba '62", in Minnion and Bolsover, p61

[31] Natasha Walter, "How my father spied for peace", *New Statesman*, 20 May 2002

[32] Caedel, Martin, "Britain's Nuclear Disarmers", in Laqueur, W., *European Peace Movements and the Future of the Western Alliance*, Transaction Publishers, 1985, p. 233, ISBN 0-88738-035-2

[33] David Cortright, *Peace: A History of Movements and Ideas*, Cambridge University Press, 2008 ISBN 0-521-85402-4

[34] Anti-war Activism in the Information Age

[35] Finlo Rohrer, "Whatever happened to CND?", *BBC News Magazine*, 5 July 2006

[36] W. P. Snyder, *The Politics of British Defense Policy, 1945-1962*, Ohio University Press, 1964.

[37] Andy Byrom, "British attitudes on nuclear weapons", *Journal of Public Affairs*, 7: 71-77, 2007.

[38] "British Attitudes to Nuclear Weapons"

[39] Kinnock wins accord on defence switch | 1980-1989 | Guardian Century

[40] Jeremy Corbyn courts new anti-nuclear row by becoming vice-president of CND - Telegraph

[41] "Lords Hansard". Hansard.millbanksystems.com. 1981-12-17. Retrieved 2011-01-09.

[42] "Commons Hansard". Hansard.millbanksystems.com. 1986-07-21. Retrieved 2011-01-09.

[43] *The Lobster*, No.3, 1984

[44]

[45] Wittner, L., *The Struggle Against the Bomb*, Volume 3, Stanford University Press, 2003.

[46] Bruce Kent, *Undiscovered Ends*, pp. 179-181.

[47] Joseph C. Goulden, "Crozier, covert acts, CIA and Cold War". *The Washington Times*, 15 May 1994

[48] Brian Crozier, Letters: Churchill, the CIA and Clinton, *The Guardian*, 3 August 1998.

[49] Tom Mills, Tom Griffin and David Miller, "The Cold War on British Muslims", Spinwatch, 2011.

[50] Bateman, D., "The Trouble With Harry: A memoir of Harry Newton, MI5 agent". *Lobster*, Issue 28, December 1994. Accessed 3 November 2011.

[51] Hudson, Kate, "Soviet funding? Rubbish". CND website

[52] Bruce Kent, *Undiscovered Ends*, Fount, 1992, pp.185-6 ISBN 0002159961

[53] Gallagher, Ian; Boffey, Daniel (2009-11-22). "EU's new 'Foreign Minister' Cathy Ashton was Treasurer of CND". London: Daily Mail. Retrieved 2011-01-09.

[54] "Myths and Misunderstandings". Mi5.gov.uk. Retrieved 2011-01-09.

[55] Charles Moore, "Was Foot a national treasure or the KGB's useful idiot?" *The Telegraph*, 5 March 2010

[56] Rhys Williams, "'Sunday Times' pays Foot damages over KGB claim". *Independent*, Sunday 23 October 2011

[57] "Secret State: Timeline". BBC News. 2002-10-17. Retrieved 2011-01-09.

[58] "Dale Campbell-Savours, MP, in Business of the House". *Hansard*. 24 July 1986. Retrieved 2011-01-09.

[59] "Domestic Intelligence Agencies: The Mixed Record of the UK's MI5" (PDF). Center for Democracy and Technology. Retrieved 2011-01-09.

[60] Department of the Official Report (Hansard), House of Commons, Westminster. "Commons Hansard". Publications.parliament.uk. Retrieved 2011-01-09.

[61] "I regret nothing, says Stasi spy". BBC News. 1999-09-20. Retrieved 2011-01-09.

19.12   Further reading

- James Aulich, *War Posters: Weapons of Mass Communication* (New York: Thames & Hudson, 2007), ISBN 9780500251416

- Ross Bradshaw, *From Protest to Resistance*, A *Peace News* pamphlet (London: Mushroom Books, 1981), ISBN 0-907123-02-3

- Paul Byrne, *Social Movements in Britain* (London: Routledge, 1997), ISBN 0-415-07123-2

- Paul Byrne, *The Campaign for Nuclear Disarmament* (Croom Helm: London, 1988), ISBN 0-7099-3260-X

- Christopher Driver, *The Disarmers: A Study in Protest* (London: Hodder and Stoughton, 1964)

- Peggy Duff, *Left, Left, Left: A personal account of six protest campaigns 1945-65* (London: Allison and Busby, 1971), ISBN 0-85031-056-3

- Kate Hudson, *CND - Now More Than Ever: The Story of a Peace Movement* (London: Vision Paperbacks, 2005), ISBN 1-904132-69-3

- John Mattausch, *A Commitment to Campaign: A Sociological Study of CND* (Manchester University Press, 1989), ISBN 0-7190-2908-2

- John Minnion and Philip Bolsover (eds), *The CND Story: The first 25 years of CND in the words of the people involved* (London: Allison & Busby, 1983), ISBN 0-85031-487-9

- Holger Nehring, "Diverging perceptions of security: NATO and the protests against nuclear weapons", in Andreas Wenger, et al. (eds), *Transforming NATO in the Cold War: Challenges beyond Deterrence in the 1960s* (London: Routledge, 2006)

- Holger Nehring, "From Gentleman's Club to Folk Festival: The Campaign for Nuclear Disarmament in Manchester, 1958-63", *North West Labour History Journal*, No. 26 (2001), pp. 18–28

- *Holger Nehring*, "National Internationalists: British and West German Protests against Nuclear Weapons, the Politics of Transnational Communications and the Social History of the Cold War, 1957–1964", *Contemporary European History*, 14, No. 4(2006)

- *Holger Nehring*, "Politics, Symbols and the Public Sphere: The Protests against Nuclear Weapons in Britain and West Germany, 1958-1963", *Zeithistorische Forschungen*, 2, No. 2 (2005)

- *Holger Nehring,* "The British and West German Protests against Nuclear Weapons and the Cultures of the Cold War, 1957–64", *Contemporary British History,* 19, No. 2 (2005)

- Frank Parkin, *Middle-class radicalism: The Social Bases of the British Campaign for Nuclear Disarmament* (Manchester University Press, 1968)

- Richard Taylor and Colin Pritchard, *The Protest Makers: The British Nuclear Disarmament of 1958-1965, Twenty Years On* (Oxford: Pergamon Press, 1980), ISBN 0-08-025211-7

- Byrne, Paul (1997). *Social Movements in Britain.* Routledge. p. 91. ISBN 0-415-07123-2.

19.13 External links

Official media pages

- Official website

- Facebook

- Twitter

- YouTube channel

News items

- McGuffin, Paddy (7 March 2007). "A new generation of CND goes on the march". *Telegraph & Argus.*

- "Anniversary demo at nuclear site". *BBC Online.* 2 January 2008.

- Campbell, Duncan; Williams, Rachel (16 February 2008). "CND veterans remain unbowed, 50 years on". *The Guardian.* London.

- Sengupta, Kim (7 January 2012). "How Thatcher's election win launched secret war on CND". *The Independent.* London.

- CND membership surge gathers pace after Jeremy Corbyn election. *The Guardian.* Published 16 October 2015. Retrieved 10 January 2017.

- Campaign for Nuclear Disarmament marches back into public arena after years of decline. *The Independent.* Published 29 January 2016. Retrieved 10 January 2017.

Historic

- "Thousands protest against H-bomb". *BBC Online.* 1960-04-18. Retrieved 8 January 2012. - Report of the 1960 Aldermaston March

- BBC Report of CND Protest in London 22 October 1983

- 20/20 Vision: *MI5's Official Secrets*

- Exhibition - CND: The story of a peace movement (LSE Archives)

- Catalogue of the CND archives, held at the Modern Records Centre, University of Warwick

- 'If at first you don't succeed...:fighting against the bomb in the 1950s and 1960s' by Rip Bulkeley, Pete Goodwin, Ian Birchall, Peter Binns and Colin Sparks, *International Socialism* journal, 2:11, (Winter 1981) - a short Marxist history of CND

Other

- Anti-Nuclear.com

- A British Museum expert's view of the CND badge

Chapter 20

International Association of Lawyers against Nuclear Arms

The **International Association of Lawyers against Nuclear Arms** (**IALANA**) is an international non-governmental organisation headquartered in The Hague. It was founded in 1988 and seeks "to build and strengthen international legal efforts to ban the use and threat of use of nuclear weapons." *[1] Its membership consists of individual lawyers and lawyer's organisations. The current President of the organisation is Christopher Weeramantry, a former Judge of the International Court of Justice. The German section of the organisation was co-founded by former German Minister of Justice Herta Däubler-Gmelin. IALANA had a central role in the process that sought an advisory ruling on the legality of nuclear arms from the International Court of Justice.*[1]

The organisation has consultative status with the United Nations Economic and Social Council.

20.1 See also

- International Court of Justice advisory opinion on the Legality of the Threat or Use of Nuclear Weapons

- Parliamentarians for Nuclear Non-Proliferation and Disarmament

20.2 References

[1] Sarah J. Diehl, James Clay Moltz, *Nuclear Weapons and Nonproliferation: A Reference Book* (p. 277), 2002

20.3 External links

- Official website

Chapter 21

International Campaign to Abolish Nuclear Weapons

The **International Campaign to Abolish Nuclear Weapons** (abbreviated to **ICAN**; pronounced /ˈaɪkæn/ *EYE-kan*) is a global civil society coalition working to mobilize people in all countries to inspire, persuade and pressure their governments to initiate and support negotiations for a nuclear-weapon-ban treaty. ICAN was launched in 2007 and today counts more than 440 partner organizations in 100 countries. ICAN calls on states, international organizations and other actors to:

·Acknowledge that any use of nuclear weapons would cause catastrophic humanitarian harm.

·Acknowledge that there exists a universal humanitarian imperative to ban nuclear weapons, even for states that do not possess these weapons.

·Acknowledge that the nuclear possessors have an obligation to eliminate their nuclear weapons.

·Take immediate action to support a multilateral process of negotiations for a treaty banning nuclear weapons.

21.1   Mission

ICAN aims to galvanize public and government support for a multilateral process for a treaty banning nuclear weapons. ICAN seeks to shift the disarmament debate to focus on the humanitarian threat posed by nuclear weapons, drawing attention to their unique destructive capacity, their catastrophic health and environmental consequences, their indiscriminate targeting, the debilitating impact of a detonation on medical infrastructure and relief measures, and the long-lasting effects of radiation on the surrounding area.[1]

Founders of ICAN were inspired by the success of the International Campaign to Ban Landmines, which was pivotal in bringing about the negotiation of the anti-personnel mine ban treaty in 1997. They sought to establish a similar campaign model.[2]

21.2   Formation

Launch of ICAN in Melbourne, Australia, in 2007

In September 2006, the Nobel Peace Prize-winning International Physicians for the Prevention of Nuclear War adopted a proposal at its biennial congress in Helsinki, Finland, to launch ICAN globally.[3] ICAN was launched publicly at two events, the first on 23 April 2007 in Melbourne, Australia, where funds had been raised to establish the campaign, and the second on 30 April 2007 in Vienna at a meeting of State parties to the Treaty on the Non-Proliferation of Nuclear Weapons. National campaigns have been organized in dozens of countries in every region of the world.

21.3   Membership and Support

ICAN is made up of more than 440 partner organizations in 100 countries. An International Steering Group (ISG) provides leadership and strategic management of the campaign, while an International Staff Team (IST) pro-

187

vides ongoing coordination of the campaign internationally. Current members of the ISG include the Acronym Institute for Disarmament Diplomacy, Article 36, International Physicians for the Prevention of Nuclear War, Norwegian Peoples Aid, PAX, Peace Boat, the Latin America Human Security Network (SEHLAC), Swedish Physicians for the Prevention of Nuclear War, the Women's International League for Peace and Freedom, and Zambian Health Professionals for Social Responsibility.

21.4   Milestones

ICAN campaigners in Mexico in 2014

October 27, 2016: UN First Committee adopts a landmark, ICAN-supported resolution to launch negotiations in 2017 on a treaty outlawing nuclear weapons. ICAN calls on all states to participate in the negotiations, stating that "every nation has an interest in ensuring that nuclear weapons are never used again, which can only be guaranteed through their complete elimination." [4]

February–August 2016: ICAN campaigns actively at UN Open-Ended Working Group in Geneva, which recommends by a large majority of 107 participating States that the General Assembly authorize negotiations on "a legally binding instrument to prohibit nuclear weapons, leading towards their total elimination." ICAN calls the OEWG recommendation "a breakthrough in the seven-decade-long global struggle to rid the world of the worst weapons of mass destruction." [5]

November 2, 2015: UN General Assembly establishes Open-Ended Working Group to review the evidence of catastrophic humanitarian impact of nuclear weapons and to make concrete recommendations for taking forward multilateral nuclear disarmament. ICAN calls on the OEWG "to begin the serious practical work of developing the elements for a treaty banning nuclear weapons."

November 2015: After mobilizing campaigners behind the Humanitarian Pledge for almost a year, ICAN takes significant credit for bringing 127 onto the Pledge as signatories;

another 23 States vote in favor of Pledge goals at General Assembly.

August 6–7, 2015: ICAN campaigners organize worldwide events to commemorate the 70th anniversaries of the atomic bombings of the Japanese cities of Hiroshima and Nagasaki.

December 2014: More than 600 ICAN campaigners gather in Vienna on the eve of the Vienna Conference on the Humanitarian Impact of Nuclear Weapons. ICAN tells conference participants "a new legal instrument prohibiting nuclear weapons would constitute a long overdue implementation of the Non-Proliferation Treaty." At the conference conclusion, Austria issues historic Humanitarian Pledge to work with all stakeholders "to fill the legal gap for the prohibition and elimination of nuclear weapons." [6]

October 26, 2014: 155 States, an increase of 30 from the previous year, submit joint humanitarian appeal for nuclear disarmament at UN General Assembly.

July 1, 2014: Beatrice Fihn is appointed ICAN Executive Director.

February 2014: Nayarit Conference on the Humanitarian Impact of Nuclear Weapons attended by 146 States and more than a hundred civil society campaigners. ICAN tells participants "the claim by some states that they continue to need these weapons to deter their adversaries has been exposed by the evidence presented at this conference···as a reckless and unsanctionable gamble with our future." At conference conclusion, Mexico calls for the start of a diplomatic process to negotiate a legally binding instrument prohibiting nuclear weapons.

August 30, 2013: UN working group highlights humanitarian concerns about the catastrophic humanitarian consequences of nuclear detonations and the need for non-nuclear nations to push forward.

March 2013: ICAN coordinates civil society participation at historic Oslo Conference on the Humanitarian Impact of Nuclear Weapons, an unprecedented gathering of States to evaluate the scientific evidence about the catastrophic consequences of nuclear weapons.

March 5, 2012: ICAN launches "Don't Bank on the Bomb" global divestment initiative. [7]

November 26, 2011: ICAN welcomes historic resolution adopted by the International Red Cross and Red Crescent movement in favor of an international agreement to prohibit nuclear weapons. [8]

June 27, 2011: P5 nations (the United States, Russia, the United Kingdom, France and China) meet in Paris to discuss ways to improve transparency in relation to their nuclear weapons. ICAN releases a video challenging them to do much more.

May 28, 2010: ICAN campaigners at the NPT Review Conference in New York call on governments to support a nuclear weapons convention. While references to a convention are included in the final document, ICAN is already considering a shift in strategy toward a new treaty banning nuclear weapons in order to empower non-nuclear-weapon states to assume more effective leadership.

April 30, 2007: ICAN is launched internationally during the Non-Proliferation Treaty preparatory committee meeting in Vienna.

September 7, 2006: International Physicians for the Prevention of Nuclear War, the 1985 Nobel Peace Laureate, adopts ICAN as top campaign priority at its world congress in Helsinki, Finland. IPPNW's Australian affiliate, MAPW, commits to fundraising and providing coordination for a campaign launch in 2007.

21.5 Supporters

Michael Douglas with ICAN executive director Beatrice Fihn

A number of prominent individuals have lent their support to the campaign, including Nobel Peace Prize laureates Desmond Tutu, the Dalai Lama and Jody Williams, the musician Herbie Hancock, the cricketer Ian Chappell, the actors Martin Sheen and Michael Douglas, and the artist Yoko Ono.

In November 2012, the Secretary-General of the United Nations, Ban Ki-moon, praised ICAN and its partners "for working with such commitment and creativity in pursuit of our shared goal of a nuclear-weapon-free world".[9] Earlier, he had provided a video message to ICAN in support of its global day of action.[10]

21.6 See also

- Nuclear disarmament

- Humanitarian Initiative

21.7 References

[1] IPPNW (2016). "The health and humanitarian case for banning and eliminating nuclear weapons." (PDF). *www.ippnw.org*. IPPNW. Retrieved January 12, 2017.

[2] "The International Campaign to Abolish Nuclear Weapons", Ronald McCoy, 30 April 2016

[3] Campaign milestones 2006| ICAN website

[4] ICAN (December 23, 2016). "UN General Assembly approves historic resolution". *www.icanw.org*. ICAN. Retrieved January 12, 2017.

[5] ICAN (August 19, 2016). "Majority of UN members declare intention to negotiate ban on nuclear weapons in 2017". *www.icanw.org*. ICAN. Retrieved January 12, 2017.

[6] "Humanitarian Pledge" (PDF). Austrian Ministry of Foreign Affairs. December 9, 2014. Retrieved January 12, 2017.

[7] "Don't Bank on the Bomb". PAX/ICAN. Retrieved January 12, 2017.

[8] "Working towards the elimination of nuclear weapons: Council of Delegates 2011: Resolution 1". International Committee of the Red Cross. November 26, 2011. Retrieved January 12, 2017.

[9] Letter from the UN Secretary-General, sent to ICAN, 2 November 2012

[10] Ban Ki-moon's message to ICAN, June 2010

•

21.8 External links

- International Campaign to Abolish Nuclear Weapons

Chapter 22

Peace Action

Peace Action is a peace organization whose focus is on preventing the deployment of nuclear weapons in space, thwarting weapons sales to countries with human rights violations, and promoting a new United States foreign policy based on common security and peaceful resolution to international conflicts.

Peace Action believes that every person has the right to live without the threat of nuclear weapons, that war is not a suitable response to conflict, and that the United States has the resources to both protect and provide for its citizens.[2] Peace Action has over 100,000 members who belong to over 70 autonomous affiliate and chapter organizations.[3]

Peace Action was formed through the merger of The Committee for a SANE Nuclear Policy and the Nuclear Weapons Freeze Campaign (also known as "The Freeze").

22.1 Campaigns

In 2003, Peace Action launched the Campaign for a New Foreign Policy, an initiative to build grassroots support and congressional pressure for a U.S. foreign policy based on human rights and democracy, nuclear disarmament and international cooperation.

Peace Action opposes the U.S. occupation of Iraq as well as any potential future action within that state to impose permanent military bases, any attempt to control Iraqi oil through U.S. government or corporate institutions, or any action on the part of the U.S. government to further influence the domestic policy of elected Iraqi officials. They lobby their activist network to demand a complete withdrawal from Iraq as soon as possible.[4]

To prevent future wars Peace Action lobbies its grassroots network to demand peaceful diplomacy with Iran. In December 2006 Peace Action began a petition to prevent war with Iran to date there are over 44,000 names.[5]

On the nuclear front Peace Action took part in a coalition lobby effort with organizations like the Arms Control Association and the Council for a Livable World to zero out funding for the Reliable Replacement Warhead and Complex 2030. Efforts of the collation helped stir the Senate Arms Services Committee to zero out the Administration's $15 million RRW request for Navy research and development.

Peace Action participated in organizing People's Climate March in September 2014. Peace Action believes war and militarism is interconnected with the climate crisis. The organization states that wars and militarism are the biggest obstacles to funding initiatives to address global warming. Both wars and climate crisis require a political solution which can only become a reality if the climate justice movement links to ending wars and militarism and the peace movement connects to justice: climate, economic and racial justice. Peace Action, as a national endorser, jumped into the organizing from the beginning launching the Peoples Climate March Peace and Justice Hub. The Hub brought together peace and faith groups to organize a No War, No Warming contingent and rally. George Martin, Peace Action Education Fund board member, Cole Harrison, executive director of Massachusetts Peace Action (MAPA), Jim Anderson, Peace Action of New York State (PANYS) Chair and Natia Bueno, PANYS Student Outreach Coordinator, led the way.[6]

22.2 Grassroots work

Peace Action has 100 chapters nationwide with a network of over 100,000 paying members. They send bi-weekly Action Alerts to almost 100,000 people worldwide, keeping them up to date on legislation regarding the Iraq war, nuclear disarmament, and preventing future wars with countries the Bush administration deems "rogue nations," like Iran. They also run a forum blog concerning issues of peace, nuclear abolition, and justice.

Their motto is "Peace Demands Action" and work on issues like Iraq, missile bases in Europe, or cutting the funding of new nuclear warheads . Peace Action's goal is to organize the nation around issues of peace and justice through protests, congressional action, and lobby days. They recently organized a petition to let our leaders know that any war with Iran, particularly one that involves nuclear weapons, should not be an option.

Peace Action initiated the Student Peace Action Network (SPAN) in 1995 to bring the voices of young activists into the forefront of the peace movement. Youth actively engaged in peace issues lacked a systematic tool to unite and organize with other young people. SPAN addresses this problem by providing advocacy tools, a nationwide network of like-minded youth, information about the issues, and support for affiliate chapters. Through coordinated direct actions, demonstrations, teach-ins, letter-writing campaigns, dissemination of materials, and other tactics, SPAN activists all over the country challenge unjust policies and work for non-violent, constructive alternatives.

22.3   History

Peace Action, was originally founded as 'SANE' in 1957 by Lenore Marshall and Norman Cousins and others in response to the nuclear arms race and the Eisenhower administration's policies on the production and testing of nuclear weapons. William Sloane Coffin, former chaplain of Yale University and political activist, retired from Riverside Church to become President of SANE/FREEZE in 1987.*[7] The name "SANE" came from the concepts put forth by Erich Fromm in his book *The Sane Society.*[8] The group's aim was to alert Americans of the threat of nuclear weapons. A full-page advertisement placed in *The New York Times* in November 1957 provoked a nationwide response, and by 1958 the membership of the organization had grown to 25,000. SANE was formally incorporated in July of that year.

Various influential people and celebrities began to get involved with the organization and show support for their cause . In 1959, Steve Allen hosted a meeting that founded the Hollywood SANE. Members included Marlon Brando, Henry Fonda, Marilyn Monroe, Arthur Miller, Harry Belafonte, and Ossie Davis. In 1960, a SANE rally was held at Madison Square Garden that attracted 20,000 to hear Eleanor Roosevelt, Norman Cousins, Norman Thomas, A. Philip Randolph, Walter Reuther, and Harry Belafonte call for an end to the arms race. International sponsors of SANE (including Martin Buber, Pablo Casals, Bertrand Russell and Albert Schweitzer) petitioned President John F. Kennedy to maintain a moratorium on testing in the atmosphere. Graphic Artists for SANE was also organized, with members that included Jules Feiffer, Ben Shahn, and Edward Sorel.

The group launched campaigns and rallies to drum up support for its cause and to put pressure on political figures. In 1961, SANE hosted an eight-day, 109-mile march from McGuire Air Force Base to the United Nations Plaza that was attended by more than 25,000 people. They organized a rally of over 10,000 people on "Cuba Sunday" to express concern and outrage over the Cuban Missile Crisis. Dr. Spock became a national sponsor and appeared in an ad stating "Dr. Spock is worried." The ad was printed in 700 papers worldwide.

22.3.1   Early political influence

As a way of seeing their goals achieved, SANE began working through its political lobbying programs. The organization began by pushing for the election of congressional candidates whose positions reflected those of the organization. In 1966, SANE formed the "Voter's Peace Pledge Campaign" to urge Congressional candidates to work for peace in Vietnam. They became one of the first national organizations to advocate removal of President Lyndon B. Johnson from office. They went on to endorse Eugene McCarthy as the Democratic presidential candidate in 1968.

SANE's Norman Cousins acted as an unofficial liaison between President Kennedy and Soviet Premier Nikita Khrushchev on the Partial Test Ban Treaty negotiations. The organization helped secure the passage of the War Powers Resolution. As the Vietnam War began to escalate, SANE organized a rally at Madison Square Garden that attracted 18,000 people opposing the war, as well as a march on Washington in November 1965 drawing 35,000. Three days after the march, Vice-president Hubert Humphrey met with SANE leaders Dr. Spock, Sanford Gottlieb, and Homer Jack "to openly, responsibly, and frankly discuss their proposals" to end the war. Many more SANE marches on Washington would occur throughout the war.

SANE would go on to criticize the Anti-Ballistic Missile Treaty and SALT agreements for ignoring offensive strategic weapons. Following Richard Nixon's re-election, SANE advocated Congressional cut-off of funds for the Vietnam war. After the end of the Vietnam War, SANE lobbied to have Congress end the bombing of Cambodia, and helped lead a successful effort to pass the War Powers Act. SANE would also take on the military budget, and produced the "America Has a Tapeworm" ad. Despite the end of the war, SANE continued actions throughout the 1970s that promoted its purpose.

22.3.2   Nuclear Weapons Freeze Campaign

During the 1980s, SANE continued to monitor the political and military actions of the U.S. government and beyond. In 1981, The Nuclear Weapons Freeze Campaign began with the purpose of pressuring the government to stop the nuclear arms build-up. The campaign was initiated by Randall Forsberg's call to "freeze and reverse the nuclear arms race".[9] Many SANE leaders participated in the creation of 'the Freeze', as it was sometimes called, which was a grassroots-based confederation of groups spanning the country. Freeze leaders included Randall Forsberg, Helen Caldicott, Pam Solo, and Randy Kehler. Elected officials such as Rep. Patricia Schroeder and Sen. Ted Kennedy helped to lead the movement in Congress. The Freeze's grassroots network pushed for nuclear reductions through ballot initiatives in towns and cities across the nation.

Specifically, the Freeze's goal was to get the U.S. and the Soviet Union to simultaneously adopt a mutual freeze on the testing, production, and deployment of nuclear weapons and of missiles, as well as new aircraft designed primarily to deliver nuclear weapons. Much emphasis was put on the MX and Pershing II missiles. Randall Forsberg was the organizer who initiated this idea of the "mutual, verifiable" Freeze.

During 1982, the SANE political action committee was formed for the political election year. Aside from working to get selected candidates elected it became a driving force behind many proposed nuclear freeze referendums. In a victory for both the Freeze campaign and SANE, Ronald Reagan proposed START I, part of a two phase treaty between the U.S. and the USSR that would reduce overall warhead counts on any missile type.

In roughly the 1983-84 period, when the Nuclear Weapons Freeze Campaign was planning expansively around mass-movement fund raising, lobbying, and Political Action Committees (PACs), SANE was merged into that entity, though local SANE chapters would continue to hold meetings for some time to come. Specific congressional races were targeted, and some of the pro-Freeze candidates credited the movement—and the grass-roots funds it raised —with their success in getting elected, or re-elected, to Congress. From 1984 on, the movement had three actual legal entities, the 'Nuclear Weapons Freeze Campaign', with both public education and lobbying arms (501.C-3 and 501.C-4 corporations), and the Freeze Voter PAC (501.C-5).

During the 1980s, SANE/FREEZE expanded its work to oppose U.S. military intervention in El Salvador and to end U.S. military aid to the Contras in Nicaragua. The organization promoted its agenda in different ways. An ad was placed in Variety magazine signed by over 250 celebrities including Jack Lemmon, Burt Lancaster, James Earl Jones, Sally Field, Shirley MacLaine, and Ed Asner supporting its causes. A weekly radio program by SANE/FREEZE,

"Consider the Alternatives", reaches 140 radio stations. Their door canvassing campaign reached 250,000 households.

22.3.3   The Gulf War and the War on Terror

Following Iraq's invasion of Kuwait, SANE/FREEZE opposed the U.S. military buildup in the Persian Gulf. Throughout the Gulf War, the organization coordinated anti-war marches in Washington, DC, helping to mobilize 500,000 protesters. Soon after, In 1993, SANE/FREEZE renamed itself **Peace Action**.

Of great concern to Peace Action in 1995 was the conference for review of the Nuclear Non-Proliferation Treaty. The signatories to the treaty decided by consensus to extend the treaty indefinitely and without conditions. The year also marked the 50th anniversary of the atomic bombing of Hiroshima and Nagasaki.

The next year Peace Action launched Peace Voter '96, the organization's largest nationally-coordinated campaign since the mid-1980s. Over one million Peace Voter Guides were distributed for the November elections. Also that year, Peace Action joined human rights groups to stop major weapons sales to Indonesia and Turkey. In 1997, Indonesia withdrew its request for U.S. fighter jets due to "unwarranted criticism" of their human rights record.

In 1999, Peace Action opposed the NATO bombing of Kosovo, which it described as "cruise missile humanitarianism", and founded the National Coalition for Peace and Justice, a body uniting most of the major peace groups in the country. Also that year, Peace Action commemorated the bombing of Nagasaki by staging a demonstration at Los Alamos National Laboratory in New Mexico. The demonstration was led by actor Martin Sheen.

Following the September 11, 2001 attacks, Peace Action responded to the war on terrorism and the bombing of Afghanistan with a call for justice, not war. The group went on to participate in two national coalitions: Win Without War and United for Peace and Justice.

22.4   See also

- Committee for Non-Violent Action

- Department of Peace

- List of anti-war organizations

- List of peace activists

- Nuclear weapons

22.5 References

[1] Staff Directory, Peace Action.

[2] Peace-Action About Peace-Action Retrieved June 19, 2007

[3] The National Network

[4] Peace Action

[5] Peace Action

[6] Peace Action Blog

[7] Mary Rourke (April 13, 2006). "William Sloane Coffin Jr, 81; Former Yale Chaplain and Civil Rights, Peace Activist" . *LA Times.*

[8] ISBN 1-199-36561-0

[9] Hugh Gusterson (30 March 2012). "The new abolitionists" . *Bulletin of the Atomic Scientists.*

22.6 Further reading

- Milton S. Katz, *Ban the Bomb: A History of SANE, 1957-1985* (New York: Greenwood Press, 1986). ISBN 0-313-24167-8

- Pam Solo, *From Protest to Policy: Beyond the Freeze to Common Security* (Ballinger, 1988). ISBN 978-0-88730-112-4

- Glen Harold Stassen and Lawrence S. Wittner, eds., *Peace Action: Past, Present and Future* (Boulder, CO: Paradigm Publishers, 2007). ISBN 978-1-59451-333-6

22.7 External links

- Official site

- Pacific Freeze

- Peace Action Blog

- Peace Action long range strategic plan

- Peace Action West official site (formerly California Peace Action)

- People's Climate Change Peace and Justice hub

- Student Peace Action

Chapter 23

History of the anti-nuclear movement

Main article: Anti-nuclear movement

The application of nuclear technology, both as a source of

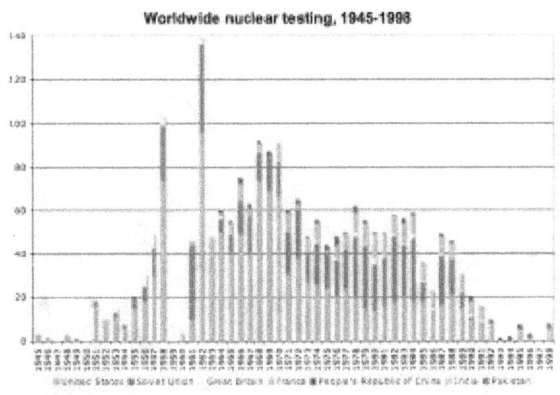

Worldwide nuclear testing totals, 1945-1998.

energy and as an instrument of war, has been controversial.[1][2][3][4][5]

Scientists and diplomats have debated nuclear weapons policy since before the atomic bombing of Hiroshima in 1945.[6] The public became concerned about nuclear weapons testing from about 1954, following extensive nuclear testing in the Pacific. In 1961, at the height of the Cold War, about 50,000 women brought together by Women Strike for Peace marched in 60 cities in the United States to demonstrate against nuclear weapons.[7][8] In 1963, many countries ratified the Partial Test Ban Treaty which prohibited atmospheric nuclear testing.[9]

Some local opposition to nuclear power emerged in the early 1960s,[10] and in the late 1960s some members of the scientific community began to express their concerns.[11] In the early 1970s, there were large protests about a proposed nuclear power plant in Wyhl, Germany. The project was cancelled in 1975 and anti-nuclear success at Wyhl inspired opposition to nuclear power in other parts of Europe and North America.[12][13] Nuclear power became an issue of major public protest in the 1970s.[14]

23.1 Early years

The 1945 Trinity explosion, 0.016 seconds after detonation. The fireball is about 200 meters (600 ft) wide. Trees may be seen as black objects in the foreground.

In 1945 in the New Mexico desert, American scientists conducted "Trinity," the first nuclear weapons test, marking the beginning of the atomic age.[15] Even before the Trinity test, national leaders debated the impact of nuclear weapons on domestic and foreign policy. Also involved in the debate about nuclear weapons policy was the scientific community, through professional associations such as the Federation of Atomic Scientists and the Pugwash Conference on Science and World Affairs.[6]

On August 6, 1945, towards the end of World War II, the Little Boy device was detonated over the Japanese military city of Hiroshima. Exploding with a yield equivalent to 12,500 tonnes of TNT, the blast and thermal wave of the bomb destroyed nearly 50,000 buildings (including the headquarters of the 2nd General Army and Fifth Division) and killed approximately 75,000 people, among them 20,000 Japanese soldiers and 20,000 Korean slave laborers.[16] Detonation of the Fat Man device exploded over the Japanese industrial city of Nagasaki three days later after Hiroshima, destroying 60% of the city and killing approximately 35,000 people, among them 23,200-28,200

The mushroom cloud over Hiroshima after the dropping of the atomic bomb nicknamed 'Little Boy' (1945).

Operation Crossroads Test Able, a 23-kiloton air-deployed nuclear weapon detonated on July 1, 1946. This bomb used, and consumed, the infamous Demon core that took the lives of two scientists in two separate criticality accidents.

Mushroom-shaped cloud and water column from the underwater nuclear explosion of July 25, 1946, which was part of Operation Crossroads.

November 1951 nuclear test at the Nevada Test Site, from Operation Buster, with a yield of 21 kilotons. It was the first U.S. nuclear field exercise conducted on land; troops shown are 6 mi (9.7 km) from the blast.

Japanese munitions workers, 2,000 Korean slave laborers, and 150 Japanese soldiers.[17] The two bombings remains the only events where nuclear weapons have been used in combat. Subsequently, the world's nuclear weapons stockpiles grew.[15]

Operation Crossroads was a series of nuclear weapon tests conducted by the United States at Bikini Atoll in the Pacific Ocean in the summer of 1946. Its purpose was to test the effect of nuclear weapons on naval ships. Pressure to cancel Operation Crossroads came from scientists and diplomats. Manhattan Project scientists argued that further nu-clear testing was unnecessary and environmentally danger-ous. A Los Alamos study warned "the water near a re-cent surface explosion will be a witch's brew" of radioac-tivity. To prepare the atoll for the nuclear tests, Bikini's na-tive residents were evicted from their homes and resettled on smaller, uninhabited islands where they were unable to sustain themselves.[18]

Radioactive fallout from nuclear weapons testing was first drawn to public attention in 1954 when a Hydrogen bomb test in the Pacific contaminated the crew of the Japanese fishing boat Lucky Dragon.[9] One of the fishermen died in Japan seven months later. The incident caused widespread concern around the world and "provided a decisive impe-tus for the emergence of the anti-nuclear weapons move-ment in many countries".[9] The anti-nuclear weapons movement grew rapidly because for many people the atomic bomb "encapsulated the very worst direction in which so-ciety was moving".[19]

Women Strike for Peace during the Cuban Missile Crisis in 1962.

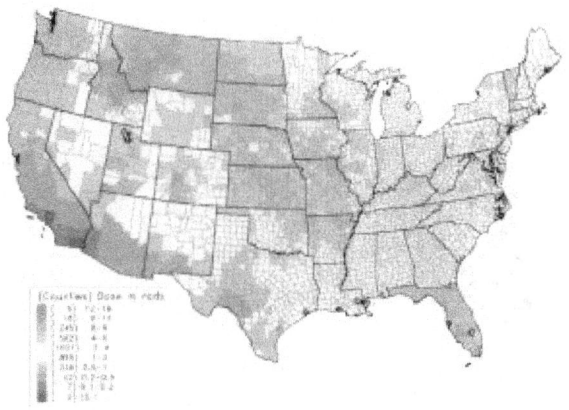

Because of concerns about worldwide fallout levels, the Partial Test Ban Treaty was signed in 1963. Above are the per capita thyroid doses (in rads) in the continental United States resulting from all exposure routes from all atmospheric nuclear tests conducted at the Nevada Test Site from 1951–1962.

Peace movements emerged in Japan and in 1954 they converged to form a unified "Japanese Council Against Atomic and Hydrogen Bombs". Japanese opposition to the Pacific nuclear weapons tests was widespread, and "an estimated 35 million signatures were collected on petitions calling for bans on nuclear weapons". [19]

German publications of the 1950s and 1960s contained criticism of some features of nuclear power including its safety.

The Phoenix of Hiroshima *(foreground) in Hong Kong Harbor in 1967, was involved in several famous protest voyages against nuclear testing in the Pacific.*

Nuclear waste disposal was widely recognized as a major problem, with concern publicly expressed as early as 1954. In 1964, one author went so far as to state "that the dangers and costs of the necessary final disposal of nuclear waste could possibly make it necessary to forego the development of nuclear energy". [20]

The Russell–Einstein Manifesto was issued in London on July 9, 1955 by Bertrand Russell in the midst of the Cold War. It highlighted the dangers posed by nuclear weapons and called for world leaders to seek peaceful resolutions to international conflict. The signatories included eleven pre-eminent intellectuals and scientists, including Albert Einstein, who signed it just days before his death on April 18, 1955. A few days after the release, philanthropist Cyrus S. Eaton offered to sponsor a conference—called for in the manifesto—in Pugwash, Nova Scotia, Eaton's birthplace. This conference was to be the first of the Pugwash Conferences on Science and World Affairs, held in July 1957.

In the United Kingdom, the first Aldermaston March organised by the Campaign for Nuclear Disarmament took place at Easter 1958, when several thousand people marched for four days from Trafalgar Square, London, to the Atomic Weapons Research Establishment close to Aldermaston in Berkshire, England, to demonstrate their opposition to nuclear weapons. [21] [22] The Aldermaston marches continued into the late 1960s when tens of thousands of people took part in the four-day marches. [19]

In 1959, a letter in the *Bulletin of Atomic Scientists* was the start of a successful campaign to stop the Atomic Energy Commission dumping radioactive waste in the sea 19 kilometres from Boston. [23]

On November 1, 1961, at the height of the Cold War, about 50,000 women brought together by Women Strike for Peace marched in 60 cities in the United States to demonstrate against nuclear weapons. It was the largest national women's peace protest of the 20th century. [7] [8]

In 1958, Linus Pauling and his wife presented the United Nations with the petition signed by more than 11,000 scientists calling for an end to nuclear-weapon testing. The "Baby Tooth Survey," headed by Dr Louise Reiss, demonstrated conclusively in 1961 that above-ground nuclear testing posed significant public health risks in the form of radioactive fallout spread primarily via milk from cows that had ingested contaminated grass.[24][25][26] Public pressure and the research results subsequently led to a moratorium on above-ground nuclear weapons testing, followed by the Partial Test Ban Treaty, signed in 1963 by John F. Kennedy and Nikita Khrushchev.[27] On the day that the treaty went into force, the Nobel Prize Committee awarded Pauling the Nobel Peace Prize, describing him as "Linus Carl Pauling, who ever since 1946 has campaigned ceaselessly, not only against nuclear weapons tests, not only against the spread of these armaments, not only against their very use, but against all warfare as a means of solving international conflicts."[6][28]

Pauling started the International League of Humanists in 1974. He was president of the scientific advisory board of the World Union for Protection of Life and also one of the signatories of the Dubrovnik-Philadelphia Statement.

23.2 After the Partial Test Ban Treaty

Radioactive materials were accidentally released from the 1970 Baneberry Nuclear Test at the Nevada Test Site.

The Shippingport Atomic Power Station was the first full-scale PWR nuclear power plant in the United States. The reactor went online December 2, 1957, and was in operation until October, 1982.

In the United States, the first commercially viable nuclear power plant was to be built at Bodega Bay, north of San Francisco, but the proposal was controversial and conflict with local citizens began in 1958.[10] The proposed plant site was close to the San Andreas Fault and close to the

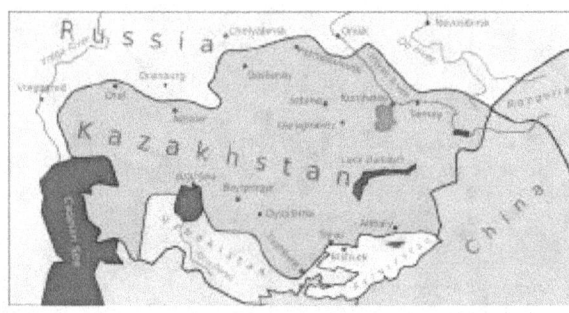

The 18,000 km² expanse of the Semipalatinsk Test Site (indicated in red), which covers an area the size of Wales. The Soviet Union conducted 456 nuclear tests at Semipalatinsk from 1949 until 1989 with little regard for their effect on the local people or environment. The full impact of radiation exposure was hidden for many years by Soviet authorities and has only come to light since the test site closed in 1991.[29]

region's environmentally sensitive fishing and dairy industries. The Sierra Club became actively involved.[31] The conflict ended in 1964, with the forced abandonment of plans for the power plant. Historian Thomas Wellock traces the birth of the anti-nuclear movement to the controversy over Bodega Bay.[10] Attempts to build a nuclear power plant in Malibu were similar to those at Bodega Bay and were also abandoned.[10]

In 1966, Larry Bogart founded the Citizens Energy Coun-

120,000 people attended an anti-nuclear protest in Bonn, Germany, on October 14, 1979, following the Three Mile Island accident.[30]

cil, a coalition of environmental groups that published the newsletters "Radiation Perils," "Watch on the A.E.C." and "Nuclear Opponents". These publications argued that "nuclear power plants were too complex, too expensive and so inherently unsafe they would one day prove to be a financial disaster and a health hazard".[32][33]

The emergence of the anti-nuclear power movement was "closely associated with the general rise in environmental consciousness which had started to materialize in the USA in the 1960s and quickly spread to other Western industrialized countries".[11] Some nuclear experts began to voice dissenting views about nuclear power in 1969, and this was a necessary precondition for broad public concern about nuclear power to emerge.[11] These scientists included Ernest Sternglass from Pittsburg, Henry Kendall from the Massachusetts Institute of Technology, Nobel laureate George Wald and radiation specialist Rosalie Bertell. These members of the scientific community "by expressing their concern over nuclear power, played a crucial role in demystifying the issue for other citizens", and nuclear power became an issue of major public protest in the 1970s.[11][34]

In 1971, 15,000 people demonstrated against French plans to locate the first light-water reactor power plant in Bugey. This was the first of a series of mass protests organized at nearly every planned nuclear site in France.[35]

Also in 1971, the town of Wyhl, in Germany, was a proposed site for a nuclear power station. In the years that followed, public opposition steadily mounted, and there were large protests. Television coverage of police dragging away farmers and their wives helped to turn nuclear power into a major issue. In 1975, an administrative court withdrew the construction licence for the plant,[12][13][36] but the Wyhl occupation generated ongoing debate. This initially centred on the state government's handling of the affair and associated police behaviour, but interest in nuclear is-

sues was also stimulated. The Wyhl experience encouraged the formation of citizen action groups near other planned nuclear sites.[12] Many other anti-nuclear groups formed elsewhere, in support of these local struggles, and some existing citizen action groups widened their aims to include the nuclear issue.[12] Anti-nuclear success at Wyhl also inspired nuclear opposition in the rest of Europe and North America.[13]

In 1972, the anti-nuclear weapons movement maintained a presence in the Pacific, largely in response to French nuclear testing there. Activists, including David McTaggart from Greenpeace, defied the French government by sailing small vessels into the test zone and interrupting the testing program.[37][38] In Australia, thousands joined protest marches in Adelaide, Melbourne, Brisbane, and Sydney.[38] Scientists issued statements demanding an end to the tests; unions refused to load French ships, service French planes, or carry French mail; and consumers boycotted French products. In Fiji, activists formed an Against Testing on Mururoa organization.[38]

In Spain, in response to a surge in nuclear power plant proposals in the 1960s, a strong anti-nuclear movement emerged in 1973, which ultimately impeded the realisation of most of the projects.[39]

In 1974, organic farmer Sam Lovejoy took a crowbar to the weather-monitoring tower which had been erected at the Montague Nuclear Power Plant site. Lovejoy felled the tower and then took himself to the local police station, where he took full responsibility for the action. Lovejoy's action galvanized local public opinion against the plant.[40][41] The Montague project was canceled in 1980,[42] after $29 million was spent on the project.[40]

By the mid-1970s anti-nuclear activism had moved beyond local protests and politics to gain a wider appeal and influence. Although it lacked a single co-ordinating organization, and did not have uniform goals, the movement's efforts gained a great deal of attention.[4] Jim Falk has suggested that popular opposition to nuclear power quickly grew into an effective anti-nuclear power movement in the 1970s.[43] In some countries, the nuclear power conflict "reached an intensity unprecedented in the history of technology controversies".[44]

In France, between 1975 and 1977, some 175,000 people protested against nuclear power in ten demonstrations.[30]

In West Germany, between February 1975 and April 1979, some 280,000 people were involved in seven demonstrations at nuclear sites. Several site occupations were also attempted. In the aftermath of the Three Mile Island accident in 1979, some 120,000 people attended a demonstration against nuclear power in Bonn.[30]

In May 1979, an estimated 70,000 people, including the

governor of California, attended a march and rally against nuclear power in Washington, D.C.[45][46]

On June 12, 1982, one million people demonstrated in New York City's Central Park against nuclear weapons and for an end to the cold war arms race. It was the largest anti-nuclear protest and the largest political demonstration in American history.[47][48] International Day of Nuclear Disarmament protests were held on June 20, 1983 at 50 sites across the United States.[49][50] In 1986, hundreds of people walked from Los Angeles to Washington DC in the Great Peace March for Global Nuclear Disarmament.[51] There were many Nevada Desert Experience protests and peace camps at the Nevada Test Site during the 1980s and 1990s.[52][53]

On May 1, 2005, 40,000 anti-nuclear/anti-war protesters marched past the United Nations in New York, 60 years after the atomic bombings of Hiroshima and Nagasaki.[54] This was the largest anti-nuclear rally in the U.S. for several decades.[55] In Britain, there were many protests about the government's proposal to replace the aging Trident weapons system with a newer model. The largest protest had 100,000 participants and, according to polls, 59 percent of the public opposed the move.[55]

The International Conference on Nuclear Disarmament took place in Oslo in February 2008, and was organized by The Government of Norway, the Nuclear Threat Initiative and the Hoover Institute. The Conference was entitled *Achieving the Vision of a World Free of Nuclear Weapons* and had the purpose of building consensus between nuclear weapon states and non-nuclear weapon states in relation to the Nuclear Non-proliferation Treaty.[56]

In May 2010, some 25,000 people, including members of peace organizations and 1945 atomic bomb survivors, marched for about two kilometers from downtown New York to the United Nations headquarters, calling for the elimination of nuclear weapons.[57]

23.3 Other issues

Early anti-nuclear advocates expressed the view that affluent lifestyles on a global scale strain the viability of the natural environment and that nuclear energy would enable those lifestyles. Examples of such expressions are:

"We can and should seize upon the energy crisis as a good excuse and a great opportunity for making some very fundamental changes that we should be making anyhow for other reasons."
—Russell E. Train, 1974.[58]

"In fact, giving society cheap, abundant energy at this point would be the moral equivalent of giving an idiot child a machine gun."
—Paul R. Ehrlich, 1975.[59]

"If you ask me, it'd be little short of disastrous for us to discover a source of clean, cheap, abundant energy because of what we would do with it. We ought to be looking for energy sources that are adequate for our needs, but that won't give us the excesses of concentrated energy with which we could do mischief to the earth or to each other."
—Amory Lovins, 1977.[60]

"Let's face it. We don't want safe nuclear power plants. We want NO nuclear power plants."
—Spokesman for the Government Accountability Project, 1985.[61]

"... we also thought that as you provide societies with more energy it enables them to do more environmental destruction. The idea of tying us to the natural forces of the wind and the sun was very appealing in that it would limit and constrain human development"
—Robert Stone (director) (of both anti-nuclear weapons and, recently, pro-nuclear power films). 2014.[62]

23.4 See also

- Debate over the atomic bombings of Hiroshima and Nagasaki

- Nuclear power debate

- Nuclear weapons debate

- The Bomb (film)

- Uranium mining debate

23.5 References

[1] "Sunday Dialogue: Nuclear Energy, Pro and Con" . *New York Times*. February 25, 2012.

[2] Robert Benford. The Anti-nuclear Movement (book review) *American Journal of Sociology*. Vol. 89, No. 6, (May 1984), pp. 1456-1458.

[3] James J. MacKenzie. Review of The Nuclear Power Controversy by Arthur W. Murphy *The Quarterly Review of Biology*, Vol. 52, No. 4 (Dec., 1977), pp. 467-468.

[4] Walker, J. Samuel (2004). *Three Mile Island: A Nuclear Crisis in Historical Perspective* (Berkeley: University of California Press), pp. 10-11.

[5] Jim Falk (1982). *Global Fission: The Battle Over Nuclear Power*, Oxford University Press.

[6] Jerry Brown and Rinaldo Brutoco (1997). *Profiles in Power: The Anti-nuclear Movement and the Dawn of the Solar Age*, Twayne Publishers, pp. 191-192.

[7] Woo, Elaine (January 30, 2011). "Dagmar Wilson dies at 94; organizer of women's disarmament protesters". *Los Angeles Times*.

[8] Hevesi, Dennis (January 23, 2011). "Dagmar Wilson, Anti-Nuclear Leader, Dies at 94". *The New York Times*.

[9] Wolfgang Rudig (1990). *Anti-nuclear Movements: A World Survey of Opposition to Nuclear Energy*, Longman, p. 54-55.

[10] Paula Garb. Review of Critical Masses, *Journal of Political Ecology*, Vol 6, 1999.

[11] Wolfgang Rudig (1990). *Anti-nuclear Movements: A World Survey of Opposition to Nuclear Energy*, Longman, p. 52.

[12] Stephen Mills and Roger Williams (1986). Public Acceptance of New Technologies Routledge, pp. 375-376.

[13] Robert Gottlieb (2005). Forcing the Spring: The Transformation of the American Environmental Movement, Revised Edition, Island Press, USA, p. 237.

[14] Jim Falk (1982). *Global Fission: The Battle Over Nuclear Power*, Oxford University Press, pp. 95-96.

[15] Mary Palevsky, Robert Futrell, and Andrew Kirk. Recollections of Nevada's Nuclear Past *UNLV FUSION*, 2005, p. 20.

[16] Emsley, John (2001). "Uranium". *Nature's Building Blocks: An A to Z Guide to the Elements*. Oxford: Oxford University Press. p. 478. ISBN 0-19-850340-7.

[17] *Nuke-Rebuke: Writers & Artists Against Nuclear Energy & Weapons (The Contemporary anthology series)*. The Spirit That Moves Us Press. May 1, 1984. pp. 22–29.

[18] Niedenthal, Jack (2008). *A Short History of the People of Bikini Atoll*, retrieved 2009-12-05

[19] Jim Falk (1982). *Global Fission: The Battle Over Nuclear Power*, Oxford University Press, pp. 96-97.

[20] Wolfgang Rudig (1990). *Anti-nuclear Movements: A World Survey of Opposition to Nuclear Energy*, Longman, p. 63.

[21] A brief history of CND

[22] "Early defections in march to Aldermaston". Guardian Unlimited. 1958-04-05.

[23] Jim Falk (1982). *Global Fission: The Battle Over Nuclear Power*, Oxford University Press, p. 93.

[24] Louise Zibold Reiss (November 24, 1961). "Strontium-90 Absorption by Deciduous Teeth: Analysis of teeth provides a practicable method of monitoring strontium-90 uptake by human populations" (PDF). Science. Retrieved October 13, 2009.

[25] Thomas Hager (November 29, 2007). "Strontium-90". Oregon State University Libraries Special Collections. Retrieved December 13, 2007.

[26] Thomas Hager (November 29, 2007). "The Right to Petition". Oregon State University Libraries Special Collections. Retrieved December 13, 2007.

[27] Jim Falk (1982). *Global Fission: The Battle Over Nuclear Power*, Oxford University Press, p. 98.

[28] Linus Pauling (October 10, 1963). "Notes by Linus Pauling. October 10, 1963.". Oregon State University Libraries Special Collections. Retrieved December 13, 2007.

[29] Togzhan Kassenova (28 September 2009). "The lasting toll of Semipalatinsk's nuclear testing". *Bulletin of the Atomic Scientists*.

[30] Herbert P. Kitschelt. Political Opportunity and Political Protest: Anti-Nuclear Movements in Four Democracies *British Journal of Political Science*, Vol. 16, No. 1, 1986, p. 71.

[31] Thomas Raymond Wellock (1998). Critical Masses: Opposition to Nuclear Power in California, 1958-1978, The University of Wisconsin Press, pp. 27–28.

[32] Keith Schneider. Larry Bogart, an Influential Critic Of Nuclear Power, Is Dead at 77 *The New York Times*, August 20, 1991.

[33] Anna Gyorgy (1980). No Nukes: Everyone's Guide to Nuclear Power South End Press, ISBN 0-89608-006-4, p. 383.

[34] Jim Falk (1982). *Global Fission: The Battle Over Nuclear Power*, Oxford University Press, p. 95.

[35] Dorothy Nelkin and Michael Pollak (1982). *The Atom Besieged: Antinuclear Movements in France and Germany*, ASIN: B0011LXE0A, p. 3.

[36] Nuclear Power in Germany: A Chronology

[37] Paul Lewis. David McTaggart, a Builder of Greenpeace, Dies at 69 *The New York Times*, March 24, 2001.

[38] Lawrence S. Wittner. Nuclear Disarmament Activism in Asia and the Pacific, 1971-1996 *The Asia-Pacific Journal*, Vol. 25-5-09, June 22, 2009.

[39] Lutz Mez, Mycle Schneider and Steve Thomas (Eds.) (2009). *International Perspectives of Energy Policy and the Role of Nuclear Power*, Multi-Science Publishing Co. Ltd. p. 371.

[40] Utilities Drop Nuclear Power Plant Plans *Ocala Star-Banner*, January 4, 1981.

[41] Anna Gyorgy (1980). No Nukes: Everyone's Guide to Nuclear PowerSouth End Press, ISBN 0-89608-006-4, pp. 393-394.

[42] Northeast Utilities System. Some of the Major Events in NU's History Since the 1966 Affiliation

[43] Jim Falk (1982). *Global Fission: The Battle Over Nuclear Power*, Oxford University Press, p. 96.

[44] Herbert P. Kitschelt. Political Opportunity and Political Protest: Anti-Nuclear Movements in Four Democracies *British Journal of Political Science*, Vol. 16, No. 1, 1986, p. 57.

[45] Jon Agnone. Amplifying Public Opinion: The Policy Impact of the U.S. Environmental Movement p. 7.

[46] Social Protest and Policy Change p. 45.

[47] Jonathan Schell. The Spirit of June 12 *The Nation*, July 2, 2007.

[48] 1982 - a million people march in New York City

[49] Harvey Klehr. Far Left of Center: The American Radical Left Today Transaction Publishers, 1988, p. 150.

[50] 1,400 Anti-nuclear protesters arrested *Miami Herald*, June 21, 1983.

[51] Hundreds of Marchers Hit Washington in Finale of Nationwaide Peace March *Gainesville Sun*, November 16, 1986.

[52] Robert Lindsey. 438 Protesters are Arrested at Nevada Nuclear Test Site *New York Times*, February 6, 1987.

[53] 493 Arrested at Nevada Nuclear Test Site *New York Times*, April 20, 1992.

[54] Anti-Nuke Protests in New York *Fox News*, May 2, 2005.

[55] Lawrence S. Wittner. A rebirth of the anti-nuclear weapons movement? Portents of an anti-nuclear upsurge *Bulletin of the Atomic Scientists*, 7 December 2007.

[56] "International Conference on Nuclear Disarmament". February 2008.

[57] A-bomb survivors join 25,000-strong anti-nuclear march through New York *Mainichi Daily News*, May 4, 2010.

[58] Train, R. E. (1974). "The Quality of Growth". *Science*. **184** (4141): 1050-3. doi:10.1126/science.184.4141.1050. PMID 17736183.

[59] "An Ecologist's Perspective on Nuclear Power", Federation of American Scientists Public Issue Report, May-June 1975

[60] Mother Earth News Nov/Dec 1977, p. 22: The Plowboy Interview with Amory Lovins

[61] The American Spectator, Vol 18, No. 11, Nov. 1985

[62] KTH Royal Institute of Technology in Stockholm, Nov. 2014: Interview with Robert Stone

Chapter 24

Göttingen Manifesto

The **Göttingen Manifesto** was a declaration of 18 leading nuclear scientists of West Germany (among them the Nobel laureates Otto Hahn, Max Born, Werner Heisenberg and Max von Laue) against arming the West German army with tactical nuclear weapons in the 1950s, the early part of the Cold War, as the West German government under chancellor Adenauer had suggested.

24.1 Historical situation

In the Second World War some of the signing scientists had been members of the Uranverein, a nuclear research project of the Nazi regime. The war ended with the nuclear destruction of the cities of Hiroshima and Nagasaki by the United States. After World War II the Cold War began. Germany was divided, and both German states were frontier states in the Cold War. After the Korean War (1950 - 1953), West Germany founded its own army, the Bundeswehr, in 1955. There were many protests against the remilitarisation of West Germany. A few months after the foundation of the West German army, the Eastern German state founded an army, too.

In 1953 the hydrogen bomb was invented. A short time later both superpowers, the United States and the Soviet Union, had a so-called overkill potential. In the whole world and especially in the frontier states of the Cold War there was a great fear of nuclear war at that time. A few years later, in the Cuban Missile Crisis, it was felt the existence of mankind was under threat.

The Göttinger 18 wrote the following manifesto on April 12, 1957:

24.2 The Manifesto

24.3 German original text

24.4 External links

- Original text (in German)
- Detailed information (University of Göttingen)

Chapter 25

International Court of Justice advisory opinion on the Legality of the Threat or Use of Nuclear Weapons

Legality of the Threat or Use of Nuclear Weapons was an advisory opinion delivered by the International Court of Justice (ICJ) on 8 July 1996.[1]

The initial request for an advisory opinion by the ICJ was presented by the World Health Organization (WHO) on 3 September 1993,[2] but the ICJ did not render an opinion on this request because the WHO was *ultra vires*, or acting outside its legal capacity. Another request was presented by the United Nations General Assembly in December 1994[3] and accepted by the Court in January 1995. The ICJ handed down an advisory opinion on 8 July 1996 the *Legality of the Threat or Use of Nuclear Weapons* case. The decision provides one of the few authoritative judicial decisions concerning the legality under international law of the use or the threatened use of nuclear weapons.

Beyond the central question, many more general issues were touched upon by the Court or raised in the pleadings. These included institutional issues such as the proper role of international judicial bodies, and the ICJ's advisory function. The main substantive issues regarded sources of international legal obligation and the interaction of various branches of international law, particularly the norms of international humanitarian law (*jus in bello*) and the rules governing the use of force (*jus ad bellum*). In addition, the proceedings explored the status of "*Lotus* approach", and employed the concept of *non liquet*. There were also strategic questions such as the legality of the practice of nuclear deterrence or the meaning of Article VI of the 1968 Treaty on the Non-Proliferation of Nuclear Weapons.

The hypothetical possibility of outlawing the use of nuclear weapons in an armed conflict was raised as early as June 30, 1950, by the Dutch representative to the International Law Commission (ILC) J.P.A. François, who suggested this "would in itself be an advance".[4] In addition, the Polish government requested this issue to be examined by the ILC as a crime against the peace of mankind.[5] However, the issue became moot due to Cold War tensions.

25.1 Request of the World Health Organization

The original advisory opinion was requested by the World Health Organization in 1993.

An advisory opinion on this issue was originally requested by the World Health Organization (WHO) on 3 September 1993:[6]

> In view of the health and environmental effects, would the use of nuclear weapons by a state in war or other armed conflict be a breach of its obligations under international law including the WHO Constitution?[7]

The ICJ considered the WHO's request, in a case known as the *Legality of the Use by a State of Nuclear Weapons in Armed Conflict* (General List No. 93), and also known as the *WHO Nuclear Weapons case*, between 1993 and 1996. The ICJ fixed 10 June 1994 as the time limit for written submissions, but after receiving many written and oral submissions, later extended this date to 20 September 1994. After considering the case the Court refused to give an advisory opinion on the WHO question. On 8 July 1996 it held, by 11 votes to three, that the question did not fall within the scope of WHO's activities, as is required by Article 96(2) of the UN Charter.[8]

25.2 Request of the UN General Assembly

UN General Assembly.

On 15 December 1994 the UN General Assembly adopted resolution A/RES/49/75K.[9] This asked the ICJ urgently to render its advisory opinion on the following question:

> Is the threat or use of nuclear weapons in any circumstances permitted under international law?
> —United Nations General Assembly[10][11]

The resolution, submitted to the Court on 19 December 1994, was adopted by 78 states voting in favour, 43 against, 38 abstaining and 26 not voting.[12]

The General Assembly had considered asking a similar question in the autumn of 1993, at the instigation of the Non-Aligned Movement (NAM), which ultimately did not that year push its request. NAM was more willing the following year, in the face of written statements submitted in the WHO proceedings from a number of nuclear-weapon states indicating strong views to the effect that the WHO

lacked competence in the matter. The Court subsequently fixed 20 June 1995 as the filing date for written statements.

Altogether forty-two states participated in the written phase of the pleadings, the largest number ever to join in proceedings before the Court. Of the five declared nuclear weapon states only the People's Republic of China did not participate. Of the three "threshold" nuclear-weapon states only India participated. Many of the participants were developing states which had not previously contributed to proceedings before the ICJ, a reflection perhaps of the unparalleled interest in this matter and the growing willingness of developing states to engage in international judicial proceedings in the "post-colonial" period.

Oral hearings were held from 30 October to 15 November 1995. Twenty-two states participated: Australia, Egypt, France, Germany, Indonesia, Mexico, Iran, Italy, Japan, Malaysia, New Zealand, Philippines, Qatar, Russian Federation, San Marino, Samoa, Marshall Islands, Solomon Islands, Costa Rica, United Kingdom, United States, Zimbabwe; as did the WHO. The secretariat of the UN did not appear, but filed with the Court a dossier explaining the history of resolution 49/75K. Each state was allocated 90 minutes to make its statement. On 8 July 1996, nearly eight months after the close of the oral phase, the ICJ rendered its opinion.

25.2.1 Composition of the Court

See also: Composition of the International Court of Justice

The ICJ is composed of fifteen judges elected to nine year terms by the UN General Assembly and the UN Security Council. The court's "advisory opinion" can be requested only by specific United Nations organisations, and is inherently non-binding under the Statute of the court.

The fifteen judges asked to give their advisory opinion regarding the legality of the threat or use of nuclear weapons were:

25.2.2 Court's analysis

Deterrence and "threat"

The court considered the matter of deterrence, which involves a *threat* to use nuclear weapons under certain circumstances on a potential enemy or an enemy. Was such a threat illegal? The court decided, with some judges dissenting, that, if a threatened retaliatory strike was consistent with military necessity and proportionality, it would not necessarily be illegal. (Judgement paragraphs 37–50)

The legality of the possession of nuclear weapons

The court then considered the legality of the possession, as opposed to actual use, of nuclear weapons. The Court looked at various **treaties**, including the UN Charter, and found no treaty language that specifically forbade the possession of nuclear weapons in a categorical way.

The UN Charter was examined in paragraphs 37-50 (paragraph 37: "The Court will now address the question of the legality or illegality of recourse to nuclear weapons in the light of the provisions of the Charter relating to the threat or use of force"). Paragraph 39 mentions: "These provisions [i.e. those of the Charter] do not refer to specific weapons. They apply to any use of force, regardless of the weapons employed. The Charter neither expressly prohibits, nor permits, the use of any specific weapon, including nuclear weapons. A weapon that is already unlawful *per se*, whether by treaty or custom, does not become lawful by reason of its being used for a legitimate purpose under the Charter."

Treaties were examined in paragraphs 53-63 (paragraph 53: "The Court must therefore now examine whether there is any prohibition of recourse to nuclear weapons as such; it will first ascertain whether there is a conventional prescription to this effect"), as part of the law applicable in situations of armed conflict (paragraph 51, first sentence: "Having dealt with the Charter provisions relating to the threat or use of force, the Court will now turn to the law applicable in situations of armed conflict"). In particular, with respect to "the argument [that] has been advanced that nuclear weapons should be treated in the same way as poisoned weapons" , the Court concluded that "it does not seem to the Court that the use of nuclear weapons can be regarded as specifically prohibited on the basis of the [...] provisions of the Second Hague Declaration of 1899, the Regulations annexed to the Hague Convention IV of 1907 or the 1925 Protocol" (paragraphs 54 and 56)". It was also argued by some that the Hague Conventions concerning the use of bacteriological or chemical weapons would also apply to nuclear weapons, but the Court was unable to adopt this argument ("The Court does not find any specific prohibition of recourse to nuclear weapons in treaties expressly prohibiting the use of certain weapons of mass destruction" , paragraph 57 *in fine*).

With respect to treaties that "deal [...] exclusively with acquisition, manufacture, possession, deployment and testing of nuclear weapons, without specifically addressing their threat or use," the Court notes that those treaties "certainly point to an increasing concern in the international community with these weapons; the Court concludes from this that these treaties could therefore be seen as foreshadowing a future general prohibition of the use of such weapons, but they do not constitute such a prohibition by themselves"

(paragraph 62). Also, regarding *regional* treaties prohibiting resource, namely those of Tlatelolco (Latin America) and Rarotonga (South Pacific) the Court notes that while those "testify to a growing awareness of the need to liberate the community of States and the international public from the dangers resulting from the existence of nuclear weapons" , "[i]t [i.e. the Court] does not, however, view these elements as amounting to a comprehensive and universal conventional prohibition on the use, or the threat of use, of those weapons as such." (paragraph 63).

Customary international law also provided insufficient evidence that the possession of nuclear weapons had come to be universally regarded as illegal.

Ultimately, the court was unable to find an *opinio juris* (that is, legal consensus) that nuclear weapons are illegal to possess. (paragraph 65) However, in practice, nuclear weapons have not been used in war since 1945 and there have been numerous UN resolutions condemning their use (however, such resolutions are not universally supported—most notably, the nuclear powers object to them).(paragraph 68-73) The ICJ did not find that these facts demonstrated a new and clear customary law absolutely forbidding nuclear weapons.

However, there are many universal humanitarian laws applying to war. For instance, it is illegal for a combatant specifically to target civilians and certain types of weapons that cause indiscriminate damage are categorically outlawed. All states seem to observe these rules, making them a part of customary international law, so the court ruled that these laws would also apply to the use of nuclear weapons.(paragraph 86) The Court decided not to pronounce on the matter of whether the use of nuclear weapons might possibly be legal, if exercised as a last resort in extreme circumstances (such as if the very existence of the state was in jeopardy).(paragraph 97)

25.2.3 Decision

The court undertook seven separate votes, all of which were passed: [14]

1. The court decided to comply with the request for an advisory opinion; [15]

2. The court replied that "There is in neither customary nor conventional international law any specific authorization of the threat or use of nuclear weapons"; [16]

3. The court replied that "There is in neither customary nor conventional international law any comprehensive and universal prohibition of the threat or use of nuclear weapons as such"; [17]

4. The court replied that "A threat or use of force by means of nuclear weapons that is contrary to Article

2, paragraph 4, of the United Nations Charter and that fails to meet all the requirements of Article 51, is unlawful";[18]

5. The court replied that "A threat or use of nuclear weapons should also be compatible with the requirements of the international law applicable in armed conflict, particularly those of the principles and rules of humanitarian law, as well as with specific obligations under treaties and other undertakings which expressly deal with nuclear weapons" [19]

6. The court replied that "the threat or use of nuclear weapons would generally be contrary to the rules of international law applicable in armed conflict, and in particular the principles and rules of humanitarian law; However, in view of the current state of international law, and of the elements of fact at its disposal, the Court cannot conclude definitively whether the threat or use of nuclear weapons would be lawful or unlawful in an extreme circumstance of self-defence, in which the very survival of a State would be at stake" [20]

7. The court replied that "There exists an obligation to pursue in good faith and bring to a conclusion negotiations leading to nuclear disarmament in all its aspects under strict and effective international control". [21]

The court voted as follows:[22]

Split decision

The only significantly split decision was on the matter of whether "the threat or use of nuclear weapons would generally be contrary to the rules of international law applicable in armed conflict", not including "in an extreme circumstance of self-defence, in which the very survival of a State would be at stake". However, three of the seven "dissenting" judges (namely, Judge Shahabuddeen of Guyana, Judge Weeramantry of Sri Lanka, and Judge Koroma of Sierra Leone) wrote separate opinions explaining that the reason they were dissenting was their view that there is no exception under any circumstances (*including* that of ensuring the survival of a State) to the general principle that use of nuclear weapons is illegal. A fourth dissenter, Judge Oda of Japan, dissented largely on the ground that the Court simply should not have taken the case.

Vice President Schwebel remarked in his dissenting opinion that

It cannot be accepted that the use of nuclear weapons on a scale which would - or could - result in the deaths of many millions in indiscriminate

inferno and by far-reaching fallout, have pernicious effects in space and time, and render uninhabitable much or all of the earth, could be lawful.

And Higgins noted that she did not

exclude the possibility that such a weapon could be unlawful by reference to the humanitarian law, if its use could never comply with its requirements.[24]

Nevertheless, the Court's opinion did not conclude definitively and categorically, under the existing state of international law at the time, whether in an extreme circumstance of self-defence in which the very survival of a State would be a stake, the threat or use of nuclear weapons would necessarily be unlawful in all possible cases. However, the court's opinion unanimously clarified that the world's states have a binding duty to negotiate in good faith, and to accomplish, nuclear disarmament.

25.3 International reaction

25.3.1 United Kingdom

Main article: Nuclear weapons and the United Kingdom

The Government of the United Kingdom has announced plans to renew Britain's only nuclear weapon, the Trident missile system.[25] They have published a white paper *The Future of the United Kingdom's Nuclear Deterrent* in which they state that the renewal is fully compatible with the United Kingdom's treaty commitments and international law.[26] These arguments are summarised in a question and answer briefing published by UK Permanent Representative to the Conference on Disarmament[27]

- Is Trident replacement legal under the Non Proliferation Treaty (NPT)? Renewal of the Trident system is fully consistent with our international obligations, including those on disarmament. ...

- Is retaining the deterrent incompatible with NPT Article VI? The NPT does not establish any timetable for nuclear disarmament. Nor does it prohibit maintenance or renewal of existing capabilities. Renewing the current Trident system is fully consistent with the NPT and with all our international legal obligations. ...

The white paper *The Future of the United Kingdom's Nuclear Deterrent* stands in contrast to two legal opinions. The first, commissioned by Peacerights,[28] was given on 19 December 2005 by Rabinder Singh QC and Professor Christine Chinkin of Matrix Chambers. It addressed

> whether Trident or a likely replacement to Trident breaches customary international law[29]

Drawing on the International Court of Justice (ICJ) opinion, Singh and Chinkin argued that:

> The use of the Trident system would breach customary international law, in particular because it would infringe the "intransgressible" [principles of international customary law] requirement that a distinction must be drawn between combatants and non-combatants.[29]

The second legal opinion was commissioned by Greenpeace[30] and given by Philippe Sands QC and Helen Law, also of Matrix Chambers, on 13 November 2006.[31] The opinion addressed

> The compatibility with international law, in particular the *jus ad bellum*, international humanitarian law ('IHL') and Article VI of the Treaty on the Non-Proliferation of Nuclear Weapons ('NPT'), of the current UK strategy on the use of Trident...The compatibility with IHL of deploying the current Trident system...[and] the compatibility with IHL and Article VI NPT of the following options for replacing or upgrading Trident: (a) Enhanced targeting capability; (b) Increased yield flexibility; (c) Renewal of the current capability over a longer period.[32]

With regards to the *jus ad bellum*, Sands and Law found that

> Given the devastating consequences inherent in the use of the UK's current nuclear weapons, we are of the view that the proportionality test is unlikely to be met except where there is a threat to the very survival of the state. In our view, the 'vital interests' of the UK as defined in the Strategic Defence Review are considerably broader than those whose destruction threaten the survival of the state. The use of nuclear weapons to protect such interests is likely to be disproportionate and therefore unlawful under Article 2(4) of the UN Charter.[33]

The phrase "very survival of the state" is a direct quote from paragraph 97 of the ICJ ruling. With regards to international humanitarian law, they found that

> it [is] hard to envisage any scenario in which the use of Trident, as currently constituted, could be consistent with the IHL prohibitions on indiscriminate attacks and unnecessary suffering. Further, such use would be highly likely to result in a violation of the principle of neutrality.[34]

Finally, with reference to the NPT, Sands and Law found that

> A broadening of the deterrence policy to incorporate prevention of nonnuclear attacks so as to justify replacing or upgrading Trident would appear to be inconsistent with Article VI; b) Attempts to justify Trident upgrade or replacement as an insurance against unascertainable future threats would appear to be inconsistent with Article VI; c) Enhancing the targeting capability or yield flexibility of the Trident system is likely to be inconsistent with Article VI; d) Renewal or replacement of Trident at the same capability is likely to be inconsistent with Article VI; and e) In each case such inconsistency could give rise to a material breach of the NPT.[35]

25.3.2   Scots law

Main article: Scots criminal law

In 1999 a legal case was put forward to attempt to use the ICJ's Opinion in establishing the illegality of nuclear weapons.

On 27 September 1999, three Trident Ploughshares activists Ulla Røder from Denmark, Angie Zelter from England, and Ellen Moxley from Scotland, were acquitted of charges of malicious damage at Greenock Sheriff Court. The three women had boarded the *Maytime* a barge moored in Loch Goil and involved in scientific work connected with of the Vanguard Class submarines berthed in the nearby Gareloch, and caused £80,000 worth of damage. As is often the case in trials relating to such actions, the defendants attempted to establish that their actions were necessary, in that they had prevented what they saw as "nuclear crime" .[36]

The acquittal of the **Trident Three** resulted in the High Court of Justiciary, the supreme criminal court in Scots law, considering a Lord Advocate's Reference, and presenting the first detailed analysis of the ICJ Opinion by another judicial body. The High Court was asked to answer four questions:[36]

1. In a trial under Scottish criminal procedure, is it competent to lead evidence as to the content of customary international law as it applies in the United Kingdom?

2. Does any rule of customary international law justify a private individual in Scotland in damaging or destroying property in pursuit of his or her objection to the United Kingdom's possession of nuclear weapons, its action in placing such weapons at locations within Scotland or its policies in relation to such weapons?

3. Does the belief of an accused person that his or her actions are justified in law constitute a defence to a charge of malicious mischief or theft?

4. Is it a general defence to a criminal charge that the offence was committed in order to prevent or bring to an end the commission of an offence by another person?

The four collective answers given by Lord Prosser, Lord Kirkwood and Lord Penrose were all negative. This did not have the effect of overturning the acquittals of Roder, Zelter and Moxley (Scots Law, like many other jurisdictions, does not allow for an acquittal to be appealed); however, it does have the effect of invalidating the *ratio decidendi* under which the three women were able to argue for their acquittal, and ensures that similar defences cannot be present in Scots Law.

25.3.3 Humanitarian Initiative

Building on the decision that states need to abide by international humanitarian law at all times, the humanitarian dimension of nuclear disarmament has gained traction. The root of the original opposition lies with the humanitarian consequences of nuclear weapons.

The Humanitarian Initiative is a group of eighty states that are calling for the humanitarian consequences to be at the core of any nuclear weapons discourse, paving the way for their gradual delegitimization and an emergent norm banning nuclear weapons in international customary law.

25.4   See also

- Global Security Institute

- International humanitarian law

- List of International Court of Justice cases

- The Martens Clause

- Mutual assured destruction

- Nuclear warfare

- Humanitarian Initiative

- Parliamentarians for Nuclear Non-Proliferation and Disarmament

25.5   References

- Sands, Philippe; and Law, Helen; *The United Kingdom's nuclear deterrent:Current and future issies of legality* (PDF) for Greenpeace.

- Singh, Rabinder; and Chinkin, Christine; *The Maintenance and Possible Replacement of the Trident Nuclear Missile System Introduction and Summary of Advice* for Peacerights

- United Kingdom Permanent Representative to the Conference on Disarmament *Britain's Nuclear Deterrent*

- United Nations General Assembly A/RES/49/75/K: Request for an advisory opinion from the International Court of Justice on the legality of the threat or use of nuclear weapons 90th plenary meeting 15 December 1994.

- Weiss, Peter; *Notes on a Misunderstood Decision: The World Court's Near Perfect Advisory Opinion in the Nuclear Weapons Case*, website of the Lawyers' Committee on Nuclear Policy (LCNP) July 22, 1996

25.6   ICJ Documents

- ICJ documents relating to the case

- *Legality of the threat or use of nuclear weapons (General List No. 95)* 8 July 1996

- Summary of the Advisory Opinion

- Declarations of individual judges:

  - Declaration of President Bedjaoui
  - Declaration of Judge Herczegh
  - Declaration of Judge Shi
  - Declaration of Judge Vereshchetin
  - Declaration of Judge Ferrari Bravo

- Separate Opinions of individual judges:

- Separate Opinion of Judge Guillaume
- Separate Opinion of Judge Ranjeva
- Separate Opinion of Judge Fleischhauer
- Dissenting Opinions of individual judges:
  - Dissenting Opinion of Vice-President Schwebel
  - Dissenting Opinion of Judge Oda
  - Dissenting Opinion of Judge Shahabuddeen
  - Dissenting Opinion of Judge Weeramantry
  - Dissenting Opinion of Judge Koroma
  - Dissenting Opinion of Judge Higgins

25.7 Further reading

- David, Eric; "The Opinion of the International Court of Justice on the Legality of the Use of Nuclear Weapons" (1997) 316 *International Review of the Red Cross* 21.

- Condorelli,Luigi; "Nuclear Weapons: A Weighty Matter for the International Court of Justice" (1997) 316 *International Review of the Red Cross* 9, 11.

- Ginger,Ann Fagan; "Looking at the United Nations through The Prism of National Peace Law," 36(2) *UN Chronicle*62 (Summer 1999).

- Greenwood, Christopher; "The Advisory Opinion on Nuclear Weapons and the Contribution of the International Court to International Humanitarian Law" (1997) 316 *International Review of the Red Cross* 65.

- Greenwood, Christopher; "Jus ad Bellum and Jus in Bello in the Nuclear Weapons Advisory Opinion" in Laurence Boisson de Chazournes and Phillipe Sands (eds), *International Law, the International Court of Justice and Nuclear Weapons* (1999) 247, 249.

- Holdstock, Dougaylas; and Waterston, Lis; "Nuclear weapons, a continuing threat to health," 355(9214) The Lancet 1544 (29 April 2000).

- McNeill, John; "The International Court of Justice Advisory Opinion in the Nuclear Weapons Cases--A First Appraisal" (1997) 316 *International Review of the Red Cross* 103, 117.

- Mohr, Manfred; "Advisory Opinion of the International Court of Justice on the Legality of the Use of Nuclear Weapons Under International Law--A Few Thoughts on its Strengths and Weaknesses" (1997) 316 *International Review of the Red Cross* 92, 94.

- Moore, Mike; "World Court says mostly no to nuclear weapons," 52(5) *Bulletin of the Atomic Scientists,* 39 (Sept-October 1996).

- Moxley, Charles J.; *Nuclear Weapons and International Law in the Post Cold War World* (Austin & Winfield 2000), ISBN 1-57292-152-8.

25.8 Footnotes

[1] "Legality of the Threat or Use of Nuclear Weapons" - Advisory Opinion of 8 July 1996 - General List No. 95 (1995-1998)

[2] "Request for advisory opinion made by the World Health Organization" (PDF). The Hague: International Court of Justice. 1993-09-03. Retrieved 2009-11-02.

[3] "Request for advisory opinion" (PDF). The Hague: International Court of Justice. 1994-12-19. Retrieved 2009-02-11.

[4] *Yearbook of the ILC, 1950,* vol. I, p. 131

[5] *Yearbook of the ILC, 1950,* vol. 1, p. 162

[6] "ICJ Press releases on the Legality of the Use by a State of Nuclear Weapons in Armed Conflict" - General List No. 93 (1993-1996)

[7] Request for an advisory opinion (on the) Legality of the Use by a State of Nuclear Weapons in Armed Conflict - General List No. 93 (1993-1996) - transmitted to the Court under a World Health Assembly resolution of 14 May 1993, paragraph 1

[8] ICJ Press release on the Legality of the threat or use of nuclear weapons - ICJ Advisory Opinion 8 July 1996, ICJ General List No. 93

[9] Resolutions adopted by the General Assembly at its 49th session A service provided by the United Nations, Dag Hammarskjöld Library

[10] "General Assembly Session 49 Meeting 90". 15 December 1994. p. 35.

[11] "General Assembly Resolution 49/75 K, Request for an advisory opinion from the International Court of Justice on the legality of the threat or use of nuclear weapons".

[12] United Nations Bibliographic Information System Dag Hammarskjold Library Voting record search: UN Symbol: A/RES/49/75K

[13] See footnote 61 to the dissenting opinion of Judge Weeramantry

[14] ICJ *Legality of the Threat or Use of Nuclear Weapons (General List No. 95),* paragraph 105.

[15] ICJ *Legality of the Threat or Use of Nuclear Weapons (General List No. 95)*, section 1.

[16] ICJ *Legality of the Threat or Use of Nuclear Weapons (General List No. 95)*, paragraph 105, section 2A.

[17] ICJ *Legality of the Threat or Use of Nuclear Weapons (General List No. 95)*, paragraph 105, section 2B.

[18] ICJ *Legality of the Threat or Use of Nuclear Weapons (General List No. 95)*, paragraph 105, section 2C.

[19] ICJ *Legality of the Threat or Use of Nuclear Weapons (General List No. 95)*, paragraph 105, section 2D.

[20] ICJ *Legality of the Threat or Use of Nuclear Weapons (General List No. 95)*, paragraph 105, section 2E.

[21] ICJ *Legality of the Threat or Use of Nuclear Weapons (General List No. 95)*, paragraph 105, section 2F.

[22] As registrar of the court Eduardo Valencia-Ospina was not entitled to vote

[23] In this instance President Bedjaoui's deciding vote carried the motion

[24] ICJ advisory opinion Dissenting opinion of Judge Higgins

[25] Memoranda on the Future of the UK's Strategic Nuclear Deterrent: the White Paper to the House of Commons Defence Committee

[26] The Future of the United Kingdom's Nuclear Deterrent(pdf) December 2006:

[27] Britain's Nuclear Deterrent by UK Permanent Representative to the Conference on Disarmament

[28] Peacerights Memorandum from Peacerights

[29] Singh, Rabinder; and Chinkin, Christine: The Maintenance and Possible Replacement of the Trident Nuclear Missile System Introduction and Summary of Advice for Peacerights (paragraph 1 and 2)

[30] Greenpeace Trident replacement may be illegal under international law

[31] Sands, Philippe; and Law, Helen: The United Kingdom's nuclear deterrent:Current and future issies of legality (see References)

[32] Sands, Philippe; and Law, Helen: References, paragraph 1

[33] Sands, Philippe; and Law, Helen: References, paragraph 4(i)

[34] Sands, Philippe; and Law, Helen: References, paragraph 4(iii)

[35] Sands, Philippe; and Law, Helen: References, paragraph 4(iv)

[36] Peter Weiss. The International Court of Justice and the Scottish High Court: Two Views of the Illegality of Nuclear Weapons, Web article states that it was first published in: Waseda Proceedings of Comparative Law, Vol.4 (2001), p. 149, Institute of Comparative Law, Waseda University, Tokyo.

Chapter 26

Mainau Declaration

The term "**Mainau Declaration**" refers to socio-political appeals by Nobel laureates who participated in the Lindau Nobel Laureate Meetings, the annual gathering with young scientists at the German town of Lindau. The name denotes that these declarations were presented on Mainau Island in Lake Constance, the traditional venue of the last day of the one-week meeting.[1]

26.1   Mainau Declaration 1955

The first Mainau Declaration was an appeal against the use of nuclear weapons. Initiated and drafted by German nuclear scientists Otto Hahn and Max Born, it was circulated at the 5th Lindau Nobel Laureate Meeting (11–15 July 1955) and presented on Mainau Island on 15 July 1955. The declaration was initially signed by 18 Nobel laureates. Within a year, the number of supporters rose to 52 Nobel laureates.

26.1.1   Full text

We, the undersigned, are scientists of different countries, different creeds, different political persuasions. Outwardly, we are bound together only by the Nobel Prize, which we have been favored to receive. With pleasure we have devoted our lives to the service of science. It is, we believe, a path to a happier life for people. We see with horror that this very science is giving mankind the means to destroy itself. By total military use of weapons feasible today, the earth can be contaminated with radioactivity to such an extent that whole peoples can be annihilated. Neutrals may die thus as well as belligerents.

If war broke out among the great powers, who could guarantee that it would not develop into a deadly conflict? A nation that engages in a total war thus signals its own destruction and imperils the whole world.

We do not deny that perhaps peace is being preserved precisely by the fear of these weapons. Nevertheless, we think it is a delusion if governments believe that they can avoid war for a long time through the fear of these weapons. Fear and tension have often engendered wars. Similarly it seems to us a delusion to believe that small conflicts could in the future always be decided by traditional weapons. In extreme danger no nation will deny itself the use of any weapon that scientific technology can produce.

All nations must come to the decision to renounce force as a final resort. If they are not prepared to do this, they will cease to exist.

—Mainau, Lake Constance, 15 July 1955[2]

26.1.2   Signatories

The initial 18 signatories were:[1]

- Kurt Alder
- Max Born
- Adolf Butenandt
- Arthur H. Compton
- Gerhard Domagk
- Hans von Euler-Chelpin
- Otto Hahn
- Werner Heisenberg
- George Hevesy
- Richard Kuhn
- Fritz Lipmann

- Hermann Joseph Muller

- Paul Hermann Müller

- Leopold Ruzicka

- Frederick Soddy

- Wendell M. Stanley

- Hermann Staudinger

- Hideki Yukawa

26.2   Mainau Declaration 2015 on Climate Change

The Mainau Declaration 2015 on Climate Change was presented on Mainau Island, Germany, on the occasion of the last day of the 65th Lindau Nobel Laureate Meeting on Friday 3 July 2015. It is an urgent warning of the consequences of climate change and was initially signed by 36 Nobel laureates. In the months thereafter, 35 additional laureates joined the group of supporters of the declaration. As of February 2016, a total of 76 Nobel laureates endorse the Mainau Declaration 2015.

The text of the declaration states that although more data needs to be analysed and further research has to be done, the climate report by the IPCC still represents the most reliable scientific assessment on anthropogenic climate change, and that it should therefore be used as a foundation upon which policymakers should discuss actions to oppose the global threat of climate change."[2]

A group photo of some of the Nobel laureates who initially signed the Mainau Declaration 2015. Photo: Christian Flemming

26.2.1   Full text

We undersigned scientists, who have been awarded Nobel Prizes, have come to the shores of

Lake Constance in southern Germany, to share insights with promising young researchers, who like us come from around the world. Nearly 60 years ago, here on Mainau, a similar gathering of Nobel Laureates in science issued a declaration of the dangers inherent in the newly found technology of nuclear weapons—a technology derived from advances in basic science. So far we have avoided nuclear war though the threat remains. We believe that our world today faces another threat of comparable magnitude.

Successive generations of scientists have helped create a more and more prosperous world. This prosperity has come at the cost of a rapid rise in the consumption of the world's resources. If left unchecked, our ever-increasing demand for food, water, and energy will eventually overwhelm the Earth's ability to satisfy humanity's needs, and will lead to wholesale human tragedy. Already, scientists who study Earth's climate are observing the impact of human activity.

In response to the possibility of human-induced climate change, the United Nations established the Intergovernmental Panel on Climate Change (IPCC) to provide the world's leaders a summary of the current state of relevant scientific knowledge. While by no means perfect, we believe that the efforts that have led to the current IPCC Fifth Assessment Report represent the best source of information regarding the present state of knowledge on climate change. We say this not as experts in the field of climate change, but rather as a diverse group of scientists who have a deep respect for and understanding of the integrity of the scientific process.

Although there remains uncertainty as to the precise extent of climate change, the conclusions of the scientific community contained in the latest IPCC report are alarming, especially in the context of the identified risks of maintaining human prosperity in the face of greater than a 2 °C rise in average global temperature. The report concludes that anthropogenic emissions of greenhouse gases are the likely cause of the current global warming of the Earth. Predictions from the range of climate models indicate that this warming will very likely increase the Earth's temperature over the coming century by more than 2 °C above its pre-industrial level unless dramatic reductions are made in anthropogenic emissions of greenhouse gases over the coming decades.

Based on the IPCC assessment, the world must make rapid progress towards lowering current and future greenhouse gas emissions to minimize the substantial risks of climate change. We believe that the nations of the world must take the opportunity at the United Nations Climate Change Conference in Paris in December 2015 to take decisive action to limit future global emissions. This endeavor will require the cooperation of all nations, whether developed or developing, and must be sustained into the future in accord with updated scientific assessments. Failure to act will subject future generations of humanity to unconscionable and unacceptable risk.

—Mainau, Germany, 3 July 2015[*][2]

Nobel laureate Brian Schmidt reading the Mainau Declaration 2015 on Climate Change on the final day of the 65th Lindau Nobel Laureate Meeting. Photo: Christian Flemming

26.2.2 Signatories and supporters

The following Nobel laureates have thus far signed the Mainau Declaration 2015 or expressed their full support after its presentation. 36 Nobel laureates (left column) signed the declaration on 3 July 2015 on the final day of the 65th Lindau Nobel Laureate Meeting; 35 agreed later on for their names to be listed as signatories.[*][2]

26.3 See also

- Lindau Nobel Laureate Meetings

26.4 References

[1] Burmester, Ralph (2015). *Science at First Hand, 65 years Lindau Nobel Laureate Meetings*. Germany: Deutsches Museum Bonn. pp. 48/49.

[2] "Mainau Declaration Official Website". Lindau Nobel Laureate Meetings. Retrieved 2015-11-24.

26.5 External links

- Mainau Declaration Official Website

- Video of Nobel laureate Brian Schmidt presenting the Mainau Declaration 2015

Chapter 27

Nuclear-Free Future Award

Since 1998 the **Nuclear-Free Future Award** (NFFA) is an award given to anti-nuclear activists, organizations and communities. The award has honored and helped facilitate the ongoing work of individuals and initiatives struggling to undo this juncture of time for the sake of the coming generations. The central message is: leave the uranium in the earth !"[1] The award is intended to promote the opposition to uranium mining, nuclear weapons and nuclear power.

The NFFA is a project of the Franz Moll Foundation for the Coming Generations and gives out awards in three categories: Resistance ($10,000 prize), Education ($10,000 prize) and Solutions ($10,000 prize). Additional optional categories are Lifetime Achievement and Special Recognition (contemporary work of art). The award ceremonies take place all around the world.

The NFFA is financed by donations, charity events, and benefit auctions.

27.1 Laureates

The Nuclear-Free Future Award Laureates:"[2]

27.2 See also

- List of Nuclear-Free Future Award recipients
- List of nuclear whistleblowers
- List of peace activists
- NAS Award for Behavior Research Relevant to the Prevention of Nuclear War
- Non-nuclear future
- Nuclear Free World Policy
- World Uranium Hearing
- Anti-nuclear movement
- Nuclear disarmament

27.3 References

[1] "Statement of Mission" .

[2] "NFFA Recipients and Locations" .

[3] Jillian Marsh

[4] Manuel Pino

[5] Recipients of the 2006 Nuclear-Free Future Awards

[6] The 2004 Nuclear-Free Future Award Recipients

[7] Jonathan Schell

[8] The 2002 Nuclear Free Future Awards

27.4 External links

- The Nuclear Free Future Award

Chapter 28

Nuclear-free zone

A **nuclear-free zone** is an area in which nuclear weapons (see nuclear-weapon-free zone) and nuclear power plants are banned. The specific ramifications of these depend on the locale in question.

Nuclear-free zones usually neither address nor prohibit radiopharmaceuticals used in nuclear medicine even though many of them are produced in nuclear reactors. They typically do not prohibit other nuclear technologies such as cyclotrons used in particle physics.

Several sub-national authorities worldwide have declared themselves "nuclear-free". However, the label is often symbolic, as nuclear policy is usually determined and regulated at higher levels of government: nuclear weapons and components may traverse nuclear-free zones via military transport without the knowledge or consent of local authorities which had declared nuclear-free zones.

Palau became the first nuclear-free nation in 1980.[1] New Zealand was the first Western-allied nation to legislate towards a national nuclear free zone by effectively renouncing the nuclear deterrent.[2]

28.1 Nuclear-free zone by geographical areas

28.1.1 Australia

Many Australian local government areas of Australia have passed anti-nuclear weaponry legislation; notable among these are Brisbane, capital of Queensland, which has been nuclear weapon free since 1983, and the South and North Sydney councils. However the passage of such legislation is generally considered just a symbolic measure.[3] The majority of councils which have passed anti-nuclear weaponry legislation are members of the Australian Nuclear Free Zones and Toxic Industries Secretariat which has 44 member councils.[4]

28.1.2 Austria

Austria is a nuclear free zone, when a nuclear power station was built during the 1970s at Zwentendorf, Austria, start-up was prevented by a popular vote in 1978. The completed power plant is now marketed as a shooting location for film and television.[5] On July 9, 1997, the Austrian Parliament voted unanimously to maintain the country's anti-nuclear policy.[6]

Ironically, the headquarters of the International Atomic Energy Agency is located in Vienna, and the IAEA maintains nuclear laboratories both in Vienna and Seibersdorf.[7] The IAEA has also established programs to assist nuclear energy projects in developing countries.

Austria's anti-nuclear stance also causes tension with its nuclear neighbors. Vienna is located close to the Czech reactor at Temelin, and four reactors are being built in neighboring Slovakia and two in neighboring Hungary.[8] Austria also draws from regional electricity grids, meaning it imports nuclear power, although chancellor Werner Faymann has pledged to eliminate Austria's reliance on foreign power by 2015.[9]

28.1.3 Canada

Vancouver is a nuclear weapons free city. Victoria, British Columbia is also a nuclear weapons free city. This has caused problems as nearby Esquimalt houses CFB Esquimalt, Canada's Pacific naval base, which is used frequently by the United States Navy. The USN routinely sends ships or aircraft carriers loaded with nuclear weapons to Esquimalt. As a result, the ships are forced to dock out of the city limits as not to violate the city by-laws. The policy does not limit operations at TRIUMF, Canada's national laboratory for nuclear and particle physics at the University of British Columbia, Vancouver. The province of British Columbia also bans mining for uranium, and the construction of nuclear power plants within its territorial limits.[10]

Nanaimo, British Columbia, Kitimat, British Columbia;

Red Deer, Alberta, and Regina, Saskatchewan are also nuclear weapons-free cities.

28.1.4   Former Soviet Union

Central Asia

All states of Central Asia have signed the Treaty on the Non-Proliferation of Nuclear Weapons and signed the Treaty on nuclear-weapon-free zones. Thus, there are no technologies to create a weapon or enrich the particles, but in Tajikistan, during the Soviet Union, such initiatives were brought to life, but with the collapse of the Soviet Union, the facilities were dismantled and moved to the Russian Federation. The Treaty came into force on 21 March 2009.

Estonia

After Estonia seceded from the Soviet Union in the early 1990s, Estonia became a nuclear-free country. The two land-based nuclear reactors at the Soviet Navy nuclear submarine training centre in Paldiski were removed when Russia finally relinquished control of the nuclear reactor facilities in September 1995. There are no nuclear power stations in Estonia.

28.1.5   Japan

Nuclear-free Kobe Port, seen from Po-ai Shiosai Park in 2011

Main article: Japan's non-nuclear policy

As a resource-poor nation, Japan is heavily reliant on nuclear power, but its unique experience in World War II has led to the wholesale rejection of nuclear weapons, holding nuclear weapons shall not be manufactured in, possessed by, or allowed entry into Japan. These tenets, known as the Three Non-Nuclear Principles, were first stated by Prime Minister Eisaku Satō in 1967, and were adopted as a parliamentary resolution in 1971, though they have never formally been entered into law. They continue to reflect the attitudes of both government and the general public, who remain staunchly opposed to the manufacture or use of nuclear weapons.

The Japan Self-Defense Forces have never made any attempt to manufacture or otherwise obtain nuclear arms, and no nuclear weapons are known to have been introduced into the Japanese Home Islands since the end of World War II. While the United States does not maintain nuclear bases within its military installations on the Home Islands, it is believed to have once stored weapons at Okinawa, which remained under US administrative jurisdiction until 1972.

28.1.6   Italy

Main article: Italian nuclear power referendum, 1987

Italy is a nuclear free zone since the Italian nuclear power referendum of November 1987. Following center-right parties' victory in the 2008 election, Italy's industry minister announced that the government scheduled the construction to start the first new Italian nuclear-powered plant by 2013. The announced project was paused in March 2011, after the Japanese earthquake, and scrapped after a referendum on 12–13 June 2011.

28.1.7   New Zealand

Main article: New Zealand's nuclear-free zone

In 1984, Prime Minister David Lange barred nuclear-powered or nuclear-armed ships from using New Zealand ports or entering New Zealand waters. Under the New Zealand Nuclear Free Zone, Disarmament, and Arms Control Act 1987,[11][12] territorial sea and land of New Zealand became nuclear weapons and nuclear-powered ship free zones. It does not ban nuclear power stations. A research reactor was operated by the University of Canterbury until 1981. Official planning for a nuclear power station continued until the 1980s. A nuclear reactor provided electricity for McMurdo Station, in the New Zealand Antarctic Territory from 1962-1972.

The Act prohibits "entry into the internal waters of New Zealand 12 miles (22.2 km) radius by any ship whose propulsion is wholly or partly dependent on nuclear power" and bans the dumping of radioactive waste within the nuclear-free zone, as well as prohibiting any New Zealand citizen or resident "to manufacture, acquire, possess, or have any control over any nuclear explosive device." [12][13] Combined with the firm policy of the United States to "neither confirm nor deny" whether particular naval vessels carry nuclear weapons, the Act effectively bars these ships from entering New Zealand waters.[14]

New Zealand's security treaty with the United States, ANZUS, did not mention nuclear deterrence and did not require unconditional port access. However, after New Zealand refused entry to USS *Buchanan*, the United States government suspended its ANZUS obligations to New Zealand, seeing New Zealand's effective rejection of United States Navy vessels as voiding the treaty. The Lange Labour government did not see their stance as incompatible with the treaty and sought a compromise for over two years before passing the Act.[14] Support for the non-nuclear policy was bolstered by the perceived over-reaction of the United States and by the sinking of the Rainbow Warrior by French spies while docked in Auckland. According to some commentators, the legislation was a milestone in New Zealand's development as a nation and seen as an important act of sovereignty, self-determination and cultural identity.[15][16] New Zealand's three decade anti-nuclear campaign is the only successful movement of its type in the world which resulted in the nation's nuclear-free zone status being enshrined in legislation.[17]

The nuclear-free zone law does not make building land-based nuclear power plants illegal. However, the relatively small electricity system, abundance of other resources to generate electricity, and public opposition has meant a nuclear power plant has never gone beyond the investigation phase – a nuclear power plant was proposed north of Auckland in the early 1970s, but the discovery of large natural gas reserves in Taranaki saw the proposal shelved.[18]

28.1.8   Nordic countries

Nuclear weapons-free Nordic (*Finn. Ydinaseeton Pohjola*) was an initiative by the President of Finland Urho Kekkonen for a nuclear weapons-free zone in the Nordic countries. The aim was to prevent the Nordic countries from becoming a nuclear battleground and a route for cruise missiles in the event of a nuclear war between the Soviet Union and NATO.

Nuclear energy, however, is used in both Finland and Sweden.

28.1.9   Palau

Palau adopted its first constitution in July 1979, stating that the Micronesian country would be "nuclear-free". The United States told the Palauan government that this constitution was likely incompatible with the Compact of Free Association. The Palauan government submitted a revised version of the constitution without the "nuclear-free" clause the following October. The Palauan people rejected the revised document and instated the original constitution in July 1980.[1] Seven years later, however, the Palauan people voted to overturn their nuclear-free status out of "economic survival".[19]

28.1.10   United Kingdom

The Nuclear Free Zone Movement in the United Kingdom was very strong in early 1980s; up to 200 local authorities including County councils, District councils and City councils such as the Greater London Council (GLC) (before its abolition) declared themselves to be 'nuclear free'. The first 'Nuclear Free Zone' in the UK was Manchester City Council in 1980 - this still exists to this day. Wales became 'nuclear free' on 23 February 1982 after Clwyd County Council declared itself 'nuclear free' and the Nuclear Free Wales Declaration was made. This policy was legally underpinned by Section 137 of the Local Government Act, which allowed local authorities to spend a small amount on whatever members considered was in the interest of their area or a part of their area.

UK nuclear-free local authorities refused to take part in civil defence exercises relating to nuclear war, which they thought were futile. The non-cooperation of the nuclear-free zone authorities was the main reason for the cancellation of the national 'Hard Rock' civil defence exercise in July 1982. In England and Wales 24 of the 54 County Councils refused to participate and seven more co-operated only in a half-hearted way.[20] This has been seen as a victory for the British Peace movement against the policies of Margaret Thatcher. Generally, nuclear-free zones were predominantly Labour Party controlled Councils but Liberal Party and even a few Conservative Party Councillors were often active in this respect too.

28.1.11   United States

A number of towns, cities and counties in the United States established themselves as Nuclear-Free Zones in the 1970s, 1980s and 1990s. The first was Missoula, Montana. In the November 1978 general election, Missoula voters overwhelmingly approved a ballot initiative in the form of a land-use ordinance establishing the entirety of Missoula County as a "'nuclear free' zoning district" banning all nuclear facilities except those for medical purposes. (In the same election, Montana voters approved a statewide initiative by a 2-1 margin barring nuclear facilities or reactors without strict state-enforced regulatory standards and ratification by popular referendum, and in a follow-up 1980 initiative, Montanans narrowly voted to ban the disposal of nuclear waste.) That Missoula's measure was originally drafted as a zoning ordinance legally enforceable by the county planning department apparently created the popular term "Nuclear Free Zone" adopted as the name of the local political action group sponsoring the initiative and later used by other jurisdictions worldwide.[21]

Subsequently, the tiny town of Garrett Park, Maryland, attracted worldwide attention with its referendum in May, 1982. The following year, Takoma Park, Maryland, was officially declared a nuclear-free zone in 1983 by then-mayor Sam Abbott. A citizen committee of the local city council continues to monitor city contracts. The city cannot hold contracts with any company associated with any aspect of nuclear weapons without a waiver from the citizen committee. In September 2005, Takoma Park took a stand against the transportation of high-level nuclear waste through the City. It voted to amend its Nuclear-Free Zone Ordinance to give its citizen committee responsibility to collect information and from this information and from consultations with individuals and organizations involved in the transportation of high-level nuclear waste, to advise the City on how to promote the safety and welfare of its citizens from harmful exposure to high-level nuclear waste.[22]

Another well-known nuclear-free community is Berkeley, California, whose citizens passed the Nuclear Free Berkeley Act in 1986 which allows the city to levy fines for nuclear weapons-related activity and to boycott companies involved in the United States nuclear infrastructure. The City of Berkeley has posted signs at city limits proclaiming its nuclear free status. The ordinance specifies possible fines for such activities within its borders. The University of California, Berkeley is deeply involved in the history of nuclear weapons, and the University of California system until recently managed operations at Los Alamos National Laboratory, a U.S. nuclear weapons design laboratory, and continues to manage the Lawrence Livermore National Laboratory. At the time of the passage of the act, the University operated a nuclear reactor for research purposes, the Etcheverry Reactor, which it continued to operate after the act went into effect. The University of California, as a state institution, is not subject to Berkeley's municipal regulations, including the ban. Berkeley also has major freeway and train lines which are used in transporting nuclear materials.

On November 14, 1984 the Davis, California City Council declared the city to be a nuclear-free zone.[23] Davis has major freeway and train arteries running through it which

are used for transporting nuclear materials. The University of California, with a campus at Davis, runs a research reactor at the nearby former McClellan Air Force Base, as well as workers who are involved with Lawrence Livermore National Laboratory.

On November 8, 1988 the city of Oakland, California passed "Measure T" with 57% of the vote, making that city a nuclear free zone. Under Ordinance No. 11062 CMS then passed on December 6, 1988, the city is restricted from doing business with *"any entity knowingly engaged in nuclear weapons work and any of its agents, subsidiaries or affiliates which are engaged in nuclear weapons work."* [24] The measure was invalidated in federal court, on the grounds that it interfered with the Federal Government's constitutional authority over national defense and atomic energy.[25][26] The issue being Oakland is a major port, and like Berkeley, and Davis, has major freeway and train arteries running through it. In 1992, the Oakland City Council unanimously reinstated modified elements of the older ordinance, reportedly bringing the total number of Nuclear Free Zones in the United States at that time to 188, with a total population of over 17 million in 27 states.[27]

Other cities, counties, and other governments within the United States passing nuclear free zone ordinances and the date of adoption, when known:

- Arcata, CA (9/15/1989)[28]
- Boulder, CO (1985)[29]
- Chicago, IL (1986)[30]
- Cleveland Heights, OH (1987)[31]
- East Windsor, CT (12/16/1992)[32]
- Eugene, OR (11/1986); revised measure defeated 5/15/1990[33]
- Garrett Park, MD (1982)[34]
- Hawaii County, HI (1981); amendment excluding military approved by referendum 11/1986[33]
- Hayward, CA (9/15/87)[35]
- Homer, AK (10/3/89)[33]
- Iowa City, IA (1985)[36]
- Marin County, CA (1986)[37]
- Oberlin, Ohio (November 1985)
- New York City, NY (11/8/1984)[38]
- Reno, NV (1996)[39]
- Sac and Fox Nation, OK (8/28/1993)[40]

- Santa Cruz, CA (11/17/1998)[41]
- Sykesville, MD (6/16/1982)[42]

28.2 See also

- African Nuclear Weapons Free Zone Treaty
- Antarctic Treaty System
- Anti-nuclear movement
- Campaign for Nuclear Disarmament
- Comprehensive Test Ban Treaty
- France and weapons of mass destruction
- Helen Caldicott
- Mongolian Nuclear-Weapons-Free Status
- Nagasaki and Hiroshima
- Non-nuclear future
- Nuclear-Free Future Award
- Nuclear testing
- Nuclear weapons and the United States
- Nuclear-Weapon-Free Zone
- Sinking of the Rainbow Warrior
- Treaty of Tlatelolco
- United States and weapons of mass destruction

28.3 References

[1] Clark, Roger; Roff, Sue Rabbitt (1984). *Micronesia: the problem of Palau* (Rev. ed.). London: Minority Rights Group. p. 13. ISBN 9780946690145.

[2] Lange, David (1990). *Nuclear Free: The New Zealand Way*. New Zealand: Penguin Books.

[3] "Nuclear-free city? Afraid no". *Brisbane Times*. March 18, 2009. Retrieved 6 Feb 2014.

[4] "[Untitled leaflet]" (PDF). Australian Nuclear Free Zones and Toxic Industries Secretariat. Archived from the original (PDF) on 22 February 2014.

[5] Zwentendorf - location

[6] "Coalition of Nuclear-Free Countries". WISE News Communique. September 26, 1997. Archived from the original on February 23, 2006. Retrieved 2006-05-19.

[7] Archived September 10, 2011, at the Wayback Machine.

[8] "Austria and Czech Republic divided over nuclear power". *BBC News*. January 4, 2012.

[9] Nucléaire: l'Autriche se débranche dès 2015

[10] EnergyBC: Nuclear Power

[11] New Zealand Nuclear Free Zone, Disarmament, and Arms Control Act 1987

[12] Nuclear Free Zone

[13] New Zealand Nuclear Free Zone Extension Bill - Green Party

[14] Pugh, Michael Charles (1989). *The ANZUS Crisis, Nuclear Visiting and Deterrence*. Cambridge: Cambridge University Press. pp. 1–2. ISBN 0-521-34355-0.

[15] "Lange's impact on NZ and world". *BBC News*. August 14, 2005. Retrieved May 20, 2010.

[16] Nuclear threat continues to grow, New Zealand warns on anniversary of anti-nuclear law - International Herald Tribune

[17] Lange, David (1990). *Nuclear Free: The New Zealand Way*. New Zealand: Penguin Books.

[18] "Nuclear Energy Prospects in New Zealand". World Nuclear Association. April 2009. Retrieved 2009-12-09.

[19] "Palau Drops Nuclear-Free Status". *The New York Times*. AP. 7 August 1987. Retrieved 22 April 2016.

[20] Bolsover, Philip. "A victory - and a new development", in Minnion, J., and Bolsover, P.. *The CND Story*. London: Alison and Busby, 1983

[21] For more on the Missoula zoning ordinance, see Missoula Independent, Nov. 30, 2000, at: http://missoulanews.bigskypress.com/missoula/reading-the-sign/Content?oid=1133059. For the zoning ordinance text as updated in 2007, see Missoula County, MT Nuclear Free Zone, at: http://greenpolicy360.net/index.php?title=Missoula_County%2C_MT_Nuclear_Free_Zone

[22] For more on Takoma Park's nuclear-free history see: http://www.takomaparkmd.gov/committees/nfz/nftpc.htm

[23] Nuclear Free Zone - Davis Wiki

[24] Schedule P, City of Oakland, rev. 7/30/01 Archived May 26, 2011, at the Wayback Machine.

[25] A Nuclear-Free Zone Is Ruled to Be Invalid, New York Times/AP. 4/28/90

[26] Guardian (US), 23 May 1990, p7, via WISE Nuclear Issues Information Service Archived June 11, 2009, at the Wayback Machine.

[27] Oakland City Council Reinstates Nuclear-Free Policy, US Newswire 7/3/92, via Highbeam

[28] " ARCATA NUCLEAR WEAPONS FREE ZONE ACT

[29] Boulder Revised Code Chapter 6-8: Nuclear Free Zone, via Colorado Code Publishing Company Archived June 14, 2009, at the Wayback Machine.

[30] The Company As Target, Ronnie Dugger, New York Times Magazine, 9/20/87

[31] Signs announcing Cleveland Heights as Nuclear Free Zone. Whatever happened to ...? I cleveland.com

[32] Town of East Windsor Nuclear Free Zone Ordinance

[33] Meiklejohn Civil Liberties Institute Archives: Human Rights and Peace Law Docket 1945-1993

[34] Town of Garrett Park: History

[35] Ordinance No. 87-024, An Ordinance Establishing Nuclear Free Hayward Archived December 29, 2010, at the Wayback Machine.

[36] press-citizen- Iowa City to replace missing signs

[37] Marin County Code, Chapter 23.12: Nuclear-Free Zone

[38] Peace Magazine. Mar 1985 The article adds that 14 of 16 ballot measures passed in the 1984 general election, and that there were 80 US NFZs at that time.

[39] Nuclear free Reno

[40] Native lands becoming nuclear free zones in US. Via WISE Nuclear Issues Information Service Archived June 11, 2009, at the Wayback Machine.

[41] COUNCIL POLICY 11.4: DECLARING THE CITY OF SANTA CRUZ A NUCLEAR FREE ZONE Archived November 18, 2008, at the Wayback Machine.

[42] UMB Langsdale Library WMAR-TV News Collection Archived September 1, 2006, at the Wayback Machine.

28.4   External links

- France's Nuclear Weapons Program at the Atomic Forum

- Mururoa protest,Time 1973

- "By-laws beat the bomb" – Commentary by Frank Johnson

- Bikini Atoll Atomic test zone

- Pictures of victims of US nuclear testing in the Marshall Islands.

- Nuclear Testing in Australia 1952-1958

- British Nuclear Test Veterans Association

- Australias Maralinga nuclear test site

- "Nuclear Free Berkeley Act" - Nuclear-free zone legislation for Berkeley, California

- Radio Nizkor International Nuclear conference

- Nuclear-free future award

Chapter 29

Russell–Einstein Manifesto

The **Russell–Einstein Manifesto** was issued in London on 9 July 1955 by Bertrand Russell in the midst of the Cold War. It highlighted the dangers posed by nuclear weapons and called for world leaders to seek peaceful resolutions to international conflict. The signatories included eleven preeminent intellectuals and scientists, including Albert Einstein, who signed it just days before his death on 18 April 1955. A few days after the release, philanthropist Cyrus S. Eaton offered to sponsor a conference—called for in the manifesto—in Pugwash, Nova Scotia, Eaton's birthplace. This conference was to be the first of the Pugwash Conferences on Science and World Affairs, held in July 1957.

29.1 Background

Main articles: History of nuclear weapons and Atomic bombings of Hiroshima and Nagasaki

The first detonation of an atomic weapon took place on 16 July 1945 in the desert north of Alamogordo, New Mexico. On 6 August 1945, the US dropped *Little Boy* on the Japanese city of Hiroshima and, three days later, *Fat Man* on Nagasaki. At least 100,000 civilians were killed outright by these two bombings.

On 18 August 1945, the *Glasgow Forward* published the first known recorded comment by Bertrand Russell on atomic weapons, which he began composing the day Nagasaki was bombed. It contained threads that would later appear in the manifesto:

> The prospect for the human race is sombre beyond all precedent. Mankind to be averted.

After learning of the bombing of Hiroshima and seeing an impending nuclear arms race, Joseph Rotblat, the only scientist to leave the Manhattan Project on moral grounds, remarked that he "became worried about the whole future of mankind".

Over the years that followed Russell and Rotblat worked on efforts to curb nuclear proliferation, collaborating with Albert Einstein and other scientists to compose what became known as the **Russell–Einstein Manifesto**.

29.2 Press conference, 9 July 1955

The manifesto was released during a press conference at Caxton Hall, London. Rotblat, who chaired the meeting, describes it as follows:

> It was thought that only a few of the Press would turn up and a small room was booked in Caxton Hall for the Press Conference. But it soon became clear that interest was increasing and the next larger room was booked. In the end the largest room was taken and on the day of the Conference this was packed to capacity with representatives of the press, radio and television from all over the world. After reading the Manifesto, Russell answered a barrage of questions from members of the press, some of whom were initially openly hostile to the ideas contained in the Manifesto. Gradually, however, they became convinced by the forcefulness of his arguments, as was evident in the excellent reporting in the Press, which in many cases gave front page coverage.

Russell had begun the conference by stating:

> I am bringing the warning pronounced by the signatories to the notice of all the powerful Governments of the world in the earnest hope that they may agree to allow their citizens to survive.

29.3   Synopsis

The manifesto called for a conference where scientists would assess the dangers posed to the survival of humanity by weapons of mass destruction. Emphasis was placed on the meeting being politically neutral. It extended the question of nuclear weapons to all people and governments. One particular phrase is quoted often, including by Rotblat upon receipt of the Nobel Peace Prize in 1995:

> Remember your humanity, and forget the rest.

29.4   The beginnings of the Pugwash Conferences

The manifesto called for an international conference, and was originally planned by Jawaharlal Nehru to be held in India. This was delayed by the outbreak of the Suez Crisis. Aristotle Onassis offered to finance a meeting in Monaco, but this was rejected. Instead, Cyrus Eaton, a Canadian industrialist who had known Russell since 1938, offered to finance the conference in his hometown of Pugwash, Nova Scotia. The **Russell–Einstein Manifesto** became the Pugwash Conferences' founding charter. The first of the conferences was held in July 1957 in London.

29.5   Signatories to the manifesto

- Max Born
- Percy W. Bridgman
- Albert Einstein
- Leopold Infeld
- Frédéric Joliot-Curie
- Hermann J. Muller
- Linus Pauling
- Cecil F. Powell
- Joseph Rotblat
- Bertrand Russell
- Hideki Yukawa

Ten of the eleven signatories of the Russell–Einstein Manifesto are Nobel Laureates, the exception being Leopold Infeld.

29.6   See also

- Mainau Declaration
- Mutually assured destruction

29.7   References

- The Origins of the Russell–Einstein Manifesto, by Sandra Ionno Butcher, May 2005.
- The First Pugwash Conference.
- Pugwash and Russell's Legacy by John R. Lenz.
- *Science and World Affairs: history of the Pugwash Conferences*, 1962, by Professor J. Rotblatt

29.8   External links

- Op-Ed: The 50-Year Shadow by Joseph Rotblat, *New York Times*, 17 May 2005.
- Meeting the Russell–Einstein Challenge to Humanity by David Krieger, October 2004.
- The Russell-Einstein Manifesto, 9 July 1955

Chapter 30

Vulnerability of nuclear plants to attack

On September 9, 1980, Daniel Berrigan (above), his brother Philip, and six others (the "Plowshares Eight") began the Plowshares Movement. They illegally trespassed onto the General Electric Nuclear Missile facility in King of Prussia, Pennsylvania, where they damaged nuclear warhead nose cones and poured blood onto documents and files. They were arrested and charged with over ten different felony and misdemeanor counts.[1]

The **vulnerability of nuclear plants to deliberate attack** is of concern in the area of nuclear safety and security. Nuclear power plants, civilian research reactors, certain naval fuel facilities, uranium enrichment plants, fuel fabrication plants, and even potentially uranium mines are vulnerable to attacks which could lead to widespread radioactive contamination. The attack threat is of several general types: commando-like ground-based attacks on equipment which if disabled could lead to a reactor core meltdown or widespread dispersal of radioactivity; and external attacks such as an aircraft crash into a reactor complex, or cyber attacks.[2]

The United States 9/11 Commission has said that nuclear power plants were potential targets originally considered for the September 11, 2001 attacks. If terrorist groups could sufficiently damage safety systems to cause a core meltdown at a nuclear power plant, and/or sufficiently damage spent fuel pools, such an attack could lead to widespread radioactive contamination. The Federation of American Scientists have said that if nuclear power use is to expand significantly, nuclear facilities will have to be made extremely safe from attacks that could release massive quantities of radioactivity into the community. New reactor designs have features of passive nuclear safety, which may help. In the United States, the NRC carries out "Force on Force" exercises at all nuclear power plant sites at least once every three years.[2]

Nuclear reactors become preferred targets during military conflict and, over the past three decades, have been repeatedly attacked during military air strikes, occupations, invasions and campaigns.[3] Various acts of civil disobedience since 1980 by the peace group Plowshares have shown how nuclear weapons facilities can be penetrated, and the group's actions represent extraordinary breaches of security at nuclear weapons plants in the United States. The National Nuclear Security Administration has acknowledged the seriousness of the 2012 Plowshares action. Non-proliferation policy experts have questioned "the use of private contractors to provide security at facilities that manufacture and store the government's most dangerous military material".[4] Nuclear weapons materials on the black market are a global concern,[5][6] and there is concern about the possible detonation of a dirty bomb by a militant group in a major city.[7][8]

The number and sophistication of cyber attacks is on the rise. *Stuxnet* is a computer worm discovered in June 2010 that is believed to have been created by the United States and Israel to attack Iran's nuclear facilities. It switched off safety devices, causing centrifuges to spin out of control.[9] The computers of South Korea's nuclear plant operator (KHNP) were hacked in December 2014. The cyber attacks involved thousands of phishing emails containing malicious code, and information was stolen.[10]

30.1 Attacks on nuclear power plants

Terrorists could target nuclear power plants in an attempt to release radioactive contamination into the community. The United States 9/11 Commission has said that nuclear power plants were potential targets originally considered for the September 11, 2001 attacks. If terrorist groups could sufficiently damage safety systems to cause a core meltdown at a nuclear power plant, and/or sufficiently damage spent fuel pools, such an attack could lead to a widespread radioactive contamination. According to a 2004 report by the U.S. Congressional Budget Office, "The human, environmental, and economic costs from a successful attack on a nuclear power plant that results in the release of substantial quantities of radioactive material to the environment could be great." [11] An attack on a reactor's spent fuel pool could also be serious, as these pools are less protected than the reactor core. The release of radioactivity could lead to thousands of near-term deaths and greater numbers of long-term fatalities.[2]

If nuclear power use is to expand significantly, nuclear facilities will have to be made extremely safe from attacks that could release massive quantities of radioactivity into the community. New reactor designs have features of passive safety, such as the flooding of the reactor core without active intervention by reactor operators. But these safety measures have generally been developed and studied with respect to accidents, not to the deliberate reactor attack by a terrorist group. However, the US Nuclear Regulatory Commission does now also require new reactor license applications to consider security during the design stage.[2]

In the United States, the NRC carries out "Force on Force" (FOF) exercises at all Nuclear Power Plant (NPP) sites at least once every three years. The FOF exercise, which is typically conducted over 3 weeks, "includes both tabletop drills and exercises that simulate combat between a mock adversary force and the licensee's security force. At an NPP, the adversary force attempts to reach and simulate damage to key safety systems and components, defined as "target sets" that protect the reactor's core or the spent fuel

pool, which could potentially cause a radioactive release to the environment. The licensee's security force, in turn, interposes itself to prevent the adversaries from reaching target sets and thus causing such a release" .[2]

In the U.S., plants are surrounded by a double row of tall fences which are electronically monitored. The plant grounds are patrolled by a sizeable force of armed guards.[12]

30.2 Military attacks

Nuclear reactors become preferred targets during military conflict and, over the past three decades, have been repeatedly attacked during military air strikes, occupations, invasions and campaigns:[3]

- In September 1980, Iran bombed the Al Tuwaitha nuclear complex in Iraq, in Operation Scorch Sword, which was a surprise IRIAF (Islamic Republic of Iran Air Force) airstrike carried out on 30 September 1980, that damaged an almost complete nuclear reactor 17 km south-east of Baghdad, Iraq.[13]

- In June 1981, an Israeli air strike completely destroyed Iraq's Osirak nuclear research facility.

- Between 1984 and 1987, Iraq bombed Iran's Bushehr nuclear plant six times.

- In 1991, the U.S. bombed three nuclear reactors and an enrichment pilot facility in Iraq.

- In 1991, Iraq launched Scud missiles at Israel's Dimona nuclear power plant.

- In September 2007, Israel bombed a Syrian reactor under construction.[3]

After several incidents in Pakistan in which terrorists attacked three of its military nuclear facilities, it became clear that there emerged a serious danger that they would gain access to the country's nuclear arsenal, according to a journal published by the US Military Academy at West Point.[14] In January 2010, it was revealed that the US army was training a specialised unit "to seal off and snatch back" Pakistani nuclear weapons in the event that militants would obtain a nuclear device or materials that could make one. Pakistan supposedly possesses about 80 nuclear warheads. US officials refused to speak on the record about the American safety plans.[15]

30.3 Nuclear terrorism

Main article: Nuclear terrorism

Amory Lovins says that the United States has for decades been running on energy that is "brittle" (easily shattered by accident or malice) and that this poses a grave and growing threat to national security, life, and liberty.[16] Lovins' claims that these vulnerabilities are increasingly being exploited. His book *Brittle Power* documents many significant assaults on energy facilities, other than during a war, in forty countries and within the United States, in some twenty-four states.[17]

Lovins further claims that in 1966, twenty natural uranium fuel rods were stolen from the Bradwell nuclear power station in England, and in 1971, five more were stolen at the Wylfa Nuclear Power Station. In 1971, an intruder wounded a night watchman at the Vermont Yankee reactor in the USA. The New York University reactor building was broken into in 1972, as was the Oconee Nuclear Station's fuel storage building in 1973. In 1975, the Kerr McGee plutonium plant had thousands of dollars worth of platinum stolen and taken home by workers. In 1975, at the Biblis Nuclear Power Plant in Germany, a Member of Parliament demonstrated the lack of security by carrying a bazooka into the plant under his coat.[18]

Nuclear plants were designed to withstand earthquakes, hurricanes, and other extreme natural events. But deliberate attacks involving large airliners loaded with fuel, such as those that crashed into the World Trade Center and Pentagon, were not considered when design requirements for today's fleet of reactors were determined. It was in 1972 when three hijackers took control of a domestic passenger flight along the east coast of the U.S. and threatened to crash the plane into a U.S. nuclear weapons plant in Oak Ridge, Tennessee. The plane got as close as 8,000 feet above the site before the hijackers' demands were met.[19][20]

In February 1993, a man drove his car past a check point the Three Mile Island Nuclear plant, then broke through an entry gate. He eventually crashed the car through a secure door and entered the Unit 1 reactor turbine building. The intruder, who had a history of mental illness, hid in a building and was not apprehended for four hours. Stephanie Cooke asks: "What if he'd been a terrorist armed with a ticking bomb?"[21]

Fissile material may be stolen from nuclear plants and this may promote the spread of nuclear weapons. Many terrorist groups are eager to acquire the fissile material needed to make a crude nuclear device, or a dirty bomb. Nuclear weapons materials on the black market are a global concern,[5][6] and there is concern about the possible detonation of a small, crude nuclear weapon by a militant group in a major city, with significant loss of life and property.[7][8] It is feared that a terrorist group could detonate a radiological or "dirty bomb", composed of any radioactive source and a conventional explosive. The radioactive material is dispersed by the detonation of the explosive. Detonation of such a weapon is not as powerful as a nuclear blast, but can produce considerable radioactive fallout. Alternatively, a terrorist group may position some of its members, or sympathisers, within the plant to sabotage it from inside.[22]

The IAEA Illicit Nuclear Trafficking Database notes 1,266 incidents reported by 99 countries over the last 12 years, including 18 incidents involving HEU or plutonium trafficking.[23]

- There have been 18 incidences of theft or loss of highly enriched uranium (HEU) and plutonium confirmed by the IAEA.[24]

- Security specialist Shaun Gregory argued in an article that terrorists have attacked Pakistani nuclear facilities three times in the recent past; twice in 2007 and once in 2008.[25]

- In November 2007, burglars with unknown intentions infiltrated the Pelindaba nuclear research facility near Pretoria, South Africa. The burglars escaped without acquiring any of the uranium held at the facility.[26][27]

- In June 2007, the Federal Bureau of Investigation released to the press the name of Adnan Gulshair el Shukrijumah, allegedly the operations leader for developing tactical plans for detonating nuclear bombs in several American cities simultaneously.[28]

- In November 2006, MI5 warned that al-Qaida were planning on using nuclear weapons against cities in the United Kingdom by obtaining the bombs via clandestine means.[29]

- In February 2006, Oleg Khinsagov of Russia was arrested in Georgia, along with three Georgian accomplices, with 79.5 grams of 89 percent enriched HEU.[24]

- The Alexander Litvinenko poisoning with radioactive polonium "represents an ominous landmark: the beginning of an era of nuclear terrorism," according to Andrew J. Patterson.[30]

- In June 2002, U.S. citizen José Padilla was arrested for allegedly planning a radiological attack on the city of Chicago; however, he was never charged with such conduct. He was instead convicted of charges that he conspired to "murder, kidnap and maim" people overseas.

30.4   Sabotage by insiders

Insider sabotage regularly occurs, because insiders can observe and work around security measures. In a study of insider crimes, the authors repeatedly said that successful insider crimes depended on the perpetrators' observation and knowledge of security vulnerabilities. Since the atomic age began, the U.S. Department of Energy's nuclear laboratories have been known for widespread violations of security rules. During the Manhattan Project, physicist Richard Feynman was barred from entering certain nuclear facilities; he would crack safes and violate other rules as pranks to reveal deficiencies in security. A better understanding of the reality of the threat will help to overcome complacency and is critical to getting countries to take stronger preventive measures.*[31]

A fire caused 5–10 million dollars worth of damage to New York's Indian Point Energy Center in 1971. The arsonist turned out to be a plant maintenance worker. Sabotage by workers has been reported at many other reactors in the United States: at Zion Nuclear Power Station (1974), Quad Cities Nuclear Generating Station, Peach Bottom Nuclear Generating Station, Fort St. Vrain Generating Station, Trojan Nuclear Power Plant (1974), Browns Ferry Nuclear Power Plant (1980), and Beaver Valley Nuclear Generating Station (1981). Many reactors overseas have also reported sabotage by workers. Suspected arson has occurred in the USA and overseas.*[18]

On 8 January 1982, the 70th anniversary of the formation of the African National Congress, Umkhonto we Sizwe, the armed wing of the ANC attacked Koeberg Nuclear Power Station while it was still under construction.*[32] Damage was estimated at R 500 million and the commissioning of the plant was put back by 18 months.*[33] In 1998 a group of workers at one of Russia's largest nuclear weapons facilities attempted to steal 18.5 kilograms of HEU—enough for a bomb.*[18]

30.5   Civil disobedience

Various acts of civil disobedience since 1980 by the peace group Plowshares have shown how nuclear weapons facilities can be penetrated, and the group's actions represent extraordinary breaches of security at nuclear weapons plants in the United States. On July 28, 2012, three members of Plowshares cut through fences at the Y-12 National Security Complex in Oak Ridge, Tennessee, which manufactures US nuclear weapons and stockpiles highly enriched uranium. The group spray-painted protest messages, hung banners, and splashed blood.*[4]

The National Nuclear Security Administration has ac-

knowledged the seriousness of the 2012 Plowshares action, which involved the protesters walking into a high-security zone of the plant, calling the security breach "unprecedented." Independent security contractor, WSI, has since had a weeklong "security stand-down," a halt to weapons production, and mandatory refresher training for all security staff.*[4]

Non-proliferation policy experts are concerned about the relative ease with which these unarmed, unsophisticated protesters could cut through a fence and walk into the center of the facility. This is further evidence that nuclear security —the securing of highly enriched uranium and plutonium— should be a top priority to prevent terrorist groups from acquiring nuclear bomb-making material. These experts have questioned "the use of private contractors to provide security at facilities that manufacture and store the government's most dangerous military material".*[4]

In 2010, there was a security breach at a Belgian air force base which possessed U.S. nuclear warheads. The incident involved six anti-nuclear activists entering Kleine Brogel Air Base. The activists stayed in the snow-covered base for about 20 minutes, before being arrested. A similar event occurred in 2009.*[34]

On December 5, 2011, two anti-nuclear campaigners breached the perimeter of the Cruas Nuclear Power Plant, escaping detection for more than 14 hours, while posting videos of their sit-in on the internet.*[35]

30.6   Cyber attacks

Stuxnet is a computer worm discovered in June 2010 that is believed to have been created by the United States and Israel to attack Iran's nuclear facilities.*[9] It switched off safety devices, causing centrifuges to spin out of control. Stuxnet initially spreads via Microsoft Windows, and targets Siemens industrial control systems. While it is not the first time that hackers have targeted industrial systems,*[36] it is the first discovered malware that spies on and subverts industrial systems,*[37] and the first to include a programmable logic controller (PLC) rootkit.*[38]*[39]

Different variants of Stuxnet targeted five Iranian organizations,*[40] with the probable target widely suspected to be uranium enrichment infrastructure in Iran;*[41]*[42] Symantec noted in August 2010 that 60% of the infected computers worldwide were in Iran.*[43] Siemens stated that the worm has not caused any damage to its customers,*[44] but the Iran nuclear program, which uses embargoed Siemens equipment procured secretly, has been damaged by Stuxnet.*[45]*[46] Kaspersky Lab concluded that the sophisticated attack could only have been conducted "with nation-state support".*[47]

Idaho National Laboratory ran the Aurora Experiment in 2007 to demonstrate how a cyber attack could destroy physical components of the electric grid.*[48] The experiment used a computer program to rapidly open and close a diesel generator's circuit breakers out of phase from the rest of the grid and explode. This vulnerability is referred to as the *Aurora Vulnerability*.

The number and sophistication of cyber attacks is on the rise. The computers of South Korea's nuclear plant operator (KHNP) were hacked in December 2014. The cyber attacks involved thousands of phishing emails containing malicious code, and information was stolen.*[10]

30.7 Population surrounding plants

Population density is one critical lens through which risks have to be assessed, says Laurent Stricker, a nuclear engineer and chairman of the World Association of Nuclear Operators:*[49]

The KANUPP plant in Karachi, Pakistan, has the most people—8.2 million—living within 30 kilometres of a nuclear plants have populations larger than 3 million within that radius.*[49] plant, although it has just one relatively small reactor with an output of 125 megawatts. Next in the league, however, are much larger plants —Taiwan's 1,933-megawatt Kuosheng plant with 5.5 million people within a 30-kilometre radius and the 1,208-megawatt Chin Shan plant with 4.7 million; both zones include the capital city of Taipei.*[49]

172,000 people living within a 30 kilometre radius of the Fukushima Daiichi nuclear power plant, have been forced or advised to evacuate the area. More generally, a 2011 analysis by *Nature* and Columbia University, New York, shows that some 21 nuclear plants have populations larger than 1 million within a 30-km radius, and six plants have populations larger than 3 million within that radius.*[49]

30.8 Implications

In his book, *Normal accidents*, Charles Perrow says that multiple and unexpected failures are built into society's complex and tightly-coupled nuclear reactor systems. Such accidents are unavoidable and cannot be designed around.*[50]

In the 2003 book, *Brittle Power*, Amory Lovins talks about the need for a resilient, secure, energy system:

The foundation of a secure energy system is to need less energy in the first place, then to get it from sources that are inherently invulnerable because they're diverse, dispersed, renewable, and mainly local. They're secure not because they're American but because of their design. Any highly centralised energy system—pipelines, nuclear plants, refineries—invite devastating attack. But invulnerable alternatives don't, and can't, fail on a large scale.*[51]

30.9 See also

- Lists of nuclear disasters and radioactive incidents
- Atomic spies
- Crimes involving radioactive substances
- Design basis accident
- Environmental impact of nuclear power
- International Nuclear Events Scale
- Nuclear and radiation accidents
- Nuclear close calls
- Nuclear criticality safety
- Nuclear fuel response to reactor accidents
- Nuclear power debate
- Nuclear power plant emergency response team
- Nuclear whistleblowers
- Nuclear weapon
- Micro nuclear reactor
- Passive nuclear safety
- Safety code (nuclear reactor)
- World Association of Nuclear Operators
- Category:Victims of radiological poisoning

30.10   Further reading

- Allison, Graham (9 August 2004). *Nuclear Terrorism: The Ultimate Preventable Catastrophe*. New York, New York: Times Books. ISBN 978-0-8050-7651-6.

- Byrne, John and Steven M. Hoffman (1996). *Governing the Atom: The Politics of Risk*, Transaction Publishers.

- Cooke, Stephanie (2009). *In Mortal Hands: A Cautionary History of the Nuclear Age*, Black Inc.

- Ferguson, Charles D., and William C. Potter, with Amy Sands, Leonard S. Spector and Fred L. Wehling (2004). *The Four Faces of Nuclear Terrorism*. Monterey, California: Center for Nonproliferation Studies. ISBN 1-885350-09-0.

- Jones, Ishmael (2010) [2008]. *The Human Factor: Inside the CIA's Dysfunctional Intelligence Culture*. Encounter Books. ISBN 978-1-59403-382-7.

- Levi, Michael (2007). *On Nuclear Terrorism*. Cambridge, Massachusetts: Harvard University Press. ISBN 978-0-674-02649-0.

- Lovins, Amory B. and John H. Price (1975). *Non-Nuclear Futures: The Case for an Ethical Energy Strategy*, Ballinger Publishing Company, 1975, ISBN 0-88410-602-0

- Schell, Jonathan (2007). *The Seventh Decade: The New Shape of Nuclear Danger*. New York, New York: Metropolitan Books.

- Kuperman, Alan J (2013). *Nuclear Terrorism and Global Security: The Challenge of Phasing out Highly Enriched Uranium*. New York, New York: Routledge.

30.11   References

[1] *Commonwealth v. Berrigan*, 501 A.2d 226, 509 Pa. 118 (1985)

[2] Charles D. Ferguson & Frank A. Settle (2012). "The Future of Nuclear Power in the United States" (PDF). *Federation of American Scientists*.

[3] Benjamin K. Sovacool (2011). *Contesting the Future of Nuclear Power: A Critical Global Assessment of Atomic Energy*. World Scientific, p. 192.

[4] Kennette Benedict (9 August 2012). "Civil disobedience". *Bulletin of the Atomic Scientists*.

[5] Jay Davis. After A Nuclear 9/11 *The Washington Post*, March 25, 2008.

[6] Brian Michael Jenkins. A Nuclear 9/11? *CNN.com*, September 11, 2008.

[7] Orde Kittrie. Averting Catastrophe: Why the Nuclear Nonproliferation Treaty is Losing its Deterrence Capacity and How to Restore It May 22, 2007, p. 338.

[8] Nicholas D. Kristof. A Nuclear 9/11 *The New York Times*, March 10, 2004.

[9] "Legal Experts: Stuxnet Attack on Iran Was Illegal 'Act of Force'". Wired. 25 March 2013.

[10] Penny Hitchin. "Cyber attacks on the nuclear industry". *Nuclear Engineering International*. 15 September 2015.

[11] "Congressional Budget Office Vulnerabilities from Attacks on Power Reactors and Spent Material".

[12] U.S. NRC: "Nuclear Security – Five Years After 9/11". Accessed 23 July 2007

[13] When Iran Bombed Iraq's Nuclear Reactor, Iraq's Osirak Destruction.

[14] Blakely, Rhys (August 11, 2009). "Terrorists 'have attacked Pakistan nuclear sites three times'", *Times Online*, London

[15] "Elite US troops ready to combat Pakistani nuclear hijacks", *Times*

[16] Brittle Power, Chapter 1, p. 1.

[17] *Brittle Power*, Chapter 1, p. 2.

[18] Amory Lovins (2001). *Brittle Power* (PDF). pp. 145–146.

[19] Threat Assessment: U.S. Nuclear Plants Near Airports May Be at Risk of Airplane Attack (Link Defunct), Global Security Newswire, June 11, 2003.

[20] Newtan, Samuel Upton (2007). *Nuclear War 1 and Other Major Nuclear Disasters of the 20th Century*, AuthorHouse, p.146.

[21] Stephanie Cooke (March 19, 2011). "Nuclear power is on trial". CNN.

[22] Frank Barnaby (2007). "Consequences of a Nuclear Renaissance" (PDF). *International Symposium*.

[23] Bunn, Matthew. "Securing the Bomb 2010: Securing All Nuclear Materials in Four Years" (PDF). President and Fellows of Harvard College. Retrieved 28 January 2013.

[24] Bunn, Matthew & Col-Gen. E.P. Maslin (2010). "All Stocks of Weapons-Usable Nuclear Materials Worldwide Must be Protected Against Global Terrorist Threats" (PDF). Belfer Center for Science and International Affairs, Harvard University. Retrieved July 26, 2012.

[25] Rhys Blakeley, "Terrorists 'have attacked Pakistan nuclear sites three times'," *Times Online* (August 11, 2009).

[26] http://www.pretorianews.co.za/?fSectionId=&fArticleId=vn20071109061218448C528585

[27] *Washington Post*, December 20, 2007, Op-Ed by Micah Zenko

[28] "Feds Hoped to Snag Bin Laden Nuke Expert in JFK Bomb Plot". Fox News. June 4, 2007.

[29] http://politics.guardian.co.uk/terrorism/story/0,,1947295,00.html

[30] "Ushering in the era of nuclear terrorism." by Patterson, Andrew J. MD, PhD, *Critical Care Medicine*, v. 35, p.953–954, 2007.

[31] Matthew Bunn and Scott Sagan (2014). "A Worst Practices Guide to Insider Threats: Lessons from Past Mistakes". The American Academy of Arts & Sciences.

[32] "History of MK". African National Congress. Archived from the original on 4 April 2007. Retrieved 14 May 2007.

[33] Helen Bamford (11 March 2006). "Koeberg: SA's ill-starred nuclear power plant". *Cape Argus*. Retrieved 14 May 2007.

[34] Kevin Dougherty (February 6, 2010). "Belgian base breach sparks nuclear worries". *Stars and Stripes*.

[35] Tara Patel (December 16, 2011). "Breaches at N-plants heighten France's debate over reactors". *Seattle Times*.

[36] "Building a Cyber Secure Plant". Siemens. 30 September 2010. Retrieved 5 December 2010.

[37] Robert McMillan (16 September 2010). "Siemens: Stuxnet worm hit industrial systems". Computerworld. Retrieved 16 September 2010.

[38] "Last-minute paper: An indepth look into Stuxnet". Virus Bulletin.

[39] "Stuxnet worm hits Iran nuclear plant staff computers". BBC News. 26 September 2010.

[40] "Stuxnet Virus Targets and Spread Revealed". BBC News. 15 February 2011. Retrieved 17 February 2011.

[41] Steven Cherry; with Ralph Langner (13 October 2010). "How Stuxnet Is Rewriting the Cyberterrorism Playbook". IEEE Spectrum.

[42] Beaumont, Claudine (23 September 2010). "Stuxnet virus: worm 'could be aimed at high-profile Iranian targets'". London: The Daily Telegraph. Retrieved 28 September 2010.

[43] MacLean, William (24 September 2010). "UPDATE 2-Cyber attack appears to target Iran-tech firms". *Reuters*.

[44] ComputerWorld (14 September 2010). "Siemens: Stuxnet worm hit industrial systems". Computerworld. Retrieved 3 October 2010.

[45] "Iran Confirms Stuxnet Worm Halted Centrifuges". *CBS News*. 29 November 2010.

[46] Ethan Bronner & William J. Broad (29 September 2010). "In a Computer Worm, a Possible Biblical Clue". *NYTimes*. Retrieved 2 October 2010. "Software smart bomb fired at Iranian nuclear plant: Experts". Economictimes.indiatimes.com. 24 September 2010. Retrieved 28 September 2010.

[47] "Kaspersky Lab provides its insights on Stuxnet worm". *Kaspersky*. Russia. 24 September 2010.

[48] "Mouse click could plunge city into darkness, experts say". *CNN*, September 27, 2007. Source: http://www.cnn.com/2007/US/09/27/power.at.risk/index.html

[49] Declan Butler (21 April 2011). "Reactors, residents and risk". *Nature*.

[50] Daniel E Whitney (2003). "Normal Accidents by Charles Perrow" (PDF). *Massachusetts Institute of Technology*.

[51] Amory B. Lovins and L. Hunter Lovins. "Terrorism and Brittle Technology" in *Technology and the Future* by Albert H. Teich, Ninth edition, Thomson, 2003, p. 169.

30.12 External links

- Annotated bibliography, Alsos Digital Library for Nuclear Issues

- Fallout: After a Nuclear Attack - slideshow by *Life magazine*

- Nuclear Emergency and Radiation Resources

- What if the terrorists go nuclear?, Center for Defense Information

- Preventing Catastrophic Nuclear Terrorism, Council on Foreign Relations

- Use of nuclear and radiological weapons by terrorists?, International Review of the Red Cross

- Nuclear-free future award

30.13 Text and image sources, contributors, and licenses

30.13.1 Text

- **Anti-nuclear movement** *Source:* https://en.wikipedia.org/wiki/Anti-nuclear_movement?oldid=763799550 *Contributors:* Ed Poor, Edward, Mac, Kingturtle, Caknuck, Alan Liefting, Redjives, Tweenk, Utcursch, Rich Farmbrough, Rama, Bender235, CanisRufus, Purplefeltangel, Karlthegreat, Rwendland, Radical Mallard, ProhibitOnions, DV8 2XL, Madmardigan53, CyrilleDunant, Jeff3000, Male1979, RichardWeiss, Drbogdan, Rjwilmsi, Hyuga, Vegaswikian, XLerate, FlaBot, Chobot, Bgwhite, Simesa, Robthebob, Wavelength, Pigman, Abhijit13, Kimchi.sg, Nirvana2013, Rax, Ospalh, Smvans7, Saudade7, Revengeofthynerd, SmackBot, Burtonpe, F, KVDP, Hmains, Chris the speller, Cadmium, Bazonka, Mark7-2, Colonies Chris, Jennica, Aces lead, Theanphibian, AdamWeeden, W Ed, Enr-v, Polonium, Will Beback, SashatoBot, Harryboyles, Vgy7ujm, Gobonobo, Biccat, Syra987, Ckatz, M@sk, Hu12, Iridescent, Joseph Solis in Australia, Ayanoa, GerryWolff, Ethnopunk, Vision Thing, Cydebot, Book M, Mombas, Thijs!bot, J.Ring, Gralo, Bobblehead, NorwegianBlue, Second Quantization, JustAGal, Borneh, Salavat, Escarbot, Mentifisto, Derzsi Elekes Andor, Superzohar, Jimothytrotter, Timwright, Afaz, NE2, Paxuscalta, RebelRobot, MaxPont, Magioladitis, Meredyth, Canyonwren, Cgingold, STBot, R'n'B, Giachen, CommonsDelinker, Petersec, Dskluz, Beyonder1, DAID, Scott Illini, Wikipedian1234, 28bytes, Johnfos, Indubitably, Dqeswn, Isultir, Wikidemon, Nsougia2, Andreas Kaganov, Cde3, NPguy, PGWG, Clamshelltvs, HybridBoy, Moonriddengirl, WereSpielChequers, Malcolmxl5, BotMultichill, Marvin Diode, Oda Mari, Nopetro, Chfriend, Lightmouse, Megansmith18, Cheapthrill, Stephen Shaw, Reginmund, Fuddle, Thatotherdude, Randy Kryn, Epthorn, De728631, ClueBot, Nailedtooth, Atom girl, Eiland, TheOldJacobite, Jackcj, Lwnf360, Niceguyedc, 718 Bot, Ktr101, Yohananw, Nukeless, Thymefromti, DumZiBoT, Fell-Gleaming, Addbot, Andunie, Chum2, Download, Freqsh0, Tassedethe, Jarble, Judit Szoleczky, Fmrauch, Dzied Bulbash, Vrinan, AnomieBOT, Bluejohntish, KatrinkaO, Gap9551, GrouchoBot, Cgnk, Nedim Ardoğa, Alexco creator11, FrescoBot, Simjonathan, Skyerise, Brian Everlasting, Larbelaitz, Nora lives, Trappist the monk, DixonDBot, LilyKitty, Robertiki, ErikvanB, Minimac, RjwilmsiBot, DASHBot, EmausBot, Dolescum, WikitanvirBot, Trofobi, Angrytoast, Boundarylayer, Dewritech, Jasonanaggie, Werieth, ZéroBot, Ehedemann, Captain Herbert, Bwikid, Gray eyes, EvenGreenerFish, Michael5770, ClueBot NG, Skjoldbro, Frietjes, Geofferybard, Dougmcdonell, Gambistics~enwiki, Eat more fruit, BG19bot, HIDECCHI001, Mark Arsten, Blaspie55, 220 of Borg, BattyBot, Unknown0106, Tgold20, Theogold20, Cookkiemonster123, Dexbot, RotlinkBot, Joeinwiki, Hermes 1900s, Lk54ui, Limnalid, Stamptrader, BegbertBiggs, Vieque, Graemem56, Ephemeratta, WC Jay, Qzd, InternetArchiveBot, Slapasentry, Jewjoo, GreenC bot, Vladdrin Ndss, Bender the Bot, Hansantinuclearwar and Anonymous: 130

- **Anti-nuclear movement in the United States** *Source:* https://en.wikipedia.org/wiki/Anti-nuclear_movement_in_the_United_States?oldid=765512772 *Contributors:* Mac, Topbanana, Arkuat, Alan Liefting, Xinoph, Rich Farmbrough, Cuppysfriend, Giraffedata, D.Holt, Sjö, Rjwilmsi, Vegaswikian, Wikiliki, Kolbasz, Simesa, RussBot, Limulus, Nirvana2013, Mikeblas, PanchoS, SmackBot, Chris the speller, Ottawakismet, Theanphibian, The PIPE, Rkmlai, E-Kartoffel, Levineps, FairuseBot, Cydebot, Ssilvers, Thijs!bot, Epbr123, Mrshaba, NE2, Jrclark, Paxuscalta, RebelRobot, R'n'B, CommonsDelinker, JMG469, Plasticup, Jamesontai, Funandtrvl, Johnfos, Oshwah, Nsougia2, Qworty, WereSpielChequers, Malcolmxl5, Nopetro, Dakinijones, Binksternet, Ktr101, SchreiberBike, DumZiBoT, CjrSellers, FellGleaming, Anaintern, Addbot, Lightbot, Yobot, Recambo51, Fmrauch, AnomieBOT, LilHelpa, BritishWatcher, Sortirdu123, Jack B108, FrescoBot, Fortdj33, DrilBot, Grufry, Guerillero, RjwilmsiBot, John of Reading, Boundarylayer, GoingBatty, Grammar conquistador, Resprinter123, ClueBot NG, Helpful Pixie Bot, WikiTryHardDieHard, Energy4All, Dexbot, Mogism, Hermes 1900s, DavidLeighEllis, Bksovacool, Stamptrader, Fer48, Graemem56, Adam9007, CAPTAIN RAJU, Sturgeontransformer, InternetArchiveBot, GreenC bot, Bender the Bot and Anonymous: 37

- **Nuclear power in the United States** *Source:* https://en.wikipedia.org/wiki/Nuclear_power_in_the_United_States?oldid=766077177 *Contributors:* Rmhermen, Fxmastermind, Mac, Julesd, Katana0182, Sander123, Munion, Alan Liefting, Wwoods, Nkocharh, Rchandra, Wmahan, Beland, Oneiros, Jayjg, Bender235, Alansohn, PaulHanson, Rwendland, Linas, Xaliqen, Tabletop, Male1979, BD2412, Rjwilmsi, Lavishluau, DonSiano, Wikiliki, Ysangkok, Kolbasz, Simesa, JWB, Limulus, Rowan O'Neill, Stephenb, WutiTheSaxon, ONEder Boy, BirgitteSB, Ospalh, Super Rad!, Ntouran, Johnpseudo, Erudy, SmackBot, Nihonjoe, Patbahn, Gilliam, Hmains, Silly rabbit, Nbarth, Andy120290, Theanphibian, Kingdon, Salamurai, Daniel.Cardenas, Acitrano, Siress, Xiaphias, Ryulong, Levineps, Iridescent, Hvatum, Samcapasso, Sahrin, Cydebot, Teratornis, Gralo, Marek69, Nick Number, Seaphoto, Tillman, Kagrenak, Mcorazao, Paxuscalta, AlmostReadytoFly, Dream Focus, AuburnPilot, Faizhaider, Nyttend, Beagel, Anaxial, Uriel8, CommonsDelinker, Hans Dunkelberg, Tyrerj, B2bomber81, Group29, Izno, Johnfos, Emeraldcrown, Ng.j, Goodmapd, Plazak, Lamro, Tungka, Noveltyghost, Grundle2600, BillTunell, Exnuke, ClueBot, Foxj, Deanlaw, Frmorrison, Mild Bill Hiccup, Eiland, Watti Renew, Lwnf360, RaymondHill101, Jusdafax, 12 Noon, Asmaybe, Sun Creator, Psinu, Keydetpiper, Jgnbr, DisambiguationGuy, DumZiBoT, XLinkBot, Maky, FellGleaming, Dthomsen8, MystBot, Gigoachef, Addbot, Neweb, Tassedethe, Lightbot, TeH nOmInAtOr, Luckas-bot, Yobot, AzureFury, House1630, AnomieBOT, Ulric1313, GB fan, Xqbot, Fionaclee, FrescoBot, Pinethicket, Hem173, Jujutacular, Full-date unlinking bot, Alex146, Vrenator, Nhorel01, Robertiki, Reaper Eternal, Tbhotch, DARTH SIDIOUS 2, RjwilmsiBot, Beyond My Ken, TGCP, Steve03Mills, John of Reading, GoingBatty, Dwalin, Dcirovic, Neun-x, H3llBot, AManWithNoPlan, ClueBot NG, Michaelmas1957, Kayz911, Dav-FL-IN-AZ-id, Drewish19, Gareth.Rose, O.Koslowski, DougieM1, Chilllls, Paulzubrinich, NuclearEnergy, Helpful Pixie Bot, Razieldavison, BG19bot, Xamnidar, Indoor-Fanatiker, Kendall-K1, Frank.esslin, Asauers, Kwaidan4, Thanatos7474, Energy4All, BattyBot, Cyberbot II, JesseAlanGordon, Wywin, Rfassbind, JMDeLuca00, Spyglasses, Limnalid, Jora8488, AndrewWyffels, TheCrimsonLegacy, LizardPatio, Sadiber, Wikideas1, Monkbot, Fer48, 13casrol, BethNaught, Trackteur, Abrowne54, ARC105, Fixwikimaps, James33035, Whensday, InternetArchiveBot, Jewjoo, GreenC bot, Bender the Bot and Anonymous: 133

- **Doomsday Clock** *Source:* https://en.wikipedia.org/wiki/Doomsday_Clock?oldid=764844286 *Contributors:* Damian Yerrick, Bryan Derksen, The Anome, XJaM, Rmhermen, Enchanter, Edward, Zanimum, Tregoweth, Ciphergoth, Evercat, GCarty, Ehn, Doradus, Jakenelson, VeryVerily, Guyster, Jeffq, Chuunen Baka, Robbot, 1984, Cingre, Auric, Doidimais Brasil, Hadal, Anthony, Netjeff, Alerante, Giftlite, Mintleaf~enwiki, Fastfission, Bfinn, Piquan, Everyking, Snowdog, Golbez, Nova77, PurpleHeather, Antandrus, Kiteinthewind, Jossi, Rdsmith4, EBB, Jokestress, Boraneer, Tsemii, Lockz, SYSS Mouse, EugeneZelenko, Bbz, Johan Elisson, Rich Farmbrough, Vsmith, Klykken, Pavel Vozenilek, MattTM, Smalljim, Viriditas, Rammer, Sam Korn, Stephen Bain, Alansohn, Atlant, Snowolf, Hessi, H2g2bob, Mattbrundage, Tainter, KTC, Mwalcoff, Dandv, Age234, Admrboltz, Encyclopedist, Bkwillwm, Mandarax, Sin-man, Graham87, Dvyost, Sjö, Drbogdan, Nightscream, Strait, Josiah Rowe, Bensin, Fred Bradstadt, Hpmons, KarlFrei, SlimXero, Jayann, Chobot, Bentisquare, Mhking, VolatileChemical, Cornellrockey, Roboto de Ajvol, Mercury McKinnon, Wavelength, I need a name, Stein, Ericorbit, Rsrikanth05, Thunderforge, Daemon8666, Abfackeln, NW036, GenestealerUK, Kvn8907, Dureo, Ragesoss, TakingUpSpace, JPMcGrath, Cosmotron, Elkman, Jrissman, Ccgrimm, SamuelRiv, Eurosong, Smkolins, Datajosh, Bosco13, Nikkimaria, Boreas231, Tim Parenti, Esprit15d, Red Jay, Dmuth, Curpsbot-unicodify, Katieh5584, Jonathan.s.kt,

Snaxe920, UltimatePyro, SmackBot, Haza-w, McGeddon, Harperbruce, Evanreyes, Crimsone, Peter Isotalo, Gilliam, Hmains, Winterheart, Psiphiorg, Thumperward, Hibernian, Stv, Victorgrigas, Ryecatcher773, Reaper X, Emurphy42, Pyreforge, Woodyteegra, UberMD, GRuban, Addshore, GrahameS, Wen D House, Thomaslau~enwiki, Gerryharrington, Gakhandal, T-borg, DoomsDayTalk, DMacks, Marcus Brute, TenPoundHammer, Synthe, Rory096, Acebrock, DickFisherVideo, KLLvr283, Calvados~enwiki, J 1982, Jrothwell, Stelio, Moogagot, Ckatz, Silvarbullet1, Timmeh, SQGibbon, Xiaphias, Zapvet, MrDolomite, AceFace905, Tony Fox, Shayonsaleh, Dlohcierekim, The Haunted Angel, SkyWalker, Wolfdog, Tere naam, Rawling, Agent Koopa, Bharnish, Strike Chaos, Borism, Karenjc, Simeon, BT14, Phatom87, Eposty, Inzy, MC10, JJC1138, Gogo Dodo, Otto4711, Gpmuscillo, UberMan5000, Taschenrechner~enwiki, Epbr123, Mercury~enwiki, Andypham3000, Mokaiba, Varavour, Nalvage, Marek69, Citizensmith, Sean William, Northumbrian, Turbo2Xs, Gczffl, Sk8a h8a, WarFighter, Canadian-Bacon, Leuko, ChrisMac, Magioladitis, Bongwarrior, VoABot II, Fusionmix, Jrg7891, Disconformist, Kevinmon, Froid, Hiplibrarianship, Allstarecho, Mike Payne, Shijualex, Medicup, DerHexer, Lenticel, The velociraptor, AVRS, Rage 0, WIKI-GUY-16, Jcronen1, Jonathan Hall, Tanarchy, Thirdright, Dinkytown, Aang-kai, J.delanoy, Trusilver, Thaurisil, BradinAlberta, SlowJog, The little neutrino, McSly, Bentonius, Certian, BeefTapa, Plasticup, Ppapadeas, Happy138, Student7, Themathkid, Potatoswatter, 2help, KylieTastic, HansRoht, Петър Петров, Nuregoftheelves, Jamesontai, Jevansen, Burbankjones, Useight, Ronbo76, Steel1943, Signalhead, Black Kite, VolkovBot, Morenooso, Thewolf37, Johnfos, RingtailedFox, AlnoktaBOT, Jomasecu, Sroc, Billiejoe26, Writing librarian, A4bot, Emilydixieson, NPrice, Charlesriver, Cody-7, Sabih omar, Melsaran, Gekritzl, Werideatdusk33, McM.bot, BeautifulToxicTears, Everything counts, Anarchangel, Crasher7, Suriel1981, TV chump, Gustav Lindwall, Andy Dingley, Jdarryl, Reo690, K. Aainsqatsi, Akitin, SieBot, 4wajzkd02, Mycophycophyta, Triwbe, Izaiahblack, Amir4it, PolarBot, Elcobbola, Tungsten92, Iameukarya, Kosack, Fratrep, Kumioko (renamed), Peulle, Philly jawn, Dabomb87, Randy Kryn, TheCatalyst31, PUMKIN81, Elassint, ClueBot, Carlos10132, SiftingJeff~enwiki, The Thing That Should Not Be, Mriya, All Hallow's Wraith, Ashwin18, Torqtorqtorq, LizardJr8, Trivialist, Jersey emt, -Midorihana-, Jusdafax, Peaches1955, Astraldust, Internet user1, Eeekster, Elephantissimo, Chrisblakemusic, CowboySpartan, The Wicked Twisted Road, Daveg123456, Mythdon, Goodvac, DumZiBoT, XLinkBot, Will-B, Garthbishop, DragonFury, Eveco, Apo-kalypso, Dmac9000, Skarebo, ZooFari, Addbot, PhilT2, Guoguo12, Friginator, Ronhjones, Laurinavicius, Bja917, Lost on Belmont, Dioxis, Favonian, Tassedethe, Bigzteve, Jarble, Rbgolfer, Luckas-bot, Yobot, Granpuff, SnoopCog, Mickey Way, AnomieBOT, Archon 2488, Floquenbeam, Jim1138, Piano non troppo, Materialscientist, Nebraska3, Xqbot, Thoranin C., Tad Lincoln, J4lambert, Anna Frodesiak, Topitsky, Mlpearc, Anonymous from the 21st century, GrouchoBot, Tngl0101, RibotBOT, Richard BB, Karl304, Nokkenbuer, HRIN, Atomicgurl00, FrescoBot, Saga650, Lukeaton, D'ohBot, Ionutzmovie, Maverick9711, BenzolBot, Drew R. Smith, AstaBOTh15, Brandon8552, Pinethicket, Prokhorovka, Loyalist Cannons, RedBot, Serols, Merrrlo, EdoDodo, Dmontoya1970, Dmsdy212, Mr.98, DCSB2005, Jerseycursed, Logical Fuzz, One15969, DRAGON BOOSTER, Shadowed Soul, NerdyScienceDude, Salvio giuliano, KinkyLipids, EmausBot, Markinsic, GoingBatty, Andromedabluesphere440, Dcirovic, K6ka, Sepguilherme, Sailsbystars, Bizzy Whizz, Anita5192, ClueBot NG, 20chances, BG19bot, Lisamccabe, Hallows AG, Michael Barera, Elisfkc, Mdy66, Danothy, Antoneeyuh, Nicno14, Lamilo1562, Black Agent, TheJJunk, Zecommie, Dexbot, SandJ-on-WP, Charles Essie, Mogism, Czech is Cyrillized, Jonny2126, YLCC23, Coxj2000, Theonlytruemathnerd, Artoria2e5, Rathnavelrat, JesseKramme, Butteredscotch, Marco.bs, Murus, Fenetosh, Bant0n12, Rider99, The Herald, Finnusertop, TheWanderer1357, SisterSalvation, Konveyor Belt, JaconaFrere, Femkemilene, Opencooper, MaverickTopGun13, Jamez42, Burklemore1, Agrade999, Ryanicus Girraticus, Alsm7253, Quaker Qweer, The 10th Doctor, Pishcal, Orduin, Mpj7, Sizeofint, Beinlichm, 211ima, Linguist111, CLCStudent, Neechy, AtGoM, Moocop, Wecassidy, Musicmaker1232, Hdjensofjfnen, AnAwesomeArticleEditor, Bender the Bot, MereTechnicality, Autumnbailey and Anonymous: 747

- **Nuclear safety in the United States** *Source:* https://en.wikipedia.org/wiki/Nuclear_safety_in_the_United_States?oldid=764865941 *Contributors:* Michael Hardy, Baylink, Wwoods, Jason Quinn, WhiteDragon, Gscshoyru, Bender235, Wtshymanski, Forteblast, Ultramarine, Tabletop, BD2412, Kolbasz, Simesa, Limulus, Rick lightburn, Petri Krohn, Johnpseudo, SmackBot, Pwt898, Yamaguchi先生, Salmar, Theanphibian, Coriolise, Slazenger, Cydebot, BillySharps, Nick Number, Calaka, Acroterion, Cgingold, Leyo, Tyrerj, AndrewTJ31, Johnfos, RingtailedFox, Philip Trueman, Masaqui, Pizzachicken, Fanra, Binksternet, Eiland, Niceguyedc, Togokill, Nuclear Menace, Zahnrad, W345thn, DumZiBoT, Addbot, Lightbot, Yobot, AnomieBOT, Ulric1313, Materialscientist, SCARECROW, Bruceschaller, HRoestBot, Chatfecter, Woody60707, Watchpup, Trappist the monk, PleaseStand, RjwilmsiBot, GoingBatty, Dcirovic, Redhanker, Lworth79, H3llBot, Δ, Ready, Frietjes, MrsEoGreen, Asauers, BattyBot, Monkbot, Fer48, InternetArchiveBot, GreenC bot, Bender the Bot and Anonymous: 26

- **Nuclear weapons debate** *Source:* https://en.wikipedia.org/wiki/Nuclear_weapons_debate?oldid=763789451 *Contributors:* Jnc, Rdsmith4, Rich Farmbrough, Bender235, Smaltjim, Arthena, Dr Gangrene, Simesa, John, Myasuda, HolyT, Natobxl, Johnfos, NPguy, Malcolmxl5, Randy Kryn, Spitfire, Addbot, AnomieBOT, Lothar von Richthofen, Pinethicket, LittleWink, Tom.Reding, Mr.98, RjwilmsiBot, Boundarylayer, Slightsmile, ZéroBot, ClueBot NG, MelbourneStar, BG19bot, BattyBot, Cyberbot II, Mysterious Whisper, XXzoonamiXX, RotlinkBot, Gabelglesia, Monkbot, Paredmon, Beebs0123, Mrhartz and Anonymous: 18

- **List of nuclear whistleblowers** *Source:* https://en.wikipedia.org/wiki/List_of_nuclear_whistleblowers?oldid=764687351 *Contributors:* Liberatus, Bgwhite, Iridescent, Cydebot, Cgingold, D.h, Misarxist, Johnfos, Cramigo, Yohananw, AnomieBOT, AManWithNoPlan, Catlemur, Netherzone, Limnalid, Achim Hering, Fer48, Adam (Wiki Ed), Bender the Bot, Yasmin Ali, Juliet Jolly and Anonymous: 6

- **Nuclear power phase-out** *Source:* https://en.wikipedia.org/wiki/Nuclear_power_phase-out?oldid=766318435 *Contributors:* AxelBoldt, Bryan Derksen, Robert Merkel, Tarquin, Taw, Maury Markowitz, Ewen, Patrick, JohnOwens, Michael Hardy, Gabbe, Stewacide, Bcrowell, Ahoerstemeier, Andrewa, Ec5618, Gestumblindi, SEWilco, Topbanana, Joy, Pstudier, Ktotam, Aleron235, Hemant~enwiki, Michael Snow, Lupo, JerryFriedman, Alan Liefting, TDC, Mboverload, Zoney, Bobblewik, Gharbeia, Andycjp, Beland, PDH, Histrion, AndrewH, Edolen1, Rich Farmbrough, ThomasK, Rama, Wikiacc, Kenb215, Bender235, Thebrid, Project2501a, Alarm, Tronno, Viriditas, Cmdrjameson, Pearle, DBrane, Jonathunder, Espoo, Orzetto, Hektor, Rosenzweig, CJ, Rwendland, ProhibitOnions, Skyring, DV8 2XL, Gene Nygaard, Alai, Nightstallion, Autrefois, Tchaika, Lkinkade, Kelly Martin, Woohookitty, Linas, JFG, Arru, Trigor, Male1979, Hillbrand, Stevenplunkett, Reisio, Minilo, Wikiliki, Ian Pitchford, Ysangkok, RexNL, Benjamin Gatti, Simesa, YurikBot, Midgley, KSmrq, DanMS, Gaius Cornelius, Sandpiper, MarcK, Albatalpa, JeremyStein, Molobo, Ospalh, PanchoS, WAS 4.250, Novasource, Brozen, Claygate, Petri Krohn, Timothy.may, Whobot, DisambigBot, DocendoDiscimus, SmackBot, Thaagenson, Direvus, Ben DeRoy, Man with two legs, IstvanWolf, Hmains, Chris the speller, Ottawakismet, Jennica, Init~enwiki, Korako, Nrcprm2026, Enr-v, RKloti, G 1, Andeggs, Salamurai, ヒ・ロ, Ajnosek, John, Carnby, Shrew, Spartanfox86, OmicronSSD, Syra987, Dave420, Ethnopunk, Dub8lad1, Dogaroon, Gralo, Nick Number, Djwork, Sbandrews, AntiVandalBot, MER-C, Rkarapin, Jgaray, Avicennasis, Engineman, Ensign beedrill, David Eppstein, ExplicitImplicity, Dellarb, Johnfos, Mátyás, Mild Bill Hiccup, Shjacks45, Manishearth, Jusdafax, Addbot, ChNPP, Legobot, Luckas-bot, Yobot, Worm That Turned, AnomieBOT, JackieBot, Materialscientist, Reing, LittleWink, Andynct, Brian Everlasting, Trappist the monk, RjwilmsiBot, John of Reading, WikitanvirBot, Boundarylayer, Moswento,

Dcirovic, Neun-x, Robbiemorrison, Ego White Tray, GermanJoe, ClueBot NG, AerobicFox, Dougmcdonell, BG19bot, LametinoWiki, Batty-Bot, Justincheng12345-bot, Guy Stewart, Cyberbot II, YFdyh-bot, Mark Bao, Athomeinkobe, Lasrick, Pinfix, Rfassbind, Moony22, Sbarnard-hbf, Bfred.lundberg, Limnalid, Eccentric.idiot, Mavericks21 ilikepie, Monkbot, Annebaader, Azureum, Jadebenn, InternetArchiveBot, Ugion, RobbieIanMorrison, GreenC bot, Bender the Bot, Juliet Jolly and Anonymous: 118

- **Anti-nuclear movement in California** *Source:* https://en.wikipedia.org/wiki/Anti-nuclear_movement_in_California?oldid=765428706 *Contributors:* Alan Liefting, Rich Farmbrough, Mutante23, Simesa, Wavelength, CWenger, SmackBot, Ottawakismet, JohnI, Dawnseeker2000, Cgingold, CommonsDelinker, Johnfos, Malcolmxl5, Energynet, Muro Bot, Graevemoore, Addbot, Mliu92, Yobot, Fmrauch, AnomieBOT, Lotje, ScottyBerg, Geofferybard, Helpful Pixie Bot, Khazar2, Mogism, Hermes 1900s, InternetArchiveBot, GreenC bot, Bender the Bot and Anonymous: 3

- **Politics of New England** *Source:* https://en.wikipedia.org/wiki/Politics_of_New_England?oldid=765001635 *Contributors:* Student7, Johnfos, De728631, This is Paul, AnomieBOT, TimothyDexter, BG19bot, OccultZone, Bender the Bot and Anonymous: 3

- **High-level radioactive waste management** *Source:* https://en.wikipedia.org/wiki/High-level_radioactive_waste_management?oldid=765259723 *Contributors:* Michael Devore, Beland, Jkl, Bender235, Gary, Rwendland, Tabletop, Teemu Leisti, Rjwilmsi, Koavf, Vegaswikian, Ground Zero, Kolbasz, Bgwhite, Simesa, WriterHound, Wavelength, Gaius Cornelius, Chrishmt0423, MaeseLeon, Twerges, Chris the speller, Mion, David ekstrand, John, Vgy7ujm, Mr Stephen, Clarityfiend, Gierszep, SamatJain, OhanaUnited, Hamiltonstone, Trusilver, Dkendr, Jarry1250, DASonnenfeld, Johnfos, Jkstark, Aymatth2, NPguy, Flyer22 Reborn, NiteSensor23, Dabomb87, ClueBot, VQuakr, Mild Bill Hiccup, Eiland, NuclearWarfare, Mitch Ames, Addbot, Yobot, Fraggle81, AnomieBOT, Danielba894, Materialscientist, Mervyn Emrys, Citation bot, Eumolpo, Fionaclee, Sophus Bie, Chongkian, FrescoBot, Citation bot 1, LittleWink, Brwass, Nhorel01, John of Reading, Hirsutism, Dcirovic, Midas02, H3llBot, Donner60, ClueBot NG, Snotbot, Frietjes, Toddrmcallister, Dougmcdonell, Helpful Pixie Bot, MusikAnimal, Mark Arsten, Blaspie55, Nebogipfel 1, BattyBot, Khazar2, Cdanchey, Frosty, Pinfix, Jodosma, Monkbot, Graemem56, Wlangstroth, Liechtenstein96, Nightmarionette, GreenC bot, Bender the Bot, Rinkesh81 and Anonymous: 50

- **Lists of nuclear disasters and radioactive incidents** *Source:* https://en.wikipedia.org/wiki/Lists_of_nuclear_disasters_and_radioactive_incidents?oldid=764026353 *Contributors:* Jpatokal, Katana0182, Alan Liefting, Sam, Jtrainor, Jkt, Survivor, ViriiK, Bgwhite, Wavelength, AndyBoal, Pegship, Sandstein, GeeCee, Serendipodous, MaeseLeon, Nickst, Stifle, Gilliam, Kasyapa, Vladislav, Danjewell, A5b, Mike1901, VoxLuna, Cydebot, Fayenatic london, Hopiakuta, JAnDbot, Magioladitis, D.h, CommonsDelinker, DASonnenfeld, ACSE, Johnfos, Oshwah, Mark v1.0, UnitedStatesian, Fasouzafreitas, Riick, Flyer22 Reborn, RW Marloe, Randy Kryn, RexxS, Addbot, Grblake, C933103, Jarbie, AnomieBOT, Jim1138, Anna Frodesiak, Hessamnia, MarkForeman, Obankston, EmausBot, Trofobi, Donner60, ClueBot NG, Pogovoreem, BG19bot, CimanyD, Cyberbot II, Figgggg, Dexbot, Mysterious Whisper, Yilku1, Kound, Limnalid, Fer48, NFSPRO, Redneck rick, *Treker, Beebs0123 and Anonymous: 39

- **Nuclear reactor accidents in the United States** *Source:* https://en.wikipedia.org/wiki/Nuclear_reactor_accidents_in_the_United_States?oldid=766077189 *Contributors:* Kku, Topbanana, Twang, Alan Liefting, Axeman89, Nihiltres, Gilliam, Cydebot, רותם, Bezking, K7aay, DagosNavy, CommonsDelinker, Jrcla2, Johnfos, RingtailedFox, Tide rolls, AnomieBOT, Jeff A. Benner, BritishWatcher, Yeaggermiester, Watchpup, PleaseStand, GoingBatty, Moswento, ClueBot NG, Daniel lightforge, Widr, Rob17072, Kizar, BattyBot, DrIngDC, Jwm353s, Fer48, KhnewKreator, Reecer1, LaughTrack07, Bender the Bot and Anonymous: 23

- **Nuclear energy policy** *Source:* https://en.wikipedia.org/wiki/Nuclear_energy_policy?oldid=734386044 *Contributors:* AxelBoldt, Mac, Andrewa, SEWilco, HarryHenryGebel, Schutz, AaronS, Ktotam, Alan Liefting, Wwoods, Michael Devore, Mboverload, Tweenk, Bobblewik, Beland, Jokestress, Esperant, Rich Farmbrough, Rama, Berkut, Bender235, Borofkin, MPerel, CJ, Rwendland, DV8 2XL, Gene Nygaard, Alai, Linas, Wikikirsc, Trigor, Male1979, JohnC, David Levy, Rjwilmsi, Bob A, Wikiliki, CooldogCongo, Bentisquare, Simesa, Noclador, Gaius Cornelius, Uni4dfx, Silverhorse, Petri Krohn, Johnpseudo, DisambigBot, ModernGeek, Victor falk, KnightRider~enwiki, SmackBot, Bluebot, Persian Poet Gal, Hibernian, Colonies Chris, A. B., DéRahier, Theanphibian, Korako, Valenciano, Enr-v, Andeggs, DavidHallett, Emcerlain, Euchiasmus, Vgy7ujm, Hu12, Tawkerbot2, CmdrObot, Mattbr, S-sully06, Chrisahn, Mattbuck, Cydebot, Mombas, Dorank, Gralo, Celuca, Seaphoto, Mrshaba, Superzohar, Corella, Alphachimpbot, JNW, Brusegadi, Engineman, Alohasoy, Fred114, Beagel, Chris G, JaGa, Gronkmeister, Ben MacDui, R'n'B, CommonsDelinker, Limongi, Antony-22, Jorfer, Alain10, Sam Blacketer, Johnfos, Kevandjess, Thadius856AWB, Staplegunther, Wikidemon, Smartguy583, Jmpenzone, NPguy, Panfakes, HybridBoy, 4wajzkd02, Moonriddengirl, Grundle2600, Nopetro, Water and Land, ClueBot, Fasettle, EoGuy, Der Golem, Eiland, Tosaka1, Lwnf360, Neomemberlands, ChesterTheWorm, JonathanCobb, Etip, DisambiguationGuy, NERIC-Security, DumZiBoT, Maky, Izmir lee, Addbot, DOI bot, Mohamed Magdy, 5 albert square, Tassedethe, I Wake Up Screaming, Isis Stafford, Yobot, Adorenarin, AnomieBOT, Ichwan Palongengi, LilHelpa, Mlpearc, Nedim Ardoğa, Full-date unlinking bot, Austeriagf, RjwilmsiBot, Orphan Wiki, Boundarylayer, Wikipelli, Dcirovic, Blahblahmh, Wo0dstock79, Kayz911, NuclearEnergy, Clara226, BG19bot, Dipankan001, BattyBot, ChrisGualtieri, Hmainsbot1, Stratoprutser, Royalcourtier, Trackteur, Jewjoo and Anonymous: 100

- **Nuclear power debate** *Source:* https://en.wikipedia.org/wiki/Nuclear_power_debate?oldid=766313480 *Contributors:* Ewen, Michael Hardy, Dante Alighieri, Dan Koehl, Topbanana, Raul654, Lowellian, Alan Liefting, Wwoods, Ketil, Arcataroger, Rich Farmbrough, Rbk, Xezbeth, Bender235, Kwamikagami, Espoo, RPaschotta, Alyeska, Rd232, Rwendland, Wtshymanski, Dr Gangrene, Kgrr, Tabletop, Wikikirsc, GregorB, Theo F, Rjwilmsi, Joffan, Josephs1, Ground Zero, Old Moonraker, Kolbasz, Bentisquare, Simesa, Chris Capoccia, Chaser, DragonHawk, Haikz, Crf, Arthur Rubin, Vicarious, Yvwv, SmackBot, Unschool, Zazaban, Bigbluefish, KVDP, Gilliam, Jushi, GoneAwayNowAndRetired, Chris the speller, Ottawakismet, RDBrown, Colonies Chris, Theanphibian, Salamurai, Daniel.Cardenas, Mion, モ~ ロ, Will Beback, Dacres, John, Vgy7ujm, AllStarZ, OmicronSSD, Ckatz, JHunterJ, Jerry-va, Scigatt, CmdrObot, Ibadibam, HappyInGeneral, Rossph1, Bjenks, AniRaptor2001, Mcorazao, MaxPont, Magioladitis, Engineman, KConWiki, Beagel, Uriel8, R'n'B, CommonsDelinker, LedgendGamer, ACSE, Johnfos, Philip Trueman, Ask123, Mishlai, Korin43, Plazak, Cde3, Doc James, Roger Jeurissen, SieBot, Malcolmxl5, Krawi, Grundle2600, RW Marloe, AWeishaupt, Asocall, Helenabella, Plastikspork, EoGuy, Eiland, Lwnf360, Blanchardb, Megiddo1013, Iohannes Animosus, 1ForThe-Money, XLinkBot, Fastily, Hiraku.n, Addbot, Roentgenium111, Chum2, TutterMouse, Amrad, Download, Roux, Tide rolls, Luckas-bot, Yobot, Enviro1, Fmrauch, KamikazeBot, Dzied Bulbash, SwisterTwister, AnomieBOT, Kingpin13, Citation bot, Jmarchn, ArthurBot, Bairdjr, Maxhotty, Felipe Schenone, AnziD, RJS29, RoodyBeep, Sheeson, Blahblahblah67, Jack B108, Biem, FrescoBot, Salfordspider, YOKOTA Kuniteru, Leightonwalter, Robert Berkshire, Jamesooders, Arlojeffrey, Citation bot 1, Gautier lebon, Jonesey95, Cnwilliams, Trappist the monk, Nhorel01, Robertiki, RjwilmsiBot, BjörnBergman, EmausBot, John of Reading, Trofobi, Clark42, Boundarylayer, Dewritech, Faceless Enemy, GoingBatty, RA0808, Klbrain, Dcirovic, Josve05a, Arbnos, H3llBot, AlexH555, SCStrikwerda, Jamesmarch, Jay-Sebastos, Robbiemorrison,

Haemetite, Ivolocy, ClueBot NG, Jankander, Widr, Dougmcdonell, Anna, Cseadmin, Helpful Pixie Bot, Athanase ch, RailstoRuin, AzureAnt, Ledjazz, Bibcode Bot, DBigXray, Anuclanus, BG19bot, Generaal klei, Wiki13, Frze, Goldenshimmer, Marcocapelle, Blaspie55, Honorsteem, Mripadkid, BattyBot, NukeEater, FlameHead96, Mr. Guye, Inqrorken, KathyErler2018251484, Thinly, Limnalid, Monkbot, Filedelinkerbot, Fer48, Woodbadger, TheQ Editor, Rsb8382, Steelersfan0817, Srednuas Lenoroc, Zupotachyon, Williamws111, Bender the Bot and Anonymous: 152

- **World Association of Nuclear Operators** *Source:* https://en.wikipedia.org/wiki/World_Association_of_Nuclear_Operators?oldid= 765978641 *Contributors:* Kku, Evolauxia, Tim!, Gaius Cornelius, Rathfelder, Droll, Cydebot, Alaibot, Lfstevens, Sadanpaamies, Nicransby, JPG-GR, Amourindian, Aspects, Abhinav, Excirial, DumZiBoT, Addbot, Lightbot, Ben Ben, DirComm, Xqbot, FrescoBot, Mean as custard, Hullernuc, Ripchip Bot, EmausBot, Dewritech, Midas02, ClueBot NG, BG19bot, M0rphzone, WanoComms, Valetude, YiFeiBot, Trixie05, Monicagellar 08, WANOHistory, Johnny higgins0, CommsWANO, C.WANO and Anonymous: 8

- **Anti-nuclear organizations** *Source:* https://en.wikipedia.org/wiki/Anti-nuclear_organizations?oldid=763799611 *Contributors:* Lquilter, Pigman, Bazonka, Cydebot, Tham153, Trusilver, JayJasper, GrahamHardy, TreasuryTag, Johnfos, Malcolmxl5, Fuddle, Randy Kryn, Addbot, Fmrauch, Loveless, Trappist the monk, Lokayat.india, George H. Harvey, BG19bot, BattyBot, BarronShores, Eplepiken, Graemem56, Amicus-Bryani, Napplicable, Person7656, Michigan12, InternetArchiveBot, GreenC bot, Bender the Bot and Anonymous: 12

- **List of anti-nuclear power groups** *Source:* https://en.wikipedia.org/wiki/List_of_anti%E2%80%93nuclear_power_groups?oldid=765937486 *Contributors:* Kku, Lquilter, IronGargoyle, Cydebot, Paxuscalta, D.h, Johnfos, Malcolmxl5, Good Olfactory, Addbot, Yobot, RoadBum1, George H. Harvey, ZéroBot, Ivolocy, Ro321, Bender the Bot, Alere spero, Juliet Jolly and Anonymous: 7

- **Campaign for Nuclear Disarmament** *Source:* https://en.wikipedia.org/wiki/Campaign_for_Nuclear_Disarmament?oldid=764418293 *Contributors:* Trelvis, Bryan Derksen, Tarquin, Ed Poor, Aldie, William Avery, Tzartzam, Soulpatch, Quercusrobur, Olivier, MartinHarper, Gabbe, Kabads, Lquilter, J'raxis, William M. Connolley, Jeandré du Toit, GCarty, Charles Matthews, Vanished user 5zariu3jisj0j4irj, Kierant, Fibonacci, Warofdreams, Timrollpickering, Saulisagenius, Wereon, Alan Liefting, Bsoffa, Bobblewik, Fys, Joeblakesley, Robert Brockway, Piotrus, Necrothesp, Thparkth, Neutrality, Klemen Kocjancic, Discospinster, TobiasTC, Rich Farmbrough, Vapour, Dave souza, Handelaar, Philip Cross, Rd232, JK the unwise, Pcpcpc, James Kemp, AshishG, DavidFarmbrough, RichardWeiss, Graham87, Rjwilmsi, Tim!, Brighterorange, MarnetteD, Leithp, SchuminWeb, Ground Zero, Str1977, Quuxplusone, DVdm, Bgwhite, Crotalus horridus, A.S. Brown, RussBot, Gaius Cornelius, Tern, Salmanazar, Logan1138, MichaelW, Caballero1967, SaveTheWhales, SmackBot, Britannicus, Mcmillen76, Hmains, Betzub, DKalkin, Greenshed, KeitHolland, Soarhead77, Salamurai, Sjeraj, BrownHairedGirl, Robofish, Molly Romanov, MTSbot~enwiki, Janetbloomfield, PedanticAI, Hu12, Seqsea, JForget, CmdrObot, Ethnopunk, WeggeBot, Cydebot, Hebrides, Bobo12345, Prof75, Mombas, PKT, Thijs!bot, Epbr123, Biruitorul, Itsmejudith, Egel, Turkeyphant, Borneh, Mentifisto, NSH001, Once in a Blue Moon, Sjamcm, RebelRobot, Marshall1946, Xoneca, Magioladitis, Unused0029, Xn4, Brother Francis, Johnbibby, Enaidmawr, CommonsDelinker, Dr Almost, EverSince, DavidB601, Paularblaster, RVJ, Andy Marchbanks, Johnfos, Pelarmian, Nicklse, Saibod, Redrocker, Brenont, Malcolmxl5, Lightmouse, RW Marloe, Ben 1982, Jza84, Randy Kryn, WikiphyteMk1, Dobermanji, Puchiko, Yorkshirian, BakerFan, John Paul Parks, DumZiBoT, Howard Clark, TFOWR, PL290, MystBot, Pyfan, Grayfell, Aimulti, Wingspeed, RetroS1mone, Ginosbot, Mattmattmattmatty, Zorrobot, Legobot, Luckas-bot, Yobot, Granpuff, Andreasmperu, Legobot II, Fmrauch, Vini 17bot5, AnomieBOT, YeshuaDavid, Hertsred, Ulric1313, Danno uk, Citation bot, Avarince, Comt Till, Julizy, Wwbread, Learner001, FrescoBot, Yickbob, Redrose64, Jonesey95, LilyKitty, Adam gardener, RjwilmsiBot, Salvio giuliano, Boundarylayer, HenryIreton1642, ClueBot NG, Proscribe, Parcly Taxel, Lainade, Widr, Wilmevans, Helpful Pixie Bot, Festermunk, BG19bot, Arithmetika, Virtualbike, Cloptonson, SirCxyrtyx, TeriEmbrey, Johnragla, Michipedian, Gravuritas, Davidcarroll, SJPreece, Sizergh, Gaylena, Helper201, KasparBot, Ceannlann gorm, Gea Jones, Godisamazingmonsterlolster123456, Lekker12, Bender the Bot, Springchickensoup and Anonymous: 104

- **International Association of Lawyers against Nuclear Arms** *Source:* https://en.wikipedia.org/wiki/International_Association_of_Lawyers_against_Nuclear_Arms?oldid=730662880 *Contributors:* Johnfos, Yobot and Maria Blaska

- **International Campaign to Abolish Nuclear Weapons** *Source:* https://en.wikipedia.org/wiki/International_Campaign_to_Abolish_Nuclear_Weapons?oldid=759794372 *Contributors:* Alan Liefting, Rjwilmsi, Jimp, Zargulon, Cydebot, Alaibot, Ebyabe, Timwright, Magioladitis, Adavidb, Jevansen, Johnfos, Randy Kryn, Addbot, Yobot, Fmrauch, Tim - ICAN, Lhoaxt, ZéroBot, Widr, JLoretz, BG19bot, Gabelglesia, Pronunn, Mu5icloverrr and Anonymous: 5

- **Peace Action** *Source:* https://en.wikipedia.org/wiki/Peace_Action?oldid=756411631 *Contributors:* SimonP, Bcorr, Alba, Fastfission, Everyking, Gamaliel, Dirthiscuit, Quadell, Kate, Pearle, Hooperbloob, RJFJR, BDD, Marudubshinki, BD2412, Rjwilmsi, TJive, Vegaswikian, SchuminWeb, Ground Zero, JdforresterBot, Common Man, RussBot, BlueZenith, DieWeisseRose, EJSawyer, SmackBot, Hmains, Colonies Chris, Cybercobra, Kukini, Outriggr (2006-2009), Cydebot, Benjiboi, Swooningdisaster, Prof75, Joetkeck, Davisxa, Cgingold, Loonymonkey, Dwalls, R'n'B, Kateshortforbob, Tagus, Johnfos, Davidwr, Someguy1221, Barkeep, Flyer22 Reborn, Barbra.Peace.Action, Fempeace, Randy Kryn, Philtheactor, ClueBot, Flotron9, Gtstricky, DumZiBoT, JohnWilmerding, Yobot, Pohick2, AnomieBOT, JoanAroma, Learner001, FrescoBot, Mean as custard, GoingBatty, Jonpatterns, SporkBot, ClueBot NG, Loriendrew, Mogism, Ausjelk, Jamesneilanderson, Filedelinkerbot, Thewhitebox, 21lima, Keeterig, Deathdragon101, Here2help and Anonymous: 30

- **History of the anti-nuclear movement** *Source:* https://en.wikipedia.org/wiki/History_of_the_anti-nuclear_movement?oldid=765776484 *Contributors:* Rich Farmbrough, Bender235, Woohookitty, Drbogdan, Simesa, Limulus, JohnI, Mombas, Nick Number, CommonsDelinker, Johnfos, Malcolmxl5, Mild Bill Hiccup, Ulric1313, Materialscientist, Widr, Helpful Pixie Bot, Seergenius, Dexbot, XXzoonamiXX, Hermes 1900s, Bender the Bot and Anonymous: 6

- **Göttingen Manifesto** *Source:* https://en.wikipedia.org/wiki/G%C3%B6ttingen_Manifesto?oldid=734243202 *Contributors:* Alan Liefting, Rich Farmbrough, Cmdrjameson, Espoo, Gene Nygaard, Lkinkade, GregorB, Doco, Olessi, Kresspahl, RussBot, Welsh, Qero, SmackBot, Njerseyguy, Writtenright, Coffeinfreak, Harej bot, Cydebot, JamesAM, TAnthony, Btiene, Pax:Vobiscum, Thaurisil, TreasuryTag, Lightbot, Bushmillsmccallan, ChrisGualtieri, Dexbot and Anonymous: 7

- **International Court of Justice advisory opinion on the Legality of the Threat or Use of Nuclear Weapons** *Source:* https://en.wikipedia.org/wiki/International_Court_of_Justice_advisory_opinion_on_the_Legality_of_the_Threat_or_Use_of_Nuclear_Weapons?oldid=749774766 *Contributors:* Trelvis, Kaihsu, Maximus Rex, Topbanana, PBS, Mattflaschen, Jacob1207, Robert Weemeyer, Get-back-world-respect, Avala, Mike Rosoft, Rich Farmbrough, YUL89YYZ, Night Gyr, Mrzaius, M. Henri Day, Yeu Ninje, Rwendland, James Kemp, Rjwilmsi,

Tim!, George Burgess, Ligulem, DanMS, Chensiyuan, Gaius Cornelius, Varlagas, Ospalh, Mais oui!, SmackBot, MyrddinEmrys, Mauls, Hmains, Rakela, Dolive21, MeekSaffron, Cybercobra, Bendzh, Eastlaw, Outriggr (2006-2009), Goatchurch, Darklilac, Wf219, Magioladitis, Robertson-Glasgow, Jeffmerkle, Johnfos, Abrichte, Fratrep, Rumping, GorillaWarfare, Vianello, Good Olfactory, Johnkatz1972, Addbot, Tassedethe, InMyHumbleOpinion, Andre Toulon, Ben Ben, Luckas-bot, Yobot, AnomieBOT, DrilBot, Full-date unlinking bot, EmausBot, Lhoaxt, Brianm358, SpikeTorontoRCP, Helpful Pixie Bot, InExcelsisDeo, Peak Player, TorbenTT and Anonymous: 27

- **Mainau Declaration** *Source:* https://en.wikipedia.org/wiki/Mainau_Declaration?oldid=765503342 *Contributors:* Danny, Piotrus, Rich Farmbrough, Zzyzx11, MidnightWolf909, Hairy Dude, RussBot, JmA, SmackBot, Illuminattile, Commander Keane bot, Outriggr (2006-2009), Myasuda, TXiKiBoT, Addbot, KamikazeBot, Bushmillsmccallan, Wbm1058, BG19bot, RichardMills65, Gutzio, Vincenzo108 and Anonymous: 7

- **Nuclear-Free Future Award** *Source:* https://en.wikipedia.org/wiki/Nuclear-Free_Future_Award?oldid=731704601 *Contributors:* Mac, Jerzy, Tweenk, Tabletop, Limulus, Bluebot, Khazar, Iridescent, ChrisCork, CmdrObot, JohnCD, Cydebot, PKT, Thijs!bot, Borneh, TheEditrix2, Recurring dreams, Martial75, Johnfos, HybridBoy, Malcolmxl5, Kjtobo, Addbot, Kman543210, Jean-Jacques Georges, Dewritech, Jasonanaggie, ZéroBot, CaroleHenson, L.Baur, Anandisu and Anonymous: 5

- **Nuclear-free zone** *Source:* https://en.wikipedia.org/wiki/Nuclear-free_zone?oldid=764865966 *Contributors:* Cyde, Mac, Jiang, Charles Matthews, Alan Liefting, Fastfission, Thomasn528, Tweenk, Gadfium, Beland, Mennonot, Kelvinc, Orlady, Apoc2400, Avenue, Rwendland, Cecil, Evil Monkey, Vuo, Kaiser matias, Dominic, LordAmeth, Gene Nygaard, LukeSurl, Firsfron, ScottDavis, Rejs, JohnC, Mutante23, Mandarax, Miq, Rjwilmsi, Dunro, Vegaswikian, Ground Zero, Steppenfox, Bgwhite, JWB, RussBot, Limulus, Epotk, DanMS, Joel7687, Lcmortensen, Matches10, Kingboyk, SmackBot, CapitalSasha, Hmains, Chris the speller, Columba livia, Samanathon, Enigma55, Theanphibian, Enr-v, Gbinal, Sjeraj, Euchiasmus, Nzgabriel, Majorclanger, Peter Horn, Midnighttonight, Andrew Hampe, ERAGON, Outriggr (2006-2009), Cydebot, 663highland, Bellerophon5685, Lossenelin, Alvesgaspar, Mombas, JamesAM, DevoutHeretic, Marek69, Jason Jones, Macmanui, Gioto, Ingolfson, Jamie Mackay, Cgingold, Magpie83, CommonsDelinker, Acalamari, Beyonder1, Aucitypops, TreasuryTag, Johnfos, Tesscass, Pelarmian, Helenalex, Mercurywoodrose, Thmazing, BenedictX, Ashtray, Lrb77, DavesPlanet, MCTales, Master of the Orichalcos, HybridBoy, Thrawn aj, Malcolmxl5, Cafeganesha, Mmales, ImageRemovalBot, Twinsday, ClueBot, John Nevard, Imavegetarian2, Certes, DumZiBoT, Dthomsen8, Good Olfactory, Addbot, Quokly, Tassedethe, Yobot, Fmrauch, AnomieBOT, Citation bot, Dr Oldekop, Tsuchida54, AndersonH.K., FrescoBot, Lotje, Drapeau, RjwilmsiBot, EmausBot, ChuispastonBot, Whoop whoop pull up, Eladingulpcmnomis, Helpful Pixie Bot, TotalFailure, JTdale, BG19bot, Mr. Joca, EdwardH, BattyBot, Cyberbot II, Dutch Ninja, Dexbot, Hmainsbot1, Mogism, Sobie002, FallingGravity, Jakec, Royalcourtier, Qwertyxp2000, Fboswell, KasparBot, GreenC bot, Bender the Bot, Fermulator and Anonymous: 69

- **Russell–Einstein Manifesto** *Source:* https://en.wikipedia.org/wiki/Russell%E2%80%93Einstein_Manifesto?oldid=766428619 *Contributors:* Edward, Schneelocke, Altenmann, Romanm, Gracefool, Toytoy, Piotrus, Gunnar Larsson, ThomasK, Bender235, Icut4you, Bastin, Graham87, Rjwilmsi, Tim!, Koavf, Daniel Collins, Pashdown, John Z, Kenmayer, Irregulargalaxies, Chobot, YurikBot, Jamesmorrison, Hairy Dude, Hede2000, Sasuke Sarutobi, CQ, David Berardan, Liujiang, SmackBot, Illuminattile, InverseHypercube, Hmains, Lapisphil, GeorgeMoney, Wybot, Ohconfucius, John, Xoán Carlos Fraga, MAG1, DI2000, Levineps, Outriggr (2006-2009), Otto4711, DBaba, Thijs!bot, Vvidetta, Ristonet, Afaz, Cgingold, Davy p, 83d40m, Mukappa, VolkovBot, Johnfos, JukoFF, SieBot, Baridiah, Mimihitam, Cyfal, KarenSutherland, Crowsnest, SilvonenBot, Good Olfactory, Addbot, Jojhutton, Martindo, Luckas-bot, Yobot, Ptbotgourou, Materialscientist, RedBot, EmausBot, WikitanvirBot, ZéroBot, Xanchester, ClueBot NG, Supasate, Dexbot, Razibot, Gassymexican, Pyrusca, Bender the Bot and Anonymous: 16

- **Vulnerability of nuclear plants to attack** *Source:* https://en.wikipedia.org/wiki/Vulnerability_of_nuclear_plants_to_attack?oldid=765690921 *Contributors:* Bender235, Art LaPella, Jprg1966, Cydebot, Johnfos, Asa Zernik, Niceguyedc, Callinus, AnomieBOT, Trappist the monk, Dcirovic, BattyBot, Cyberbot II, ChrisGualtieri, Yilku1, Bksovacool, Filedelinkerbot, Fer48, Graemem56, Elmeter, CaptainCarlosdeCorona, *Treker and Anonymous: 6

30.13.2 Images

- **File:"End_Nuclear_War_Today".png** *Source:* https://upload.wikimedia.org/wikipedia/commons/d/db/%22End_Nuclear_War_Today%22.png *License:* CC BY-SA 3.0 *Contributors:* Own work *Original artist:* Cookkiemonster123

- **File:12-05-08_AS1.JPG** *Source:* https://upload.wikimedia.org/wikipedia/commons/e/e5/12-05-08_AS1.JPG *License:* GFDL *Contributors:* Own work *Original artist:* BSMPS

- **File:1983_Easter_CND_demo_Aldermaston.jpg** *Source:* https://upload.wikimedia.org/wikipedia/commons/e/e2/1983_Easter_CND_demo_Aldermaston.jpg *License:* CC BY-SA 3.0 *Contributors:* Own work *Original artist:* Johnragla

- **File:2005_Energy_Policy_Act.jpg** *Source:* https://upload.wikimedia.org/wikipedia/commons/5/52/2005_Energy_Policy_Act.jpg *License:* Public domain *Contributors:* http://georgewbush-whitehouse.archives.gov/news/releases/2005/08/images/20050808-6_f1g3456-515h.html *Original artist:* Eric Draper

- **File:ANTIAKW.jpg** *Source:* https://upload.wikimedia.org/wikipedia/commons/4/40/ANTIAKW.jpg *License:* CC BY-SA 2.0 de *Contributors:* Own work *Original artist:* Hans Weingartz (Leonce49 at de.wikipedia)

- **File:Able_crossroads.jpg** *Source:* https://upload.wikimedia.org/wikipedia/commons/f/f4/Able_crossroads.jpg *License:* Public domain *Contributors:* ? *Original artist:* ?

- **File:Aegopodium_podagraria1_ies.jpg** *Source:* https://upload.wikimedia.org/wikipedia/commons/b/bf/Aegopodium_podagraria1_ies.jpg *License:* CC-BY-SA-3.0 *Contributors:* Own work *Original artist:* Frank Vincentz

- **File:All-Atomic_Comics_(1st_edition_front_cover).jpg** *Source:* https://upload.wikimedia.org/wikipedia/en/1/10/All-Atomic_Comics_%281st_edition_front_cover%29.jpg *License:* Fair use *Contributors:* **Original publication**: 1976, USA

  Immediate http://2.bp.blogspot.com/-IE4e-pz4ohk/T0kZ8DYWWII/AAAAAAAAj0/43aZCHUxeAo/s1600/atomic.jpg via http://comicsclasicosinglesespanol.blogspot.com/2012/02/all-atomic-comics-usa-1976.html *Original artist:* Leonard Rifas

- **File:Alternative_Energies.jpg** *Source:* https://upload.wikimedia.org/wikipedia/commons/b/bb/Alternative_Energies.jpg *License:* CC BY 2.0 *Contributors:* Flickr *Original artist:* Jürgen from Sandesneben, Germany

- **File:Ambox_current_red.svg** *Source:* https://upload.wikimedia.org/wikipedia/commons/9/98/Ambox_current_red.svg *License:* CC0 *Contributors:* self-made, inspired by Gnome globe current event.svg, using Information icon3.svg and Earth clip art.svg *Original artist:* Vipersnake151, penubag, Tkgd2007 (clock)

- **File:Ambox_important.svg** *Source:* https://upload.wikimedia.org/wikipedia/commons/b/b4/Ambox_important.svg *License:* Public domain *Contributors:* Own work, based off of Image:Ambox scales.svg *Original artist:* Dsmurat (talk · contribs)

- **File:Anti-EPR_demonstration_in_Toulouse_0166_2007-03-17.jpg** *Source:* https://upload.wikimedia.org/wikipedia/commons/e/ef/Anti-EPR_demonstration_in_Toulouse_0166_2007-03-17.jpg *License:* CC BY-SA 2.5 *Contributors:* Own work *Original artist:* Guillaume Paumier

- **File:Anti-Nuclear_Power_Plant_Rally_on_19_September_2011_at_Meiji_Shrine_Outer_Garden_03.JPG** *Source:* https://upload.wikimedia.org/wikipedia/commons/1/18/Anti-Nuclear_Power_Plant_Rally_on_19_September_2011_at_Meiji_Shrine_Outer_Garden_03.JPG *License:* Public domain *Contributors:* Own work *Original artist:* 伏丁

- **File:Anti-nuclear_protest,_USA,_1977_(1).jpg** *Source:* https://upload.wikimedia.org/wikipedia/commons/1/1c/Anti-nuclear_protest%2C_USA%2C_1977_%281%29.jpg *License:* CC BY-SA 3.0 *Contributors:* Own work *Original artist:* Derzsi Elekes Andor

- **File:Anti-nuclear_protest_at_the_NTS_3.jpg** *Source:* https://upload.wikimedia.org/wikipedia/commons/3/34/Anti-nuclear_protest_at_the_NTS_3.jpg *License:* Public domain *Contributors:* This image is available from the National Nuclear Security Administration Nevada Site Office Photo Library under ID 858. *Original artist:* National Nuclear Security Administration / Nevada Site Office

- **File:Anti-nuclear_protesters_sprayed_by_water_cannons_in_Taipei,_Taiwan.jpg** *Source:* https://upload.wikimedia.org/wikipedia/commons/5/5f/Anti-nuclear_protesters_sprayed_by_water_cannons_in_Taipei%2C_Taiwan.jpg *License:* CC BY-SA 3.0 *Contributors:* Own work *Original artist:* MrWiki321

- **File:Anti-nuke_rally_in_Harrisburg_USA.jpg** *Source:* https://upload.wikimedia.org/wikipedia/commons/5/53/Anti-nuke_rally_in_Harrisburg_USA.jpg *License:* Public domain *Contributors:* http://arcweb.archives.gov ARC Identifier 540016, Item from Record Group 220: Records of Temporary Committees, Commissions, and Boards, 1893-1999 *Original artist:* unknown, National Archives and Records Administration (NARA)

- **File:Antinuclear_Walk_Geneva-Brussels_2009_Geneva.jpg** *Source:* https://upload.wikimedia.org/wikipedia/commons/e/e9/Antinuclear_Walk_Geneva-Brussels_2009_Geneva.jpg *License:* GFDL *Contributors:* Own work *Original artist:* Yann (<a href='//commons.wikimedia.org/wiki/User_talk:Yann' title='User talk:Yann'>talk</a>)

- **File:Antinulear_Demonstration_in_Colmar,_October_3,_2009.jpg** *Source:* https://upload.wikimedia.org/wikipedia/commons/4/4e/Antinulear_Demonstration_in_Colmar%2C_October_3%2C_2009.jpg *License:* CC BY 2.0 *Contributors:* originally posted to **Flickr** as COLMAR-14H36 *Original artist:* sortirdunucleaire

- **File:Atom-Moratorium.svg** *Source:* https://upload.wikimedia.org/wikipedia/commons/5/58/Atom-Moratorium.svg *License:* CC BY-SA 2.5 *Contributors:* Kernkraftwerke in Deutschland.svg *Original artist:* Kernkraftwerke in Deutschland.svg: **Lencer**

- **File:Atomic_cloud_over_Hiroshima.jpg** *Source:* https://upload.wikimedia.org/wikipedia/commons/b/b7/Atomic_cloud_over_Hiroshima.jpg *License:* Public domain *Contributors:* This media is available in the holdings of the National Archives and Records Administration, cataloged under the ARC Identifier (National Archives Identifier) **542192**. *Original artist:* Enola Gay Tail Gunner S/Sgt. George R. (Bob) Caron

- **File:Bedrijfsafval.jpg** *Source:* https://upload.wikimedia.org/wikipedia/commons/3/3a/Bedrijfsafval.jpg *License:* Public domain *Contributors:* No machine-readable source provided. Own work assumed (based on copyright claims). *Original artist:* No machine-readable author provided. Fun4life.nl assumed (based on copyright claims).

- **File:Bertrand_Russell_leads_anti-nuclear_march_in_London,_Feb_1961.jpg** *Source:* https://upload.wikimedia.org/wikipedia/commons/4/4e/Bertrand_Russell_leads_anti-nuclear_march_in_London%2C_Feb_1961.jpg *License:* CC BY-SA 3.0 *Contributors:* Own work *Original artist:* Tony French

- **File:Beznau_-_emergency_switch.jpeg** *Source:* https://upload.wikimedia.org/wikipedia/commons/9/90/Beznau_-_emergency_switch.jpeg *License:* Public domain *Contributors:* www.swissinfo.ch *Original artist:* Thomas Kern

- **File:Browns_Ferry_Unit_1_under_construction.jpg** *Source:* https://upload.wikimedia.org/wikipedia/commons/3/36/Browns_Ferry_Unit_1_under_construction.jpg *License:* Public domain *Contributors:* TVA's 75th Anniversary webpage, picture from [1] *Original artist:* Tennessee Valley Authority

- **File:Bulletin_Atomic_Scientists_Cover.jpg** *Source:* https://upload.wikimedia.org/wikipedia/en/3/39/Bulletin_Atomic_Scientists_Cover.jpg *License:* ? *Contributors:*
  The Bulletin of the Atomic Scientists
  Original artist: ?

- **File:CND_rally,_Aberystwyth_(5184388447).jpg** *Source:* https://upload.wikimedia.org/wikipedia/commons/7/78/CND_rally%2C_Aberystwyth_%285184388447%29.jpg *License:* CC0 *Contributors:* CND rally, Aberystwyth *Original artist:* Geoff Charles

- **File:Carter_leaving_Three_Mile_Island.jpg** *Source:* https://upload.wikimedia.org/wikipedia/commons/3/3b/Carter_leaving_Three_Mile_Island.jpg *License:* Public domain *Contributors:* This media is available in the holdings of the National Archives and Records Administration, cataloged under the ARC Identifier (National Archives Identifier) **540021**. *Original artist:* President's Commission on the Accident at Three Mile Island

- **File:Castor_2011_-_Demonstration_in_Dannenberg_(9).jpg** *Source:* https://upload.wikimedia.org/wikipedia/commons/a/ab/Castor_2011_-_Demonstration_in_Dannenberg_%289%29.jpg *License:* CC BY 2.0 *Contributors:* Großdemo auf der Essowiese in Dannenberg *Original artist:* BUNDjugend

- **File:Exercise_Desert_Rock_I_(Buster-Jangle_Dog)_002.jpg** *Source:* https://upload.wikimedia.org/wikipedia/commons/5/50/Exercise_Desert_Rock_I_%28Buster-Jangle_Dog%29_002.jpg *License:* Public domain *Contributors:* http://www.dtra.mil/press_resources/photo_library/CS/CS-3.cfm *Original artist:* Federal Government of the United States

- **File:Fallout_shelter.jpg** *Source:* https://upload.wikimedia.org/wikipedia/commons/e/e6/Fallout_shelter.jpg *License:* CC-BY-SA-3.0 *Contributors:* Transferred from en.wikipedia to Commons. *Original artist:* The original uploader was Ex11e at English Wikipedia

- **File:Fernald_Production-era_Aerial.jpg** *Source:* https://upload.wikimedia.org/wikipedia/commons/d/d0/Fernald_Production-era_Aerial.jpg *License:* Public domain *Contributors:* Fernald Production-era Aerial *Original artist:* ENERGY.GOV

- **File:Flag_of_Algeria.svg** *Source:* https://upload.wikimedia.org/wikipedia/commons/7/77/Flag_of_Algeria.svg *License:* Public domain *Contributors:* SVG implementation of the 63-145 Algerian law *"on Characteristics of the Algerian national emblem"* ("Caractéristiques du Drapeau Algérien", in English). *Original artist:* This graphic was originaly drawn by User:SKopp.

- **File:Flag_of_Colombia.svg** *Source:* https://upload.wikimedia.org/wikipedia/commons/2/21/Flag_of_Colombia.svg *License:* Public domain *Contributors:* Drawn by User:SKopp *Original artist:* SKopp

- **File:Flag_of_France.svg** *Source:* https://upload.wikimedia.org/wikipedia/en/c/c3/Flag_of_France.svg *License:* PD *Contributors:* ? *Original artist:* ?

- **File:Flag_of_Germany.svg** *Source:* https://upload.wikimedia.org/wikipedia/en/b/ba/Flag_of_Germany.svg *License:* PD *Contributors:* ? *Original artist:* ?

- **File:Flag_of_Guyana.svg** *Source:* https://upload.wikimedia.org/wikipedia/commons/9/99/Flag_of_Guyana.svg *License:* Public domain *Contributors:* ? *Original artist:* ?

- **File:Flag_of_Hungary.svg** *Source:* https://upload.wikimedia.org/wikipedia/commons/c/c1/Flag_of_Hungary.svg *License:* Public domain *Contributors:*

- Flags of the World – Hungary *Original artist:* SKopp

- **File:Flag_of_Italy.svg** *Source:* https://upload.wikimedia.org/wikipedia/en/0/03/Flag_of_Italy.svg *License:* PD *Contributors:* ? *Original artist:* ?

- **File:Flag_of_Japan.svg** *Source:* https://upload.wikimedia.org/wikipedia/en/9/9e/Flag_of_Japan.svg *License:* PD *Contributors:* ? *Original artist:* ?

- **File:Flag_of_Madagascar.svg** *Source:* https://upload.wikimedia.org/wikipedia/commons/b/bc/Flag_of_Madagascar.svg *License:* Public domain *Contributors:* ? *Original artist:* ?

- **File:Flag_of_Russia.svg** *Source:* https://upload.wikimedia.org/wikipedia/en/f/f3/Flag_of_Russia.svg *License:* PD *Contributors:* ? *Original artist:* ?

- **File:Flag_of_Sierra_Leone.svg** *Source:* https://upload.wikimedia.org/wikipedia/commons/1/17/Flag_of_Sierra_Leone.svg *License:* Public domain *Contributors:* ? *Original artist:* Zscout370

- **File:Flag_of_Sri_Lanka.svg** *Source:* https://upload.wikimedia.org/wikipedia/commons/1/11/Flag_of_Sri_Lanka.svg *License:* Public domain *Contributors:* SLS 693 - National flag of Sri Lanka *Original artist:* Zscout370

- **File:Flag_of_Venezuela.svg** *Source:* https://upload.wikimedia.org/wikipedia/commons/0/06/Flag_of_Venezuela.svg *License:* Public domain *Contributors:* official websites *Original artist:* Zscout370

- **File:Flag_of_WHO.svg** *Source:* https://upload.wikimedia.org/wikipedia/commons/8/89/Flag_of_WHO.svg *License:* Public domain *Contributors:*

- Open Clip Art *Original artist:* WHO

- **File:Flag_of_the_People'{}s_Republic_of_China.svg** *Source:* https://upload.wikimedia.org/wikipedia/commons/f/fa/Flag_of_the_People%27s_Republic_of_China.svg *License:* Public domain *Contributors:* Own work, http://www.protocol.gov.hk/flags/eng/n_flag/design.html *Original artist:* Drawn by User:SKopp, redrawn by User:Denelson83 and User:Zscout370

- **File:Flag_of_the_United_Kingdom.svg** *Source:* https://upload.wikimedia.org/wikipedia/en/a/ae/Flag_of_the_United_Kingdom.svg *License:* PD *Contributors:* ? *Original artist:* ?

- **File:Flag_of_the_United_States.svg** *Source:* https://upload.wikimedia.org/wikipedia/en/a/a4/Flag_of_the_United_States.svg *License:* PD *Contributors:* ? *Original artist:* ?

- **File:Flamanville-3_2010-07-15.jpg** *Source:* https://upload.wikimedia.org/wikipedia/commons/3/3c/Flamanville-3_2010-07-15.jpg *License:* CC BY 3.0 *Contributors:* panoramio *Original artist:* schoella

- **File:Folder_Hexagonal_Icon.svg** *Source:* https://upload.wikimedia.org/wikipedia/en/4/48/Folder_Hexagonal_Icon.svg *License:* Cc-by-sa-3.0 *Contributors:* ? *Original artist:* ?

- **File:Fukushima_I_by_Digital_Globe.jpg** *Source:* https://upload.wikimedia.org/wikipedia/commons/7/7d/Fukushima_I_by_Digital_Globe.jpg *License:* CC BY-SA 3.0 *Contributors:* Earthquake and Tsunami damage-Dai Ichi Power Plant, Japan *Original artist:* Digital Globe

- **File:Fukushima_I_by_Digital_Globe_crop.jpg** *Source:* https://upload.wikimedia.org/wikipedia/commons/1/16/Fukushima_I_by_Digital_Globe_crop.jpg *License:* CC BY-SA 3.0 *Contributors:* Earthquake and Tsunami damage-Dai Ichi Power Plant, Japan *Original artist:* Digital Globe

- **File:Gerald_w_brown.jpg** *Source:* https://upload.wikimedia.org/wikipedia/commons/5/59/Gerald_w_brown.jpg *License:* Public domain *Contributors:* Own work *Original artist:* Achim Hering

- **File:Global_public_support_for_energy_sources_(Ipsos_2011).png** *Source:* https://upload.wikimedia.org/wikipedia/commons/6/6e/Global_public_support_for_energy_sources_%28Ipsos_2011%29.png *License:* CC0 *Contributors:* Own work *Original artist:* Enescot

- **File:Trojan1.jpg** *Source:* https://upload.wikimedia.org/wikipedia/commons/b/b6/Trojan1.jpg *License:* CC BY-SA 2.5 *Contributors:* Transferred from en.wikipedia to Commons by LeaW. *Original artist:* The original uploader was Erikpatt at English Wikipedia
- **File:Tschernobyl-Fukushima-Gedenken_Wien2011-04-25_2000_Kerzen.jpg** *Source:* https://upload.wikimedia.org/wikipedia/commons/a/a7/Tschernobyl-Fukushima-Gedenken_Wien2011-04-25_2000_Kerzen.jpg *License:* CC BY-SA 3.0 *Contributors:* Own work *Original artist:* Manfred Werner - Tsui
- **File:U.S._2014_Electricity_Generation_By_Type.png** *Source:* https://upload.wikimedia.org/wikipedia/commons/5/54/U.S._2014_Electricity_Generation_By_Type.png *License:* CC BY 4.0 *Contributors:* Own work https://docs.google.com/spreadsheets/d/1YS4Po6RgMp65HfzCBlVfpJFlghmxfQQ_MHXwGWZyraQ/edit#gid=416614765 *Original artist:* Daniel.Cardenas
- **File:UN_General_Assembly_hall.jpg** *Source:* https://upload.wikimedia.org/wikipedia/commons/0/05/UN_General_Assembly_hall.jpg *License:* CC BY-SA 2.0 *Contributors:* originally posted to **Flickr** as UN General Assembly *Original artist:* Patrick Gruban, cropped and downsampled by Pine
- **File:USA_California_location_map.svg** *Source:* https://upload.wikimedia.org/wikipedia/commons/1/f9/USA_California_location_map.svg *License:* CC BY 3.0 *Contributors:* own work, using

  - United States National Imagery and Mapping Agency data
  - World Data Base II data
  - U.S. Geological Survey (USGS) data

Original artist: NordNordWest
- **File:US_Electrical_Generation_1949-2011.png** *Source:* https://upload.wikimedia.org/wikipedia/commons/d/d4/US_Electrical_Generation_1949-2011.png *License:* Public domain *Contributors:* http://www.eia.gov/totalenergy/data/annual/showtext.cfm?t=ptb0802a *Original artist:* US Energy Information Administration
- **File:US_Nuclear_Capacity_Factor.png** *Source:* https://upload.wikimedia.org/wikipedia/commons/a/a8/US_Nuclear_Capacity_Factor.png *License:* Public domain *Contributors:* http://www.eia.gov/totalenergy/data/annual/showtext.cfm?t=ptb0902 *Original artist:* US Energy Information Administration
- **File:US_Nuclear_Electricity_1949-2011.png** *Source:* https://upload.wikimedia.org/wikipedia/commons/0/09/US_Nuclear_Electricity_1949-2011.png *License:* Public domain *Contributors:* http://www.eia.gov/totalenergy/data/annual/showtext.cfm?t=ptb0902 *Original artist:* US Energy Information Administration
- **File:US_Nuclear_Power_Plant_Status_9-2013.PNG** *Source:* https://upload.wikimedia.org/wikipedia/commons/d/de/US_Nuclear_Power_Plant_Status_9-2013.PNG *License:* Public domain *Contributors:* http://www.eia.gov/electricity/monthly/update/ Electric Monthly Update 23 Sept. 2013 *Original artist:* US Energy Information Administration
- **File:US_Nuclear_Power_Reactors_1955-2011.png** *Source:* https://upload.wikimedia.org/wikipedia/commons/4/48/US_Nuclear_Power_Reactors_1955-2011.png *License:* CC BY-SA 3.0 *Contributors:* Own work *Original artist:* Plazak
- **File:US_Nuclear_Summer_Capacity.png** *Source:* https://upload.wikimedia.org/wikipedia/commons/0/02/US_Nuclear_Summer_Capacity.png *License:* Public domain *Contributors:* http://www.eia.gov/totalenergy/data/annual/showtext.cfm?t=ptb0902 *Original artist:* US Energy Information Administration
- **File:US_Uranium_Imports_2012.png** *Source:* https://upload.wikimedia.org/wikipedia/commons/d/d9/US_Uranium_Imports_2012.png *License:* Public domain *Contributors:* http://www.eia.gov/todayinenergy/detail.cfm?id=12731 *Original artist:* US Energy Information Administration
- **File:US_and_USSR_nuclear_stockpiles.svg** *Source:* https://upload.wikimedia.org/wikipedia/commons/b/bb/US_and_USSR_nuclear_stockpiles.svg *License:* Public domain *Contributors:* Own work Source data from: Robert S. Norris and Hans M. Kristensen, "Global nuclear stockpiles, 1945-2006," *Bulletin of the Atomic Scientists* 62, no. 4 (July/August 2006), 64-66. Online at http://thebulletin.metapress.com/content/c4120650912x74k7/fulltext.pdf *Original artist:* Created by User:Fastfission first by mapping the lines using OpenOffice.org's Calc program, then exporting a graph to SVG, and the performing substantial aesthetic modifications in Inkscape.
- **File:US_fallout_exposure.png** *Source:* https://upload.wikimedia.org/wikipedia/commons/3/37/US_fallout_exposure.png *License:* Public domain *Contributors:* Slightly modified (whitespace made transparent, converted to PNG) from Figure 1 in *Study Estimating Thyroid Doses of I-131 Received by Americans From Nevada Atmospheric Nuclear Bomb Test*, National Cancer Institute (1997) [1] *Original artist:* National Cancer Institute
- **File:US_nuclear_sites_map.svg** *Source:* https://upload.wikimedia.org/wikipedia/commons/2/28/US_nuclear_sites_map.svg *License:* CC BY 2.5 *Contributors:* ? *Original artist:* ?
- **File:View_of_Chernobyl_taken_from_Pripyat.JPG** *Source:* https://upload.wikimedia.org/wikipedia/commons/6/6e/View_of_Chernobyl_taken_from_Pripyat.JPG *License:* Public domain *Contributors:* This photo is the author's own work *Original artist:* Jason Minshull
- **File:WANO_SYMBOL.jpg** *Source:* https://upload.wikimedia.org/wikipedia/en/3/39/WANO_SYMBOL.jpg *License:* Fair use *Contributors:* The logo may be obtained from World Association of Nuclear Operators. *Original artist:* ?
- **File:Wfm_sts_overview.png** *Source:* https://upload.wikimedia.org/wikipedia/commons/c/cb/Wfm_sts_overview.png *License:* CC-BY-SA-3.0 *Contributors:* ? *Original artist:* ?
- **File:Wiki_letter_w_cropped.svg** *Source:* https://upload.wikimedia.org/wikipedia/commons/1/1c/Wiki_letter_w_cropped.svg *License:* CC-BY-SA-3.0 *Contributors:* This file was derived from Wiki letter w.svg: <a href='//commons.wikimedia.org/wiki/File:Wiki_letter_w.svg' class='image'><img alt='Wiki letter w.svg' src='https://upload.wikimedia.org/wikipedia/commons/thumb/6/6c/Wiki_letter_w.svg/50px-Wiki_letter_w.svg.png' width='50' height='50' srcset='https://upload.wikimedia.org/wikipedia/commons/thumb/6/6c/Wiki_letter_w.svg/75px-Wiki_letter_w.svg.png 1.5x, https://upload.wikimedia.org/wikipedia/commons/thumb/6/6c/Wiki_letter_w.svg/100px-Wiki_letter_w.svg.png 2x' data-file-width='44' data-file-height='44' /></a> *Original artist:* Derivative work by Thumperward

30.13.3 Content license